ON KLAUDER'S PATH:

A Field Trip

ON KLAUDER'S PATH:

A Field Trip

Essays in Honor of John R Klauder

Editors

G G Emch
Department of Mathematics, University of Florida
USA

G C Hegerfeldt
Institut für Theoretische Physik, Universität Göttingen
Germany

L Streit
Universidade da Madeira, Portugal; and
BiBos, Universität Bielefeld, Germany

World Scientific
Singapore • New Jersey • London • Hong Kong

Published by

World Scientific Publishing Co. Pte. Ltd.

P O Box 128, Farrer Road, Singapore 9128

USA office: Suite 1B, 1060 Main Street, River Edge, NJ 07661

UK office: 73 Lynton Mead, Totteridge, London N20 8DH

ON KLAUDER'S PATH: A FIELD TRIP

ISBN 981-02-1687-4

Printed in Singapore.

Dedicated to John R. Klauder
on the occasion of his 60th birthday

PREFACE

At a seminar held on 22 January 1992, two days before his sixtieth (!) birthday, John R. Klauder was presented with a pre-edition of a collection of essays to be published under the title *On Klauder's Path: A Field Trip*. For that purpose, the editors had obtained the backbone of a body of contributions; subsequently — and then with John's knowing participation — this list was fleshed out and garnished to become the present volume. Authors have been sought who would reflect the special brand of mathematical physics John helped define, both through the repercussions of his own scientific work and through his actions on the governing bodies of several professional societies and governmental agencies: Member of the Executive Committee and then President of the International Association of Mathematical Physics, Associate Secretary–General of the Executive Council of the International Union of Pure and Applied Physics, Member of the National Science Foundation Physics Advisory Panel, to mention only a few, and not to forget that he was also the Editor of the Journal of Mathematical Physics (1979–85).

A Princeton Ph.D. (1959), John spent some thirty–five years (1953–1988) of his career as a member of the technical staff of the AT&T Bell Laboratories at Murray Hill NJ, where he also served as head of two of their departments (Theoretical Physics, 1966–67 and 1969–71; and Solid State Spectroscopy, 1971–76). In 1988 he was persuaded to move from there to the University of Florida in response to an offer of a joint appointment as Professor of Physics and of Mathematics. Over the years, John also held Visiting Professorships at several American, Japanese and European Universities. All over the world, his lectures have the deserved reputation of being clear, enthusiatic and seminal; he has had a profound impact on the career orientation of many Ph.D. students and postdoctoral fellows.

John's scientific scope embraces vast territories from Quantum Theories to General Relativity, Optics, and Chaotic Dynamics. A recurrent theme in his research is the role(s) played by *coherent states*, in particular in connection with path integral formulations of quantization; perhaps at a less lofty level, this concept has had at least two spectacular applications: as a powerful investigative tool in quantum optics, and as a precursor to wavelets. In a different vein, John also attacked specific, non-renormalizable but exactly soluble, hard-core models in field theory, where he uncovered what has since been called the *'Klauder phenomenon'*.

All of these and many others of John's interests are represented in the diverse papers through which this volume attempts to suggest his imprint on our science. Also, the current trend in his publication list indicates that we should watch as his tracks continue to lead to wider vistas on fertile fields.

G.G.E.
G.C.H.
L.S.

CONTENTS

THE PUBLICATIONS OF JOHN R. KLAUDER

Books

1. (with E.C.G. Sudarshan), *Fundamentals of Quantum Optics*, W.A. Benjamin, New York, 1968.
2. J.R. Klauder, *Magic Without Magic: John Archibald Wheeler*, W.H. Freeman and Co., San Francisco (editor), 1972.
3. (with Bo-Sture Skagerstam), *A Coherent-State Primer," Introductory Chapter of a reprint volume "Coherent States: Applications to Physics and Mathematical Physics*, World Scientific, Singapore, 1985.

Papers

1. (with J.A. Wheeler), *On the Question of a Neutrino Analog to Electric Charge*, Rev. Mod. Phys. **29** (1957), 516.
2. J.R. Klauder, *The Action Option and the Feynman Quantization of Spinor Fields in Terms of Ordinary C-Numbers*, Ann. of Physics **11** (1960), 123.
3. (with A.C. Price, S. Darlington and W.J. Albersheim), *The Theory and Design of Chirp Radars*, Bell Syst. Tech. J. **39** (1960), 745.
4. J.R. Klauder, *The Design of Radar Signals Having Both High Range Resolution and High Velocity Resolution*, Bell Syst. Tech. J. **39** (1960), 809.
5. (with J.E. Kunzler and G.F. Brennert), *Anisotropy in the Hall Effect and Magnetoresistance of Single Crystals of High Purity Copper*, Proc. VIIth Int. Conf. on Low Temp. Phys., G.M. Graham and A.C. Hollis Hallett, Univ. of Toronto Press (1961), 210.
6. (with J.E. Kunzler), *Open Oribits and the Fermi Surface of Copper from Magnetoresistance and Hall Effect Data*, Proc. VIIth Int. Conf. on Low Temp. Phys., G.M. Graham and A.C. Hollis Hallett, Univ. of Toronto Press (1961), 213.
7. ______, *Concerning the Fermi Surface Topology of Metals and Open Orbits from Magnetoresistance and Hall Effect Data*, J. Phys. Chem. Solids **18** (1961), 256.
8. ______, *Higher Order Open Orbits and the Interpretation of Magnetoresistance and Hall Effect Data for Copper*, in "Fermi Surface Conference", W.A. Harrison and M.B. Webb, Wiley (1961), 125.
9. J.R. Klauder, *Modification of Electron Energy Levels by Impurity Atoms*, Ann. of Physics **14** (1961), 43.
10. ______, *Covariant Quantization of the Gravitational Field*, Nuovo Cimento **19** (1961), 1059.
11. (with J.E. Kunzler), *Transverse-Even Voltage: A High-Field Galvanomagnetic Effect Associated with Open Orbits in Metals*, Phys. Rev. Lett. **6** (1961), 179.

12. ———, *The Observation of Four Types of Hall Constant Anisotropy in Copper and Their Role in the Determination of the Fermi Surface*, Phil. Mag. **6** (1961), 1045.
13. J.R. Klauder, *The Calculuation of Metallic Hall Constants: Topological Aspects and Applications to Copper*, Bell Syst. Tech. J. **40** (1961), 1349.
14. (with P.W. Anderson), *Spectral Diffusion Decay in Spin Resonance Experiments*, Phys. Rev. **125** (1962), 912.
15. J.R. Klauder, *Restricted Variation of the Quantum Mechanical Action Functional, and Their Relation to Classical Dynamics*, Helv. Phys. Acta **35** (1962), 333.
16. (with H. Leutwyler and J. Schaer), γ_5 - *Ambiguity in the Generally Covariant Spin Affine Connection*, Nuovo Cimento **23** (1962), 1099.
17. J.R. Klauder, *Covariant Quantization of the Gravitational Field*, Nuovo Cimento **25** (1962), 542.
18. ———, *The Sum Over Histories: Formalism and Some Applications*, Lecture notes, University of Bern, Bern, Switzerland (1962), unpublished.
19. ———, *Continuous-Representation Theory I. Postulates of Continuous-Representation Theory*, J. Math. Phys. **4** (1963), 1055.
20. ———, *Continuous-Representation Theory II. Generalized Relation between Quantum and Classical Dynamics*, J. Math. Phys. **4** (1963), 1058.
21. ———, *Continuous-Representation Theory III. On Functional Quantization of Classical Systems*, J. Math. Phys. **5** (1964), 177.
22. (with J. McKenna), *Continuous-Representation Theory IV. Structure of a Class of Function Spaces Arising from Quantum Mechanics*, J. Math. Phys. **5** (1964), 878.
23. J.R. Klauder, *Linear Representation of Spinor Fields by Antisymmetric Tensors*, J. Math. Phys. **5** (1964), 1204.
24. (with J. McKenna), *Continuous-Representation Theory V. Construction of a Class of Scalar Boson Field Continuous Representations*, J. Math. Phys. **6** (1965), 68.
25. (with J. McKenna and D.G. Currie), *On 'Diagonal' Coherent-State Representations for Quantum-Mechanical Density Matrices*, J. Math. Phys. **6** (1965), 734.
26. (with J.K. Galt and F.R. Merritt), *Cyclotron Resonance in Cadmium*, Phys. Rev.**139** (1965) A823.
27. (with H. Leutwyler), *Some Remarks on the Boundary Value Problem in General Relativity*, Helv. Phys. Acta **38** (1965), 241.
28. J.R. Klauder, *Rotationally-Symmetric Model Field Theories*, J. Math. Phys. **6** (1965), 1666.
29. (with W.A. Reed, G.F. Brennert and J.E. Kunzler), *Study of the Fine Structure in the High-Field Galvanomagnetic Properties and the Fermi Surface of Copper*, Phys. Rev. **141** (1966), 592.
30. J.R. Klauder, *Canonical Quantization and the Weak Correspondence Principle*, Proc. Sem. on Unified Theories of Elementary Particles, Munich (1965).
31. ———, *Improved Version of Optical Equivalence Theorem*, Phys. Rev. Lett. **16** (1966), 534.
32. (with J. McKenna and E.J. Woods), *Direct-Product Representation of the Canonical Commutation Relations*, J. Math. Phys. **7** (1966), 822.

33. (with H.D.I Abarbanel and J.G. Taylor), *Green's Functions for Rotationally symmetric Models*, Phys. Rev. **152** (1966), 1198.
34. J.R. Klauder, *Weak Correspondence Principle*, J. Math. Phys. **8** (1967), 2392.
35. ______, *Coherence, Correlation and Quantum Field Theory*, in "Brandeis University Summer Institute in Theoretical Physics (1967)," Gordon and Breach, NY (1970).
36. (with E.W. Aslaksen), *Unitary Representations of the Affine Group*, J. Math. Phys. **9** (1968), 206.
37. (with M.A. Duguay, J.A. Giormaine and S.L. Shapiro), *Correlation Effects in the Display of Picosecond Pulses by Two-Photon Techniques*, Appl. Phys. Lett. **13** (1968), 174.
38. J.R. Klauder, *The Application of the Affine Group to the Quantum Theory of Gravitation*, Proc. 5th Int. Conf. on Gravitation and the Theory of Relativity, Tbilisi, USSR (1968).
39. ______, *Spectral Criterion for Mode Locked Laser Signals*, Appl. Phys. Lett. **14** (1969), 147.
40. (with E.I. Blount), *Recovery of Laser Intensity from Correlation Data*, Appl. Phys. **40** (1969), 2874.
41. (with L. Streit), *Properties of 'Quadratic' Canonical Commutation Relation Representations*, J. Math. Phys. **10** (1969), 1661.
42. (with E.W. Aslaksen), *Continuous Representation Theory Using the Affine Group*, J. Math. Phys. **10** (1969), 2267.
43. J.R. Klauder, *Hamiltonian Approach to Quantum Field Theory*, Acta Physica Austriaca, Suppl., VI (1969), 167.
44. ______, *Exponential Hilbert Space: Fock Space Revisited*, J. Math. Phys. **11** (1970), 609.
45. ______, *Soluble Models of Quantum Gravitation*, in "Relativity," L. Witten, Plenum Press (1970), 5.
46. (with G.C. Hegerfeldt), *Metrics on Test Function Space for Canonical Field Operators*, Comm. Math. Phys. **16** (1970), 329.
47. (with H. Leutwyler and L. Streit), *Quantum Field Theory on Lightlike Slabs*, Nuovo Cimento **66** (1970), 536.
48. (with E.W. Aslaksen), *Elemenatary Model for Quantum Gravity*, Phys. Rev. **D2** (1970) 272.
49. J.R. Klauder, *Ultralocal Scalar Field Models*, Comm. Math. Phys. **18** (1970), 307.
50. ______, *Ultralocal Quantum Field Theory*, Acta Physica Austriaca, Suppl., VIII (1971), 227.
51. (with V. Bargmann, P. Butera and L. Girardello), *On the Completeness of the Coherent States*, Reports on Math. Phys. **2** (1971), 221.
52. J.R. Klauder, *Heisenberg Operator Equations of Motion*, Proc. Symp. on Basic Questions in Elementary Particle Physics (1971), 82.
53. ______, *Functional Techniques and Their Application in Quantum Field Theory*, in "Math. Methods Theoret. Phys.", W.E. Brittin, Colorado Assoc. Univ. Press (1973), 329.

54. (with L. Streit), *Tachyon Quantization*, in "Math. Methods Theoret. Phys.," W.E. Brittin, Colorado Assoc. Univ. Press (1973), 450.
55. (with G.C. Hegerfeldt), *Fields Without Partners*, Nuovo Cimento **10** (1972), 723.
56. J.R. Klauder, *Covariant Diastrophic Quantum Field Theory*, Phys. Rev. Lett. **28** (1972), 769.
57. (with G.C. Hegerfeldt), *The Ultralocal Model: Fields Without Partners*, Acta. Physica Austriaca, Suppl., IX (1972), 856.
58. J.R. Klauder, *Ultralocal Spinor Field Models*, Ann. Phys. **79** (1973), 111.
59. (with M.B. Ruskai), *Local Products as Operators*, J. Math. Phys. **14** (1973), 1199.
60. J.R. Klauder, *Remarks on Nonrenormalizable Interactions*, Phys. Lett **47B** (1973), 523.
61. ———, *Ultralocal Approach to Quantum Field Theory: A Survey of Present Results*, Trudy Math. Inst. Steklov **135** (1975), 163.
62. ———, *Field Theory as Noise Theory*, Phys. Lett., B **56** (1975), 93.
63. ———, *Field Structure Through Model Studies: Aspects of Nonrenormalizable Theories*, Acta Physica Austriaca, Suppl., XI (1973), 341.
64. (with G.C. Hegerfeldt), *Field Product Renormalization and Wilson-Zimmermann Expansion in a Class of Model Field Theories*, Nuovo Cimento **19** (1974), 153.
65. (with L.A. Shepp and H. Ezawa), *On the Divergence of Certain Integrals of the Wiener process*, Ann. Inst. Fourier, Grenoble **24** (1974), 189.
66. (with L. Streit), *Optical Equivalence Theorem for Unbounded Osbervables*, J. Math. Phys. **15** (1974), 760.
67. (with H. Ezawa and L.A. Shepp), *Vestigial Effects of Singular Potentials in Diffusion Theory and Quantum Mechanics*, J. Math. Phys. **16** (1975), 783.
68. ———, *A Path Space Picture for Feynman-Kac Averages*, Ann. of Physics **88** (1974), 588.
69. J.R. Klauder, *The Value and the Scope of Quantum Field Theory*, in "Math. Phys. and Phys. Math.," K. Maurin and R. Raczka, Reidel, Dortrecht (1976), 3.
70. ———, *On the Meaning of a Nonrenormalizable Theory of Gravitation*, GRG **6** (1975), 13.
71. (with L.C. Detwiler), *Supersingular Quantum Perturbations*, Phys. Rev. **D11** (1975), 1436.
72. J.R. Klauder, *On Model Fields with Independent Values at Every Space-Time Point*, Acta Physica Austriaca **41** (1975), 237.
73. ———, *Soluble Models and the Meaning of Nonrenormalizability*, in "International Symposium on Mathematical Problems in Theoretical Physics.," H. Araki, Springer, NY (1975), 160.
74. (with H. Narnhofer), *Multicomponent, Independent-Value Quantum Field Models*, Acta Physica Austriaca **44** (1976), 161.
75. J.R. Klauder, *Stochastic Processes and Quantum Theory*, Acta Physica Austriaca, Suppl., XIV (1975), 581.
76. ———, *Soluble Covariant Nonrenormalizable Quantum Field Theory*, Phys. Rev. **D12** (1975), 1590.

77. (with H. Narnhofer), *Large-N Behavior for Independent-Value Models*, Phys. Rev. **D13** (1976) 257.
78. J.R. Klauder, *A Characteristic Glimpse of the Renormalization Group*, J. Math. Phys. **17** (1976), 826.
79. (with H. Narnhofer), *Boundary Conditions and Singular Potentials in Diffusion Theory*, J. Math. Phys **17** (1976), 1201.
80. J.R. Klauder, *Challenging Problems in Singular Dynamical Systems*, Acta Physica Austriaca, Suppl., XVI (1976), 185.
81. ______, *Continuous and Discontinuous Perturbations*, Science **199** (1978), 735.
82. ______, *Augmented Quantum Field Theory: A Proposal to Extend Conventional Formulations*, Phys. Rev. **D14** (1976), 1952.
83. ______, *Quantum Optics*, Encyclopedia of Physics (R.G. Lerner and G.L. Trigg, eds.), Addison–Wesley Publ. Co., 1981, pp. 819.
84. ______, *Functional Equations and Green's Functions for Augmented Scalar Fields*, Phys. Rev. **D15** (1977) 2830.
85. ______, *Affine Fields and the Operator Formulation of Augmented Scalar Fields*, J. Math. Phys. **18** (1977), 1711.
86. ______, *Classical Concepts in Quantum Contexts*, Acta Physica Austriaca, Suppl., XVIII (1977), 1.
87. ______, *Continuous Representations and Path Integrals, Revisited*, in "Path Integrals," G.J. Papadopoulos and J.T. Devreese, Plenum Publ. Corp. (1978), 5.
88. ______, *Nonrenormalizability and New Measures for Quantum Field Theory*, Proc. III School of Elementary Particles and High Energy Physics, Primorsko, Bulgaria (1977).
89. ______, *Scale-Covariant Quantization*, Phys. Lett. **73B** (1978), 152.
90. ______, *New Measures for Nonrenormalizable Quantum Field Theory*, Ann. of Physics **117** (1979), 19.
91. ______, *Scale-Covariant Field Theory*, Proc. XVth Annual Winter School for Theoretical Physics, Karpacz, Poland (1978).
92. ______, *Path Integrals and Stationary Phase Approximations*, Phys. Rev. **D19** (1979) 2349.
93. ______, *Path Integrals, Stationary Phase Approximations, and Complex Histories*, in "Quantum Fields – Algebras, Processes," L. Streit, Springer-Verlag (1980), 65.
94. ______, *Interaction Picture for Stochastic Differential Equations*, in "Quantum Fields – Algebras, Processes," L. Streit, Springer-Verlag (1980), 53.
95. ______, *Bosons Without Bosons*, Proc. Conf. on Quantum Theory and the Structures of Time and Space 3, Tutzing, West Germany (1978).
96. ______, *Stochastic Differential Equations for Euclidean Quantum Theory*, Colloquia Mathematica Societatis Janos Bolyai 27, Random Fields, Esztergom, Hungary, 1979, 617.
97. ______, *Path Integrals for Affine Variables*, in "Functional Integration," J.–P. Antoine and E. Tirapequi, Plenum Publ. Corp. (1980), 101.
98. ______, *Path Integrals*, Acta Physica Austriaca, Suppl., XXII (1980), 3.

99. ______, *Quantum Theory I*, Lecture Notes for Introductory Quantum Mechanics Course, University of Bern, Bern, Switzerland (1980), 172 pages (unpublished)..
100. ______, *The Fusion of Classical and Quantum Theory*, Proc. of the Conference on Quantum Theory and the Structures of Time and Space 4, Tutzing, West Germany (1980).
101. (with I. Daubechies), *Measures for Path Integrals*, Phys. Rev. Lett. **48** (1982), 117.
102. J.R. Klauder, *Scale Covariance, Nonrenormalizable Interactions, and High-Temperature Expansions*, Phys. Rev. **D24** (1982), 2599.
103. ______, *High-Temperature Expansion for Nonrenormalizable Quantum Field Theories*, Czech. J. Phys. **B32** (1982), 494.
104. ______, *Applications of Stochastic Differential Equations in Physics*, Proc. of the 5-Tage-Kurs on Stochastic Differential Equations, Bielefeld, West Germany (1981).
105. (with I. Daubechies), *Constructing Measures for Path Integrals*, J. Math. Phys. **23** (1982), 1806.
106. J.R. Klauder, *Extension of the Ito Calculus*, (unpublished).
107. ______, *Constructing Measures for Spin-Variable Path Integrals*, J. Math. Phys. **23** (1982), 1797.
108. (with H. Ezawa), *Remarks on a Stochastic Quantization of Scalar Fields*, Prog. Theoret. Phys. **69** (1983), 664.
109. J.R. Klauder, *A Langevin Approach to Fermion and Quantum Spin Correlation Functions*, J. Phys. A: Math. Gen. **16** (1983) L317.
110. (with I. Daubechies), *Measures for More Quadratic Path Integrals*, Lett. in Math. Phys. **7** (1983), 229.
111. ______, *Wiener Measures for Quantum Mechanical Path Integrals*, in "Stochastic Processes in Quantum Theory and Statistical Physics" S. Albeverio, Ph. Combe, and M. Sirugue-Collin, Springer (1982), 245.
112. J.R. Klauder, *Stochastic Quantization*, Acta Physica Austriaca, Suppl., XXV (1983), 251.
113. ______, *Coherent-State Langevin Equations for Canonical Quantum Systems with Applications to the Quantized Hall Effect*, Phys. Rev. **A29** (1984), 2036.
114. (with C.B. Lang, P. Salomonsson and B.–S. Skagerstam), *Universality of the Continuum Limit of Lattice Quantum Theories*, Z. Phys. C–Particles and Fields **26** (1984), 149.
115. J.R. Klauder, *Stochastic Phenomena in Physics*, Acta. Appl. Math. **1** (1983), 151.
116. (with I. Daubechies), *Quantum Mechanical Path Integrals with Wiener Measures for all Polynomial Hamiltonians*, Phys. Rev. Lett. **52** (1984), 1161.
117. (with W.P. Petersen), *Numerical Integration of Multiplicative-Noise Stochastic Differential Equations*, SIAM J. Numer. Anal. **22** (1985), 1153.
118. (with H. Ezawa), *Fermions Without Fermions: The Nicolai Map Revisited*, Prog. Theoret. Phys. **74** (1985), 904.
119. (with W.P. Petersen), *Spectrum of Certain Nonself-Adjoint Operators and Solutions of Langevin Equations with Complex Drift*, J. Stat. Phys. **39** (1985), 53.
120. (with I. Daubechies), *Quantum Mechanical Path Integrals with Wiener Measures for All Polynomial Hamiltonians. II*, J. Math. Phys. **26** (1985), 2239.

121. (with H. Hasegawa and M. Lakshmanan), *Towards a Satisfactory Formulation of Quantum Langevin Equation*, J. Phys. A: Math. Gen. **18** (1985) L123.
122. J.R. Klauder, *Stochastic Quantization: Theme and Variation*, Festschrift, honoring E. Fradkin's 60th birthday (1988), 577.
123. ______, *Measure and Support in Functional Integration*, in "Progress in Quantum Field Theory," H. Ezawa and S. Kamefuchi, Elsevier, B.V. (1986), 31.
124. ______, *Wave Theory of Imaging Systems*, in "Lie Methods in Optics", J. Sanchez Mondragon and K.B. Wolf, Springer, Berlin (1986), 183.
125. (with F. Cooper, D.J.E. Callaway and H.A. Rose), *Langevin Simulations in Minkowski Space*, Nuclear Physics **B262** (1985), 19.
126. J.R. Klauder, *Stochastic Quantization: Application to Field Theory and General Relativity*, in "Quantum Field Theory," F. Mancini, Elsevier, B. V. (1986), 205.
127. (with H. Gausterer), *Complex Langevin Equations for Fermion Models*, Phys. Lett., B164 (1985), 127.
128. J.R. Klauder, *Path Integrals and Semiclassical Approximations to Wave Equations*, in "Path Integrals from meV to MeV," M.C. Gutzwiller, A. Inomata, J.R. Klauder and L. Streit, World Scientific, Singapore (1986), 85.
129. ______, *Coherent-State Path Integrals for Unitary Group Representations*, in "Proc. of 14th ICGTMP," Y.M. Cho, World Scientific, Singapore (1986), 14.
130. ______, *Are Coherent States the Natural Language of Quantum Theory?*, in "Fundamental Aspects of Quantum Theory," V. Gorini and A. Frigerio, Plenum, New York (1986), 1.
131. (with B. Yurke and S.L. McCall), *SU(2) and SU(1,1) Interferometers*, Phys. Rev. **A33** (1986), 4033.
132. (with S.L. McCall and B. Yurke), *Squeezed States from Nondegenerate, Four-wave Mixers*, Phys. Rev. **A33** (1986), 3204.
133. (with H. Gausterer), *Complex Langevin Solution of the Schwinger Model*, Phys. Rev. Lett. **56** (1986), 306.
134. J.R. Klauder, *Global Uniform Semiclassical Approximation to Wave Equations*, Phys. Rev. Lett. **56** (1986), 897; (1986), 1513(E).
135. ______, *Some Recent Results on Wave Equations, Path Integrals, and Semiclassical Approximations*, in "Random Media," G. Papanicolaou, Springer, New York (1987), 163.
136. ______, *An Introduction to Squeezed States*, in "Symmetries of Science II," B. Gruber, Plenum, New York (1987), 271.
137. (with H. Gausterer), *Complex Langevin Equations and their Application to Quantum Statistical and Lattice Field Models*, Phys. Rev. **D33** (1986), 3678.
138. J.R. Klauder, *Coherent States and the Global, Uniform Approximation of Wave Equation Solutions*, in "The Phys. of Phase Space," H. Kim and W. Zachary, Plenum, New York (1987), 290.
139. (with H. Gausterer), *Two-Dimensional Model for Strongly Interacting Matter at High Particle Density*, Phys. Letters **B177** (1986) 391.
140. (with I. Daubechies and T. Paul), *Wiener Measures for Path Integrals with Affine Kinematic Variables*, J. Math. Phys. **28** (1987), 85.

141. J.R. Klauder, *Global, Uniform Semiclassical Approximations for Quantum Systems on the Half Line*, Phys. Rev. **A34** (1986), 4486.
142. ______, *Coherence and Quantum Optics*, Walter C. Michels Memorial Lectures, Bryn Mawr College (1986) unpublished.
143. ______, *Wiener Measure Regularization for Quantum Mechanical Path Integrals*, Acta Physica Austriaca, Suppl., XXVI (1987), 19.
144. ______, *Semiclassical Quantization of Classically Chaotic Systems*, Phys. Rev. Lett. **59** (1987), 748.
145. ______, *Semiclassical Quantization of Classically Chaotic Systems: Pedagogical Aspects*, Proc. of the Int. Conf. on Chaos in Education, Koszeg, Hungary (1987).
146. (with H.-Y. Fan and H.R. Zaidi), *New Approach for Calculating the Normally Ordered Form of Squeeze Operators*, Phys. Rev. **D35** (1987), 1831.
147. J.R. Klauder, *Global, Uniform, Asymptotic Wave-Equation Solutions for Large Wave Numbers*, Ann. of Physics **180** (1987), 108.
148. ______, *New Asymptotics for Old Wave Equations*, Science **239** (1988), 760.
149. (with H.-Y. Fan), *Canonical Coherent-State Representation of Some Squeeze Operators*, J. Phys. A: Math. Gen. **21** (1988) L725.
150. J.R. Klauder, *Quantization Is Geometry after All*, Ann. of Physics **188** (1988), 120.
151. (with C.J. Isham), *Affine Fields and Operator Representations for the Nonlinear σ-Model*, J. Math. Phys **31** (1990), 699.
152. (with E. Onofri), *Landau Levels and Geometric Quantization*, Int. Jour. of Mod. Phys. **A4** (1989), 3939.
153. J.R. Klauder, *The Regularized Feynman Path Integral*, in "Path Integrals from meV to MeV", eds. V. Sa-yakami et.al., World Scientific, Singapore (1989), 48.
154. ______, *Quantization=Geometry+Probability*, in "Probabilistic Methods in Quatum Field Theory and Quantum Gravity," P.H. Damgaard, H. Hüffel, and A. Rosenblum, North-Holland (1990).
155. (with J. Navarro-Salas), *The Affine Group Model for Gravity, Revisited*, Class. Quant. Gravity (in press).
156. (with C.J. Isham), *Coherent States for n-Dimensional Euclidean Groups E(n) and their Application*, J. Math. Phys **32** (1991), 607.
157. (with V. Aldaya, J. Navarro-Salas), *The Affine Field Model for Gravity: A Zero Central Change Realization of the SU(1,1) Loops Groups*, Proc. of the "9th Italian Conference on General Relativity and Gravitational Physics", Eds. R. Cianci et al, World Scientific, Singapore, 1991.
158. (with R. Streater), *A Wavelet Transform for the Poincaré Groups*, J. Math. Phys. **32** (1991), 1609.
159. (with V. Aldaya and J. Navarro-Salas), *Group Quantization of the Affine Model for Gravity and non-Standard Representations of the SU(1,1) - Kac-Moody Group*, Class. Quant. Gravity **8** (1991), 413.
160. (with I. Daubechies), *Squeezed States and Path Integrals*, in "Workshop on Squeezed States & Uncertainty Relations", University of Maryland (1990) (to be published).

161. (with E. Onofri), *The Quantum Hall Effect and Coordinate-Free Quantization*, in "The Foundations of Quantum Mechanics," M. Scully, M.M Nieto, World Scientific, Singapore (to be published).
162. J.R. Klauder, *Coherent States and Coordinate-Free Quantization*, Proc. of the Workshop in honor of the 60th birthday of E.C.G. Sudarshan, A. Gleason, World Scientific, Singapore (to be published).
163. (with S. Lee), *Improved Complex Langevin Method for d=2+1 Dimensional Lattices*, Phys. Rev. D. **45** (1992), 2101.
164. , *A Stochastic Primer*, Computational Methods in Field Theory, H. Gausterer and C.B. Lang (eds.), Springer, Vienna, 1992, p.1.
165. (with B.F. Whiting), *Extended Coherent States and Path Integrals with Auxiliary Variables*, J. Phys. A: Math. Gen. (in press).

DIRICHLET OPERATORS AND GIBBS MEASURES

S. Albeverio, Yu. G. Kondratiev and M. Röckner

Abstract. We prove the essential self–adjointness on a natural domain for the Dirichlet operators H associated with Gibbs measures describing statistical mechanical lattice models with continuous–valued variables having polynomial self–interaction and short range quadratic coupling. This is used to prove that H has a gap at the lower end of its spectrum and the logarithmic Sobolev inequality holds. As a consequence we have the ergodic property of the process solving the corresponding stochastic differential equation (stochastic quantization equation).

Contribution in honor of the 60th birthday of John R. Klauder

1. Introduction

The theory of Dirichlet forms over infinite dimensional spaces is a powerful tool for the construction of infinite dimensional stochastic processes, the study of the associated potential theory and infinite dimensional analysis, see e.g. [2,3,4,13,14,15,17]. Of particular interest are the infinite dimensional processes with state spaces over $\mathbf{R}^d$ or $\mathbf{Z}^d$ in which natural invariance groups (the Euclidean group and its subgroups) act. One can construct processes associated with Dirichlet forms in such a way that these processes have as invariant probability measures Gibbs measures associated with Euclidean quantum fields over $\mathbf{R}^d$ [2,3,4], resp. lattice models of classical statistical mechanics over $\mathbf{Z}^d$ [3,6,21,34]. Once this association is made, natural questions about the nature of the generator of the processes [2,3,4,14], its spectrum and the time ergodic properties of the processes resp. the ergodic properties with respect to translations in path space [3] arise. Recently these questions have regained a lot of attention in connection with the establishment of logarithmic Sobolev inequalities, hypercontractive estimates and the relations betwen phase transitions in the sense of the processes versus phase transitions in the sense of the associated Gibbs measures, see e.g. [22,38,41] and references therein.

As for the theory of Dirichlet forms and associated processes a recently discussed problem, formulated originally in the finite dimensional case in [23] and in the infinite dimensional case in [2], is the one of uniqueness (in the sense of Markov uniqueness, see [35], or in the sense of strong uniqueness, i.e., essential self adjointness of the associated generator on a natural domain of functions, see [2,7,8,19,29,30,39,40]). The solution of the uniqueness problem is also an ingredient in the existing proofs of log–Sobolev inequalities and hypercontractive estimates.

In the present paper we describe the results we have obtained in our study of the uniqueness problem for infinite dimensional processes associated with lattice systems, refering to [9] for more details. To be specific, we study stochastic processes having as invariance measures Gibbs measures formally given by

$$d\mu_\epsilon(x) = Z_\epsilon^{-1} \exp\{-\epsilon \sum_{k\in\mathbf{Z}^d} P(x_k)\}\, d\mu_0(x) \ ,$$

where ϵ is a small parameter, P is a polynomial, $x = \{x_k | k \in \mathbf{Z}^d\} \in \mathbf{R}^{\mathbf{Z}^d}$, μ_0 is a Gaussian measure with mean zero and covariance given by a strictly positive translation invariant operator on $l_2(\mathbf{Z}^d)$ of finite support. These models describe thus classical mechanical anharmonic crystals, of the type of those studied in [5] (see also [21]) (and in particular those arising in lattice approximations of (Euclidean) quantum fields, see e.g. [36,24,1]). One can also look at these processes as the stochastic quantization processes associated with the above Gibbs measures (see [28,27,14,35] for corresponding continuous cases).

In section 2 we define the processes as the Hunt processes associated in the sense of the theory of (infinite dimensional) Dirichlet forms with a Dirichlet form given by μ_ϵ. We also describe the associated generator $\tilde{H}_{\mu_\epsilon}$ and show that it is essentially self–adjoint on the natural domain of smooth bounded cylinder functions. This then solves, in the strong sense, the uniqueness problem for these models. The method of proof is based on previous results of [19], as well as a new method of estimates developed in [8]. In section 3 we show that μ_ϵ is log–concave in the sense of [11], 0 is a simple eigenvalue as the bottom of the spectrum of $\tilde{H}_{\mu_\epsilon}$ and that there exists a gap of finite length at the bottom of the spectrum of $\tilde{H}_{\mu_\epsilon}$ (Theorem 2). This then implies that the logarithmic Sobolev inequality holds (Theorem 3), and the process is ergodic. We also briefly describe extensions of our model to the case of other types of interactions.

2. Essential self–adjointness of Dirichlet operators associated with Gibbs measures

Let us introduce some notation. We denote by $\mathbf{Z}^d \subset \mathbf{R}^d$ the d–dimensional integer lattice and by $\mathbf{R}^{\mathbf{Z}^d}$ the set of all real sequences $x = \{x_k | k \in \mathbf{Z}^d\}$; $\mathbf{R}_0^{\mathbf{Z}^d} \subset \mathbf{R}^{\mathbf{Z}^d}$ denotes the subset of finite sequences. The real Hilbert space $l_2(\mathbf{Z}^d) = \mathcal{H}_0$ is the completion of $\mathbf{R}_0^{\mathbf{Z}^d}$ with respect to the norm $|\cdot|$ generated by the scalar product

$$<\phi, \psi> = \sum_{k\in\mathbf{Z}^d} \phi_k \psi_k \quad , \quad \phi, \psi \in \mathbf{R}_0^{\mathbf{Z}^d} .$$

Note that in the chain $\mathbf{R}^{\mathbf{Z}^d} \supset \mathcal{H}_0 \supset \mathbf{R}_0^{\mathbf{Z}^d}$ the scalar product $< \cdot, \cdot >$ gives the duality between $\mathbf{R}^{\mathbf{Z}^d}$ and $\mathbf{R}_0^{\mathbf{Z}^d}$: for any $x \in \mathbf{R}^{\mathbf{Z}^d}$, $\phi \in \mathbf{R}_0^{\mathbf{Z}^d}$ the pairing is thus defined by $< x, \phi > = \sum_{k\in\mathbf{Z}^d} x_k \phi_k$.

Let A be a bounded positive operator in $\mathcal{H}_0$ which commutes with all lattice shifts $S_k : \mathcal{H}_0 \to \mathcal{H}_0, k \in \mathbf{Z}^d$, where

$$(S_k x)_j = x_{k+j} \quad , \quad x \in \mathcal{H}_0 \, , \, j \in \mathbf{Z}^d .$$

Then for the matrix $(a_{kj})_{k,j\in\mathbf{Z}^d}$ of the operator A (in the natural basis of the space $\mathcal{H}_0$) we have $a_{kj} = a(k-j)$, where $a(\cdot) : \mathbf{Z}^d \to \mathbf{R}$ is some function on the lattice such that $a(-k) = a(k)$, $k \in \mathbf{Z}^d$.

We assume that the operator A satisfies the following assumptions
A_1) A is strictly positive, i.e.,

$$\exists\, C_A > 0 : \forall h \in \mathcal{H}_0 \,, < Ah, h > \geq C_A \, |h|^2 \tag{1}$$

A_2) the function $a : \mathbf{Z}^d \to \mathbf{R}$ has finite support, i.e.,

$$\exists\, N_0 \in N : \forall k \in \mathbf{Z}^d \,, |k| > N_0 \;\; a(k) = 0 \tag{2}$$

(here $|k| = \|k\|_{\mathbf{R}^d}$).

Note that a sufficient condition for A_1) is the following:

$$a(0) - \sum_{k\neq 0} |a(k)| \geq C_A > 0 \,. \tag{3}$$

From the statistical physics point of view property A_2) means that we consider a system with local interaction.

Let μ_0 be the Gaussian measure on $\mathbf{R}^{\mathbf{Z}^d}$ with zero mean and correlation operator $A^{-1} \in \mathcal{L}(\mathcal{H}_0, \mathcal{H}_0)$. The measure μ_0 can be defined by its Fourier transform:

$$\tilde{\mu}_0(\phi) = \int_{\mathbf{R}^{\mathbf{Z}^d}} \exp(i < x, \phi >)\, d\mu_0(x) = \exp(-\frac{1}{2} < A^{-1}\phi, \phi >) \quad (\phi \in \mathbf{R}_0^{\mathbf{Z}^d})$$

due to Minlos' theorem (e.g. [19, Ch. 2], in this book also a direct construction of μ_0 is given).

For any finite set $\Lambda \subset \mathbf{Z}^d$ and $\epsilon > 0$ we introduce the following Gibbsian modification μ_ϵ^Λ of the Gaussian measure μ_0:

$$d\mu_\epsilon^\Lambda(x) = \frac{1}{Z_{\Lambda,\epsilon}} \exp\{-\epsilon \sum_{k\in\Lambda} P(x_k)\}\, d\mu_0(x) \,, \tag{4}$$

where $Z_{\Lambda,\epsilon}$ is the normalizing factor (making μ_ϵ^Λ into a probability measure), P is any polynomial of the form

$$P(s) = a_{2m} s^{2m} + \cdots + a_1 s + a_0 \quad , \quad s \in \mathbf{R}\,, \, a_{2m} > 0.$$

It is well known that in the case of sufficiently small $\epsilon > 0$, $\epsilon < \epsilon_0$, the limit $\lim_{\Lambda\to\mathbf{Z}^d} \mu_\epsilon^\Lambda$ exists in the sense of weak convergence of finite dimensional distributions [24,32]. In the following we consider only $\epsilon \in [0, \epsilon_0)$.

Note that the measure μ_ϵ has the property of invariance with respect to the shifts $S_k, k \in \mathbf{Z}^d$:

$$\forall f \in L^1(\mu_\epsilon) : \int f(S_k x) d\mu_\epsilon(x) = \int f(x) d\mu_\epsilon(x) \,. \tag{5}$$

Let us denote by $\mathcal{F}C_b^\infty$ the set of all smooth cylinder functions on $\mathbf{R}^{\mathbf{Z}^d}$ with all derivatives bounded. A function f belongs to $\mathcal{F}C_b^\infty$ if there exist $N \in \mathbf{N}$, $\{k_1, \ldots, k_N\} \subset \mathbf{Z}^d$ and $f_N \in C_b^\infty(\mathbf{R}^N)$ such that

$$f(x) = f_N(x_{k_1}, \ldots, x_{k_N}),\ x \in \mathbf{R}^{\mathbf{Z}^d}.$$

For any such f we define gradient ∇f as the mapping

$$\mathbf{R}^{\mathbf{Z}^d} \ni x \mapsto \nabla f(x) = \{\nabla_k f(x) | k \in \mathbf{Z}^d\} \in \mathbf{R}_0^{\mathbf{Z}^d},$$

where $\nabla_k f = \frac{\partial}{\partial x_k} f \in \mathcal{F}C_b^\infty$. Therefore, for all $f, g \in \mathcal{F}C_b^\infty$ the following pre–Dirichlet form

$$\varepsilon_{\mu_\epsilon}(f,g) = \frac{1}{2} \int < \nabla f(x), \nabla \bar{g}(x) > d\mu_\epsilon(x) \tag{6}$$

is well defined. We want to introduce a differential operator which is associated with the form (6). To find this operator we look at the integration by parts formula associated with (6).

Due to the condition A_2) the linear operator A can be extended to the measurable mapping

$$\mathbf{R}^{\mathbf{Z}^d} \ni x \mapsto \beta^0(x) = -Ax \in \mathbf{R}^{\mathbf{Z}^d} \tag{7}$$

and the following integration by parts formula holds (e.g. [4;19,Ch.6]):

$$\begin{aligned} \forall f, g \in \mathcal{F}C_b^\infty, \forall \phi \in \mathbf{R}_0^{\mathbf{Z}^d} : \quad & \int f(x) < \nabla g(x), \phi > d\mu_0(x) = \\ - \int < \nabla f(x), \phi > g(x)\, d\mu_0(x) - & \int f(x) g(x) < \beta^0(x), \phi > d\mu_0(x). \end{aligned} \tag{8}$$

By using (4) and the weak convergence $\mu_\epsilon^\Lambda \to \mu_\epsilon$, $\Lambda \to \mathbf{Z}^d$, it is not hard to prove the integration by parts formula also in the case of the measure μ_ϵ:

$$\begin{aligned} \forall f, g \in \mathcal{F}C_b^\infty, \forall \phi \in \mathbf{R}_0^{\mathbf{Z}^d} : \quad & \int f(x) < \nabla g(x), \phi > d\mu_\epsilon(x) = \\ - \int < \nabla f(x), \phi > g(x)\, d\mu_\epsilon(x) - & \int f(x) g(x) < \beta^\epsilon(x), \phi > d\mu_\epsilon(x), \end{aligned} \tag{9}$$

where the mapping $\beta^\epsilon : \mathbf{R}^{\mathbf{Z}^d} \to \mathbf{R}^{\mathbf{Z}^d}$ is defined as

$$\begin{aligned} \beta^\epsilon(x) = \beta^0(x) + \epsilon\, b(x) \quad &; \quad b(x) = \{b_k(x) | k \in \mathbf{Z}^d\} \in \mathbf{R}^{\mathbf{Z}^d} \\ b_k(x) = -P'(x_k) \quad &, \quad k \in \mathbf{Z}^d,\, x \in \mathbf{R}^{\mathbf{Z}^d}. \end{aligned} \tag{10}$$

The relation (9) shows that β^ϵ is the logarithmic derivative of the measure μ_ϵ [3,37]. Let us remark that because

$$\begin{aligned} \beta_k^\epsilon(x) = \beta_k^0(x) + \epsilon\, b_k(x) = (-Ax)_k + \epsilon\, b_k(x) = \\ - \sum_{j:|k-j| \le N_0} a(k-j) x_j - \epsilon P'(x_k) \end{aligned}$$

and $x_k, P'(x_k) \in L^p(\mu_\epsilon), p \geq 1, k \in \mathbf{Z}^d$ (see [32]), we have for any $\phi \in \mathbf{R}_0^{\mathbf{Z}^d}$

$$< \beta^\epsilon, \phi > \in L^p(\mu_\epsilon), \ \forall p \geq 1 . \tag{11}$$

Note that $< \beta^\epsilon(x), \phi >$ can be defined as the following limit in $L^2(\mu_\epsilon)$

$$< \beta^\epsilon(x), \phi >= \lim_{t\to 0} \frac{1}{t} \left\{ \frac{d\mu_\epsilon(x+t\phi)}{d\mu_\epsilon(x)} - 1 \right\}.$$

Let us introduce the following differential operator in $L^2(\mu_\epsilon)$ on the domain $\mathcal{F}C_b^\infty \subset L^2(\mu_\epsilon)$:

$$(H_{\mu_\epsilon} u)(x) = -\frac{1}{2}\Delta u(x) - \frac{1}{2} < \beta^\epsilon(x), \nabla u(x) > . \tag{12}$$

Here

$$\left.\begin{aligned} \Delta u(x) &= \sum_{k\in\mathbf{Z}^d} \frac{\partial^2 u(x)}{\partial x_k^2} \in \mathcal{F}C_b^\infty \subset L^2(\mu_\epsilon), \\ < \beta^\epsilon(x), \nabla u(x) > &= \sum_{k\in\mathbf{Z}^d} \beta_k^\epsilon(x) \frac{\partial u(x)}{\partial x_k} \in L^2(\mu_\epsilon) , \end{aligned}\right. \qquad u \in \mathcal{F}C_b^\infty.$$

The integration by parts formula (9) gives immediately the equality

$$\forall u, v \in \mathcal{F}C_b^\infty : \quad \varepsilon_{\mu_\epsilon}(u,v) = (H_{\mu_\epsilon} u, v)_{L^2(\mu_\epsilon)} , \tag{13}$$

therefore H_{μ_ϵ} is the pre–Dirichlet operator which corresponds to the pre–Dirichlet form $\varepsilon_{\mu_\epsilon}$ (see e.g. [2,13;19,Ch.5]).

Theorem 1. *The operator H_{μ_ϵ} is essentially self–adjoint on $\mathcal{F}C_b^\infty$, i.e., its closure $\tilde{H}_{\mu_\epsilon}$ in $L^2(\mu_\epsilon)$ is a self–adjoint operator.*

Proof. The argument is based on a modification of the approximate criterium of essential self–adjointness which is proved in [8]. For more details we refer to [9].

For any $n \in \mathbf{N}$ we denote by Λ_n the set

$$\Lambda_n = \{k \in \mathbf{Z}^d \| k| \leq n\}$$

and introduce the mapping $\beta^{\epsilon,n} : \mathbf{R}^{\mathbf{Z}^d} \to \mathbf{R}^{\mathbf{Z}^d}$ which is defined as

$$\begin{aligned} \beta^{\epsilon,n}(x) &= \beta^0(x) + \epsilon\, b^{\Lambda_n}(x) , \ x \in \mathbf{R}^{\mathbf{Z}^d}, \\ b_k^{\Lambda_n}(x) &= \begin{cases} b_k(x), & k \in \Lambda_n \\ 0, & k \in \mathbf{Z}^d \backslash \Lambda_n \end{cases} \end{aligned}$$

Let us define the differential operator H_{Λ_n} on the domain $\mathcal{F}C_b^\infty \subset L^2(\mu_\epsilon)$ as

$$(H_{\Lambda_n} u)(x) = -\frac{1}{2}\Delta u(x) - \frac{1}{2} < \beta^{\epsilon,n}(x), \nabla u(x) > \quad , \quad u \in \mathcal{F}C_b^\infty.$$

To prove Theorem 1 we use the following parabolic criterium for essential self–adjointness (see [19, Ch. 6]):

Let us consider the sequence of Cauchy problems

$$\begin{cases} \frac{d}{dt}u_n(t) + \tilde{H}_{\Lambda_n}u_n(t) = 0, & t \in [0,1] \\ u_n(0) = f, & n \in \mathbf{N}, \end{cases} \tag{14}$$

where the initial condition f is any function from $\mathcal{F}C_b^\infty$, $\tilde{H}_{\Lambda_n}$ denote the closure in $L^2(\mu_\epsilon)$ of the operator H_{Λ_n}. We need to prove the following two facts:

i) The existence of solutions.

For any $n \in \mathbf{N}$ the system (14) has a strong solution such that $u_n(t) \in Dom(\tilde{H}_{\Lambda_n})$ for all $t \in [0,1]$, i.e.

$$[0,1] \ni t \mapsto u_n(t) \in L^2(\mu_\epsilon)$$

are continuously differentiable vector–valued functions satisfing (14).

ii) Approximating property.

For all $n \in \mathbf{N}, t \in [0,1]$, we have the inclusions $u_n(t) \in Dom(\tilde{H}_{\mu_\epsilon})$ and

$$\int_0^1 \|(\tilde{H}_{\mu_\epsilon} - \tilde{H}_{\Lambda_n})u_n(t)\|_{L^2(\mu_\epsilon)}dt \to 0 \quad , \quad n \to \infty. \tag{15}$$

If i) and ii) are satisfied, then H_{μ_ϵ} is an essentially self–adjoint operator (see [19, Ch. 5]).

For the construction of the solutions u_n we use the well known representation by solutions of stochastic differential equations of the following form:

$$\begin{cases} d\xi_x^{(n)}(t) = \frac{1}{2}\beta^{\epsilon,n}(\xi_x^{(n)}(t))dt + dw(t) \\ \xi_x^{(n)}(0) = x \in \mathbf{R}^{\mathbf{Z}^d}, \end{cases} \tag{16}$$

where $w(t) = \{w_k(t) | k \in \mathbf{Z}^d\}$, $t \geq 0$, is a family of independent standard one–dimensional Wiener processes, $\xi_x^{(n)}(t) = \{\xi_{x,k}^{(n)}(t) | k \in \mathbf{Z}^d\} \in \mathbf{R}^{\mathbf{Z}^d}$. The next step consists in proving i) for the functions $u_n(t,x) = E\{f(\xi_x^{(n)}(t))\}$. In the proof of ii) an important role is played by a uniform estimate (with respect to $n \in \mathbf{N}$) for ∇u_n. Such an estimate can be obtained in a way similar to the one used in [8]. □

3. Some properties of Dirichlet operators. Stochastic quantization.

We shall consider the derivative of the logarithmic derivative of the measure μ_ϵ, understood in the following sense. For $\phi, \psi \in \mathbf{R}_0^{\mathbf{Z}^d}$ we consider the limit in $L^2(\mu_\epsilon)$:

$$\lim_{t \to 0} t^{-1}(< \beta^\epsilon(x + t\psi), \phi > - < \beta^\epsilon(x), \phi >) = < \nabla\beta^\epsilon(x)\phi, \psi > . \tag{17}$$

Due to (10) the limit in (17) exists and for the matrix elements of the linear operator $R_\epsilon(x) = -\nabla\beta^\epsilon(x) = (r_{kj}(x))_{k,j\in\mathbf{Z}^d}$ in the natural basis of the space $\mathcal{H}_0$ we have

$$r_{kj}(x) = a(k-j) + \epsilon P''(x_k)\delta_{kj} \quad , \quad k,j \in \mathbf{Z}^d \, , \, x \in \mathbf{R}^{\mathbf{Z}^d}. \tag{18}$$

Therefore, the operator $R_\epsilon(x)$ is well defined as a linear operator in $\mathbf{R}_0^{\mathbf{Z}^d}$ for any $x \in \mathbf{R}^{\mathbf{Z}^d}$. We remark that it is not hard to see that $R_\epsilon(x)$ is an unbounded linear operator in $\mathcal{H}_0$ for μ_ϵ-a.e. $x \in \mathbf{R}^{\mathbf{Z}^d}$. By (18) we have for any $\phi \in \mathbf{R}_0^{\mathbf{Z}^d}$

$$< R_\epsilon(x)\phi, \phi > = < A\phi, \phi > + \sum_{k \in \mathbf{Z}^d} P''(x_k)\,\phi_k^2 \;\geq (C_A + \epsilon C_P)\,|\phi|^2\,, \tag{19}$$

where

$$C_P = \inf_{s \in \mathbf{R}} P''(s) > -\infty$$

So we can choose $\epsilon_1 > 0$ such that for any $\epsilon \in [0, \epsilon_1)$

$$< R_\epsilon(x)\phi, \phi > \geq \frac{1}{2} C_A |\phi|^2 \quad , \quad x \in \mathbf{R}^{\mathbf{Z}^d}\,,\, \phi \in \mathbf{R}_0^{\mathbf{Z}^d} \tag{20}$$

(if $C_P \geq 0$ we put $\epsilon_1 = +\infty$). The inequality (20) shows that for $\epsilon \in [0, \epsilon_1)$ the measure μ_ϵ has the property of log–concavity in the sense of [11,12] (for another approach to the definition of log–concave measures see [10]).

Theorem 2. *For any $\epsilon \in [0, \epsilon_1)$ the point $0 \in \mathbf{R}$ is a simple eigenvalue of $\tilde{H}_{\mu_\epsilon}$ with the eigenfunction $f_0(x) \equiv 1$ and $\tilde{H}_{\mu_\epsilon}$ has a gap in the spectrum $\sigma(\tilde{H}_{\mu_\epsilon})$ of a size not less that $\frac{1}{4}C_A$, i.e., $\sigma(\tilde{H}_{\mu_\epsilon}) \cap (0, \frac{1}{4}C_A) = \emptyset$.*

The proof of Theorem 2 is based on the property of log–concavity for the measure μ_ϵ and the following very useful identity ([19,Ch.6;29,30]):

$$4\int |H_{\mu_\epsilon} u|^2 d\mu_\epsilon = \int Tr_{\mathcal{H}_0}(u''\bar{u}'')d\mu_\epsilon + \int < R_\epsilon \nabla u, \nabla \bar{u} > d\mu_\epsilon \quad , \quad u \in \mathcal{F}C_b^\infty \tag{21}$$

For the case of smooth log–concave measures on Hilbert spaces see also [11,12].

Note that the existence of a spectral gap implies by the spectral theorem the following ergodic property for the semigroup $P_t^\epsilon = \exp(-t\tilde{H}_{\mu_\epsilon})$, $t > 0$, in $L^2(\mu_\epsilon)$:

$$\forall f \in L^2(\mu_\epsilon): \quad \|P_t^\epsilon f - \int f d\mu_\epsilon\|_{L^2(\mu_\epsilon)} \to 0\,,\, t \to \infty\,. \tag{22}$$

An important property of log–concave measures is that they satisfy the logarithmic Sobolev inequality. This is proved in the case of smooth measures on Hilbert spaces in [11] by using the Bakry–Emery approach [16]. The log–Sobolev inequality for the Gibbs measures on infinite products of compact Riemannian manifolds is studied in [22,41,12]. In our present situation we have the following result.

Theorem 3. *For any $\epsilon \in [0, \epsilon_1)$ and all $f \in Dom(\tilde{H}_{\mu_\epsilon})$*

$$\int |f|^2 \log |f|^2 \, d\mu_\epsilon \leq \frac{8}{C_A}(f, \tilde{H}_{\mu_\epsilon} f)_{L^2(\mu_\epsilon)} + \|f\|^2_{L^2(\mu_\epsilon)} \log \|f\|^2_{L^2(\mu_\epsilon)}\,. \tag{23}$$

By Gross' theorem [25] the log–Sobolev inequality (23) implies hypercontractivity for the semigroup T_t^ϵ , $t \geq 0$ (for the definition of hypercontractivity, see e.g. [25;19,Ch. 6]).

Let us write the system of Langevin's equations for the measure μ_ϵ [26]:

$$\begin{aligned} d\xi_{x,k}(t) &= -\frac{1}{2} \sum_{j:|k-j|\leq N_0} a(k-j)\xi_{x,j}(t)dt - \frac{\epsilon}{2}P'(\xi_{x,k}(t))dt + dw_k(t) \\ \xi_{x,k}(0) &= x_k\,, \quad k \in \mathbf{Z}^d. \end{aligned} \tag{24}$$

The system (24) can be obtained by using the (formal) limit in (16) as $n \to \infty$. It has a strong solution in some Hilbert subspace $\mathcal{H}_- \subset \mathbf{R}^{\mathbf{Z}^d}$ for any initial date $x \in \mathcal{H}_-$ [21,31,33,34]. It is not hard to prove that the measure μ_ϵ is invariant for the system (24). Moreover, for some class of initial distributions we have the convergence to μ_ϵ under the time evolution generated by (24) (cf. [26]). By using Theorems 2 and 3 we obtain the following ergodicity property of the solution of the system (24).

Theorem 4. *For any* $f \in \mathcal{F}C_b^\infty$ *and* μ_ϵ–a.e. $x \in \mathbf{R}^{\mathbf{Z}^d}$

$$E\{f(\xi_x(t))\} \to \int f(y)d\mu_\epsilon(y) \quad , \quad t \to \infty,$$

with $\xi_x = \{\xi_{x,k} | k \in \mathbf{Z}^d\}$.

Remark 1. Our results can be extended to the case of non–local interactions where condition A_2 is replaced by the condition

$$\sum_{k \in \mathbf{Z}^d} |a(k)| < +\infty.$$

In this case we must consider the solutions of the stochastic differential equations (16) and (24) in a special Hilbert space $X \subset \mathbf{R}^{\mathbf{Z}^d}$ of full μ_ϵ–measure. The space X is constructed by the operator A similarly as in [31].

Remark 2. The polynomial P in the definition of the measure μ_ϵ can be replaced by any potential $V \in C^3(\mathbf{R})$ with polynomially bounded derivatives and such that the following conditions are fulfilled:

V_1) for some $a, b \in \mathbf{R}$

$$sV'(s) \geq as^2 + b, \ s \in \mathbf{R}.$$

V_2) $\inf_{s \in \mathbf{R}} V''(s) > -\infty$.

In this case we also have analogues of Thereoms 1-4 (modulo additional conditions for the existence and uniqueness of the measure μ_ϵ, see e.g. [18,20]). A detailed description of such cases (and also the case of Gibbs measures of a more general form) is given in [9].

Acknowledgment. Financial support by BiBoS, SFB 237 Essen–Bochum–Düsseldorf, SFB 256 Bonn and Alexander von Humboldt Foundation is gratefully acknowledged.

References

1. Albeverio S., Fenstad J.E., Høegh–Krohn R., Lindstrøm T., *Nonstandard methods in stochastic analysis and mathematical physics*, Academic Press, 1986.
2. Albeverio S., Høegh–Krohn R., *Quasi invariant measures, symmetric diffusion processes and quantum fields*, in Proceedings of the International Colloquium on Mathematical Methods of Quantum Field Theory, Editions du CNRS, No 248, 1976, pp. 11-59.
3. Albeverio S., Høegh–Krohn R., *Hunt processes and analytic potential theory on rigged Hilbert spaces*, Ann. Inst. H. Poincaré **B 13** (1977), 269-291.
4. Albeverio S., Høegh–Krohn R., *Dirichlet forms and diffusion processes on rigged Hilbert spaces*, Z. Wahrsch. und verw. Gebiete **40** (1977), 1-57.
5. Albeverio S., Høegh–Krohn R., *Homogeneous random fields and statistical mechanics*, J. Funct. Anal. **19** (1975), 242-272.
6. Albeverio S., Høegh–Krohn R., Olsen G., *The global Markov property for lattice fields*, J. Multiv. Anal. **11** (1981), 599-607.
7. Albeverio S., Høegh–Krohn R., Streit L., *Energy forms, Hamiltonians and distorted Brownian paths*, J. Math. Phys. **18** (1977), 907-917.
8. Albeverio S., Kondratiev Yu. G., Röckner M., *An approximate criterium of essential self–adjointness of Dirichlet operators*, (to appear in Potential Anal.) BiBoS Preprint Nr. 502 (1992).
9. Albeverio S., Kondratiev Yu. G., Röckner M., *Dirichlet operators via Stochastic Analysis*, (in preparation).
10. Albeverio S., Kusuoka S., *Maximality of infinite dimensional Dirichlet forms and Høegh–Krohn's model of quantum fields*, in Memorial Volume for R. Høegh–Krohn "Ideas and Methods in Quantum and Statistical Physics" (S. Albeverio, J.E. Fenstad, H. Holden, T. Linstrøm, eds.), Cambridge University Press, 1992, pp. 301-330.
11. Antonjuk A. V., Kondratiev Yu. G., *Log–concave smooth measure on Hilbert spaces and some properties of corresponding Dirichlet operators*, BiboS Preprint Nr. 478 (1991).
12. Antonjuk A. V., Kondratiev Yu. G., *Log–Sobolev inequality for Dirichlet operators on Riemannian manifolds and its applications*, BiboS Preprint Nr. 460 (1991).
13. Albeverio S., Röckner M., *Classical Dirichlet forms on topological vector spaces - closability and a Cameron–Martin formula*, J. Func. Anal. **88** (1990), 395-436.
14. Albeverio S., Röckner M., *Classical Dirichlet forms on topological vector spaces - the construction of the associated diffusion process*, Prob. Theory and Rel. Fields **83** (1989), 405-434.
15. Albeverio S., Röckner M., *Stochastic differential equations in infinite dimension: solution via Dirichlet forms*, Prob. Theory and Rel. Fields **89** (1991), 347-386.
16. Bakry D., Emery M, *Hypercontractivité de semi–groupes de diffusion*, C. R. Acad. Sci., Ser **I 299** (1984), 775-777.
17. Bouleau N., Hirsch F., *Dirichlet Forms and Analysis on Wiener Space*, Walter de Gruyter, Berlin-New York, 1991.
18. Bellissard J., Høegh–Krohn R., *Compactness and the maximal Gibbs state for random Gibbs fields on a lattice*, Comm. Math. Phys. **84** (1982), 297-327.
19. Berezansky Yu. M., Kondratiev Yu. G., *Spectral methods in infinite–dimensional analysis*, Naukova Dumka, Kiev, 1986 (English translation to appear in Kluwer Academic Publishers (1992).
20. Cassandro M., Olivieri E., Pellegrinotti A., Presutti E., *Existence and uniqueness of DLR measures for unbounded spin system*, Z. Wahrsch. u. verw. Gebiete **41** (1978), 313-334.
21. Doss H., Royer G., *Processus de diffusion associés aux mesures de Gibbs sur* $R^{\mathbf{Z}^d}$, Z. Wahrsch. u. verw. Gebiete **46** (1979), 125-158.
22. Deuschel J.-D., Stroock D., *Hypercontractivity and spectral gap of symmetric diffusion with application to the stochastic Ising model*, J. Func. Anal. **92** (1990), 30-48.
23. Fukushima M, *Dirichlet forms and Markov Processes*, North–Holland, Amsterdam, 1980.
24. Glimm J., Jaffe A., *Quantum Physics*, Springer-Verlag, Berlin-Heidelberg-New York, 1984.
25. Gross L., *Logarithmic Sobolev inequalities*, Amer. J. Math. **97** (1976), 1061-1083.

26. Ignatiuk I. A., Malishev V. A., Sidoravicius V., *Convergence of the stochastic quantization method*, in Prob. Theory and Math. Stat. , Vol.1, Mokslas, Vilnius, 1990, pp. 526-538.
27. Jona–Lasinio G., Mitter P. K., *On the stochastic quantization of field theory*, Comm. Math. Phys. **101** (1985), 409-436.
28. Klauder J., *Stochastic quantization: theme and variation*, in Quantum field theory and quantum statistics (I. A. Batalin et al., eds.), A. Hilger, Bristol, 1987, pp. 577-586.
29. Kondratiev Yu. G., *Dirichlet operators and smoothness of solution of elliptic equation*, Soviet Math. Dokl. **282** (1985), 269-273.
30. Kondratiev Yu. G., Tsicalenko T. V., *Dirichlet operators and associated differential equations*, Selecta Math. Sov. **10** (1992), 345-397.
31. Leha G., Ritter G., *On solutions to stochastic differential equations with discontinuous drift in Hilbert space*, Math. Ann. **270** (1985), 109-123.
32. Malishev V. A., Minlos R. A., *Gibbsian random fields*, Nauka, Moscow, 1985.
33. Rozovsky B. R., *Evolutional stochastic systems*, Nauka, Moscow, 1981.
34. Royer G., *Processus de diffusion associé à certains modèles d'Ising à spins continus*, Z. Wahrsch. u. verw. Gebiete **46** (1979), 165-176.
35. Röckner M., Zhang T., *On uniqueness of generalized Schrödinger operators and applications*, J. Funct. Anal. **105** (1992), 187-231.
36. Simon B., *The* $P(\phi)_2$ *Euclidean (Quantum) Field Theory*, Princeton Univ. Press, Princeton, 1974.
37. Skorohod A. V., *Integration on Hilbert space*, Springer-Verlag, Berlin-Heidelberg-New York, 1974.
38. Stroock D., Zegarlinski B., *The equivalence of the logarithmic Sobolev inequality and the Dobrushin-Shlosman mixing condition*, Comm. Math. Phys. **144** (1992), 303-323.
39. Takeda M., *On the uniqueness of Markovian self–adjoint extension of diffusion operators on infinite dimensional spaces*, Osaka J. Math. **22** (1985), 733-742.
40. Wielens N., *On the essential self–adjointness of generalized Schrödinger operators*, J. Funct. Anal. **61** (1985), 98-115.
41. Zegarlinski B., *On log–Sobolev inequalities for infinite lattice systems*, Lett. Math. Phys. **20** (1990), 173-182.

Fakultät für Mathematik Ruhr–Universität, D 4630 Bochum 1; BiBoS(Bielefeld); SFB 237(Essen-Bochum-Düsseldorf); CERFIM (Locarno) / BiBoS Research Centre, Bielefeld; Institute of Mathematics, Kiev, Ukraine; Supported by the Alexander von Humboldt Foundation / Institut für Angewandte Mathematik, Universität Bonn; SFB 256(Bonn).

A REMARK ON A CONNECTION OF RETURN TO EQUILIBRIUM AND MULTIPLE GROUND STATES IN SOME PERTURBED XY–MODEL

HUZIHIRO ARAKI

ABSTRACT. We consider a class of perturbed XY–models which are related to quasifree Fermion system via Jordan–Wigner Transsformation, recently treated by Carey and Hume. In contrast to their examples, in which failure of return to equilibrium under perturbed dynamics and multiple ground states of the unperturbed dynamics occur in the same range of parameters, we present a similar and general class of examples in which return to equilibrium under perturbed dynamics holds and multiple ground states occur under unperturbed as well as perturbed dynamics.

Dedicated to 60th anniversary of Professor John Klauder "Kanreki", or return of the birth year according to Japanese tradition

1. Introduction.

In a recent paper, Carey and Hume [3] have studied return to equilibrium of a perturbed XY–model and made an interesting observation that there is a fixed perturbation such that return to equilibrium fails for the perturbed system exactly when the parameters of the unperturbed XY–model are in the range of double ground states.

In this note, we point out that this coincidence might be fortuitous to some extent because there are a class of examples in which multiple ground states occur for unperturbed as well as perturbed dynamics by the same mechanism as the case of Carey and Hume, and yet return equilibrium holds for perturbed (and unperturbed) dynamics.

2. Model.

The XY–model we will consider is defined on a one–dimensional spin chain by Hamiltonian

$$H = -J\sum_j\{(1+r)\sigma_x^{(j)}\sigma_x^{(j+1)} + (1-r)\sigma_y^{(j)}\sigma_y^{(j+1)}\} + 2\lambda\sum_j\sigma_z^{(j)}. \tag{2.1}$$

The time evolution on the C^*–algebra $\mathcal{A}^S$ generated by Pauli spins $\sigma_x^{(j)}, \sigma_y^{(j)}, \sigma_z^{(j)}, (j = 0, \pm 1, \dots)$ is defined by

$$\tau_t(A) = \lim_{N\to\infty} e^{itH_N} A e^{-itH_N} \quad , \quad A \in \mathcal{A}^S, \tag{2.2}$$

where H_N is the Hamiltonian (2.1) in which the sum extends over only those terms involving $\sigma's$ with $-N \leq j \leq N$.

To describe the special class of perturbation we are going to use, we first explain the CAR algebra associated with the spin algebra $\mathcal{A}^S$. We expand the algebra $\mathcal{A}^S$ to a bigger algebra

$$\hat{\mathcal{A}} = \mathcal{A}^S + T\mathcal{A}^S \tag{2.3}$$

by introducting an additional element T satisfying

$$T^2 = 1 \quad , \quad T^* = T \quad , \quad TAT = \theta_-(A) \quad (A \in \mathcal{A}^S), \tag{2.4}$$

where θ_- is an involutive automorphism of $\mathcal{A}^S$ given by

$$\theta_-(A) = \lim_{N\to\infty} S_N A S_N^* \quad , \quad S_N = \Pi_{j=0}^{-N} \sigma_z^{(j)}. \tag{2.5}$$

$\hat{\mathcal{A}}$ is a C^*-algebra, being a crossed product of $\mathcal{A}^S$ by θ_- action of the two elements group $\mathbf{Z}_2$. The time evolution τ_t is extended to $\hat{\mathcal{A}}$ by (2.1) which converges also for $A \in \hat{\mathcal{A}}$.

The annihilation and creation operators defined by

$$c_j^* = TS_j(\sigma_x^{(j)} + i\sigma_y^{(j)})/2 \tag{2.6a}$$

$$c_j = TS_j(\sigma_x^{(j)} - i\sigma_y^{(j)})/2 \tag{2.6b}$$

satisfy the canonical anticommutation relations and generate a C^*-subalgebra $\mathcal{A}^{CAR}$ of $\hat{\mathcal{A}}$. The C^*-subalgebra of $\mathcal{A}^{CAR}$ generated by jointly even polynomials of c's and c^*'s will be denoted by $\mathcal{A}_+^{CAR}$. It is is also generated by those polynomials of $\sigma's$ which are jointly even in σ_x's and $\sigma_y's$ and

$$\mathcal{A}_+^{CAR} = \mathcal{A}^S \cap \mathcal{A}^{CAR}. \tag{2.7}$$

The finite interval Hamiltonian H_N belongs to this subalgebra and is a homogeneous second degree polynomial of c's and c^*'s. As stated in (2.7), such a P belongs to $\mathcal{A}^S$ (more specifically to $\mathcal{A}_+^{CAR}$) and hence can be considered as a perturbation in the original spin system. The perturbed time evolution is denoted by τ_t^P. It is given by

$$\tau_t^P(A) = \lim_{N\to\infty} e^{it(H_N+P)} A e^{-it(H_N+P)}. \tag{2.8}$$

3. Results of Carey and Hume.

There is a unique equilibrium state (characterized by the KMS condition) for τ_t and τ_t^P, for any given inverse temperature β, denoted by ω_β and ω_β^P, respectively. Return to equilibrium discussed in Theorem 7.2 of [3] is whether the following holds for all A in $\mathcal{A}^S$

$$\lim_{t\to\pm\infty} \omega_\beta(\tau_t^P(A)) = \omega_\beta^P(A). \tag{3.1}$$

The algebra $\mathcal{A}^S$ can be split into even and odd parts:

$$\mathcal{A}^S = \mathcal{A}_+^S + \mathcal{A}_-^S \tag{3.2}$$

where $\mathcal{A}_+^S = \mathcal{A}_+^{CAR}$, and $\mathcal{A}_-^S$ is the closed linear span of polynomials of σ's, jointly odd in σ_x's and σ_y's. The equilibrium states ω_β and ω_β^P are even in the sense

$$\omega_\beta(A) = \omega_\beta^P(A) = 0 \quad (A \in \mathcal{A}_-^S) \tag{3.3}$$

and τ_t^P leaves $\mathcal{A}_-^S$ invariant as a set. Therefore we have to check (3.1) only for $A \in \mathcal{A}_+^S = \mathcal{A}_+^{CAR}$, which can be done within $\mathcal{A}^{CAR}$.

We introduce the following notation for $f = (f_i)$ and $g = (g_j)$ in $l_2(\mathbf{Z})$:

$$B(h) = c^*(f) + c(g) \quad , \quad h = \begin{pmatrix} f \\ g \end{pmatrix} \tag{3.4a}$$

$$c^*(f) = \sum_j c_j^* f_j \quad , \quad c(f) = \sum_j c_j f_j . \tag{3.4b}$$

The B's generate $\mathcal{A}^{CAR}$ and the time evolution of $B(h)$ can be described as

$$\tau_t(B(h)) = B(e^{iKt}h) \quad , \tag{3.5}$$

$$\tau_t^P(B(h)) = B(e^{i(K+V)t}h) \quad . \tag{3.6}$$

Carey and Hume [3] prove that the spectrum of $K + V$ consists of an absolutely continuous part and possibly a finite number of eigenvalues with finite multiplicity, without any singular continuous spectrum, except for the Ising case of $\lambda = 0$, $\gamma = \pm 1$ where the spectrum is a purely point spectrum. By quoting an earlier result of Hume and Robinson [4], they then obtain the following result (Theorem 7.2 in [3]):

Theorem (Carey and Hume). *The return to equilibrium (3.1) holds if and only if the spectrum of $K + V$ is purely absolutely continuous.*

In other words, the return to equilibrium fails if and only if $K + V$ has an eigenvalue. Carey and Hume then show that, for a specific (decoupling) perturbation P which makes time evolution of the right half and left half of the chain independent, return to equilibrium fails due to the appearance of an eigenvalue 0 exactly for those parameter values leading to double ground states of the unperturbed dynamics.

4. Main result and its proof.

We now consider any P described above and consider the perturbed dynamics $K+\mu P$ for the parameter values (λ, γ) leading to double ground states of the unperturbed dynamics (namely $|\lambda| < 1$, $\gamma \neq 0$ in our notation). Our main result in this note is the following proposition, which contrasts with the above quoted result of Carey and Hume.

Proposition. *For sufficiently small μ, return to equilibrium holds generically while multiple ground states occur both for perturbed and unperturbed dynamics.*

Proof. From the above quoted theorem of Carey and Hume, and Proposition 4.1 (a) in [3], it follows that $K + \mu P$ has a point spectrum and return to equilibrium (3.1) fails only if either μ satisfies a polynomial equation (depending on λ, γ and P) or μP is decoupling (which happens only for a specific P and a specific value of μ). Therefore return to equilibrium holds generically.

A complete general criterion for the number of ground states is given in Theorem 5 (assisted by Theorems 3 and 4) in [2]. In particular, we claim that case (iii) in Theorem 5 (1) holds for the ground state of A_+^{CAR} given by the Fock vacuum state of $\mathcal{A}^{CAR}$ in the present situation and hence the number of pure ground states (with non–equivalent representations) is 2 or more.

To obtain this conclusion, we use the combination of Theorem 5 (5), Lemma 4.1 (1) and Corollary 4.4 in [2], where π_+ and π_- in Lemma 4.1 (1) should be identified with π_{11} and π_{22} in Corollary 4.4.

For the parameter values under consideration, the spectrum of K is separated away from 0. Then the spectrum of $K + \mu V$ is uniformly separated away from 0 for small μ. Thus the spectral projection E_+^μ of the perturbed operator $K + \mu V$ for $(0, \infty)$ is obtained by

$$E_+^\mu = \frac{1}{2\pi i}\int_\Gamma \frac{dz}{z - K - \mu V} \tag{4.1}$$

where the integration path Γ encircles the positive spectrum of $K + \mu' V$ for $0 \leq \mu' \leq \mu$. Since V is of finite rank, the equality

$$\frac{1}{Z - K - \mu V} - \frac{1}{Z - K} = \frac{1}{Z - K}\mu V \frac{1}{Z - K - \mu V} \tag{4.2}$$

shows that $E_+^\mu - E_+$ is of Hilbert–Schmidt class. Thus E_+^μ satisfies the first criterion of (4.10) in [2].

The quantity in the second equation in (4.10) in [2] is known as Z_2–index and is norm continuous in the projections by Theorem 3 in [1]. Since the second equation of (4.10) in [2] holds for $\mu = 0$, the condition [1] is satisfied for all μ for which the spectrum of $K + \mu' V$ remains away from 0 for $0 \leq \mu' \leq \mu$ because the norm continuity of (4.1) in μ can be proven by an equality similar to (4.2). For such a μ, the number of ground states (with non–equivalent cyclic representations) is 2 or more by lemma 4.1 (1) and Corollary 4.4 in [2].

Q.E.D.

REFERENCES

1. H. Araki and D. E. Evans, *On a C^*–algebra approach to phase transition in the two–dimensional Ising model*, Commun. Math. Phys. **91** (1983), 489–503.
2. H. Araki and T. Matsui, *Ground states of the XY–model*, Commun. Math. Phys. **101** (1985), 213-245.
3. A. L. Carey and L. R. Hume, *Multiple phases and return to equilibrium*, preprint.
4. L. R. Hume and D. W. Robinson, *Return to equilibrium in the XY–model*, J. Statistical Phys. **44** (1986), 829-848.

RESEARCH INSTITUTE FOR MATHEMATICAL SCIENCES KYOTO UNIVERSITY, KYOTO 606-01, JAPAN

THE CHARACTERISTIC FUNCTIONAL FOR A WICK SQUARE

Klaus Baumann

Abstract. We construct the characteristic functional $E(f)$ of $\frac{1}{2} :j^2: (f)$, where j^μ is a free conserved current in $1+1$ dimensions, and discuss its relevance for conformal quantum field theories in two dimensions.

Dedicated to John Klauder at the occasion of his sixtieth birthday

1. Introduction

In the following we construct the characteristic functional

$$E(f) := (\Omega, e^{\frac{i}{2}:j^2:(f)}\Omega),$$

where $j^\mu(x)$ is a free, conserved current in two dimensional space-time.

Our interest in this functional is twofold:

i) As an example - because explicitely known is only the characteristic functional of a generalized free field φ

$$E_0(f) = (\Omega, e^{i\varphi(f)}\Omega) = e^{-\frac{1}{2}\mathcal{W}_2(f\otimes f)} .$$

Together with the commutation relations

$$e^{i\varphi(f)}e^{i\varphi(g)} = e^{i\varphi(f+g)}e^{-\Delta(f,g)}$$

where $\Delta(f,g) = \frac{1}{2}(\Omega, [\varphi(f), \varphi(g)]\Omega)$ is half the commutator, it gives all the information which is contained in the Wightman functional $\mathcal{W}$. This construction goes through in any space-time dimension.

ii) In $1+1$-dimensions a free conserved current $j^\mu(x)$ – i.e. $\Box j^\mu = 0\,, \partial_\mu j^\mu = 0$ – decomposes into two commuting fields j_+ and j_-, which depend only on the light cone coordinates $x_+ = x^0 + x'$ and $x_- = x^0 - x'$ respectively, - i.e. $j_+ \equiv j_+(x_+), j_- \equiv j_-(x_-)$ and $[j_+(x_+), j_-(x_-)] = 0$.

Let $j(x)$ denote either $j_+(x_+)$ or $j_-(x_-)$. The commutation relations are given by

$$[j(x), j(y)] = -i\delta'(y-x)$$

if we normalize the two point function $\mathcal{W}_2(x,y) = (\Omega, j(x)j(y)\Omega) = W_2(y-x)$ according to $W_2(\xi) = -2\pi(\xi + i0)^{-2}$. The Wick square $\Theta(x) := \frac{1}{2} :j^2: (x)$ fulfills the commutation relation

$$[\Theta(x), \Theta(y)] = \frac{i}{24\pi}\delta'''(y-x) - i\delta'(y-x)\{\Theta(x) + \Theta(y)\}.$$

More generally we consider fields $\Theta_c(x)$ fulfilling

$$[\Theta_c(x), \Theta_c(y)] = \frac{i}{24\pi} \cdot c \cdot \delta'''(y-x) - i\delta'(y-x)\{\Theta_c(x) + \Theta_c(y)\}.$$

This class contains all building blocks for energy momentum tensors $\Theta_{\mu\nu}$ of conformal quantum field theories as has been discussed by several authors [1,2]. Friedan, Qui and Shenker [3] proved that the central charge c can only have the values

$$c = 1 - \frac{1}{(n+1)(n+2)} \quad , \quad n \in \mathbf{N}_0 \quad ; \quad \text{or } c \geq 1$$

if the corresponding Wightman functional $\mathcal{W}^{(c)}$ is positive definite. With a suitable product definition for functionals (see [4] for details) we can write

$$\mathcal{W} = \exp \mathcal{W}^T.$$

For $r \in \mathbf{N}$ the functional $\mathcal{W}^r := \exp r\,\mathcal{W}^T$ fulfills all the Wightman axioms. The same is true for $r \in (0, \infty)$ but positivity may be violated. The commutation relations for Θ_c can be equivalently expressed by the truncated Wightman functions

$$\mathcal{W}_2^{(c)}([x,y])^T = \frac{i}{24} \cdot c \cdot \delta'''(y-x)$$

and for $m + n \geq 1$

$$\mathcal{W}_{m+n+2}^{(c)}(x_1, ..., x_m, [x,y], y_1, ..., y_n)^T = -i\delta'(y-x)\mathcal{W}_{m+n+1}^{(c)}(x_1, ..., x_m, x+y, y_1, ..., y_n)^T.$$

From these formulae we see, that the truncated Wightman functions of Θ_c and $\Theta \equiv \Theta_1 \equiv \frac{1}{2} :j^2:$ are related by $\mathcal{W}^{(c)T} = c \cdot \mathcal{W}^T$ and we have

$$\mathcal{W}^{(c)} = \exp c\,\mathcal{W}^T \equiv \mathcal{W}^c.$$

A good knowledge of the Wightman functions of $\frac{1}{2} :j^2:$ is therefore of great importance – e.g. if we could show that $\mathcal{W}^T$ is conditionally positive definite then $\mathcal{W}^{(c)} = \exp c\,\mathcal{W}^T$ would fulfill positivity for all $c > 0$! Of course $\mathcal{W}^T$ is not conditional positive definite because only for certain values of c positivity of $\mathcal{W}^{(c)}$ is true! The characteristic functional of Θ_c is defined by

$$\begin{aligned}
E_c(f) := (\Omega, e^{i\Theta_c(f)}\Omega) &= \sum_{n=0}^{\infty} \frac{(i)^n}{n!} \mathcal{W}^{(c)}(f^{\otimes n}) \\
&= \exp \sum_{n=0}^{\infty} \frac{(i)^n}{n!} \mathcal{W}^{(c)}(f^{\otimes n})^T \\
&= \exp c \sum_{n=0}^{\infty} \frac{(i)^n}{n!} \mathcal{W}(f^{\otimes n})^T \\
&= [\exp \sum_{n=0}^{\infty} \frac{(i)^n}{n!} \mathcal{W}(f^{\otimes n})^T]^c \\
&= [E(f)]^c
\end{aligned}$$

where $E(f)$ denotes the characteristic functional of $\Theta_1 \equiv \frac{1}{2} :j^2:$.

At first glance the characteristic functional $E(f) = (\Omega, e^{i\varphi(f)}\Omega)$ seems to contain less information than the Wightman functional $\mathcal{W}$ itself, because we only get symmetrized n-point functions. But if the underlying field φ fulfills the Wightman axioms the n-point functions are uniquely fixed by their values at totally spacelike points.

Let $f_\alpha(x) = \sum_{k=1}^n \alpha_k f_k(x)$ with suppf_k spacelike to suppf_l for all $k \neq l$ then we have

$$\partial_{\alpha_1} \cdots \partial_{\alpha_n} E(f_\alpha) \mid_{\alpha=0} = i^n\, \mathcal{W}_n(f_1 \otimes \cdots \otimes f_n)$$

and therefore via analytic continuation all the n-point functions can be constructed starting from $E(f)$.

Also an explicit formula for $(\Omega, e^{\frac{i}{2}:j^2:}\Omega) \equiv E(f)$ may offer a new approach to determine the special values of c where $\mathcal{W}^{(c)}$ is positive definite. For example in [5] it has been shown that the characteristic functional belonging to the Wick square of a free euclidean, massive scalar field is infinitely divisible, in contrast to the relativistic case, where this is not true!

2. A discrete approximation of E(f)

In analogy to the Fourier transform in N dimensions

$$e^{-t\,p\cdot fp} = \frac{1}{\mid det(4\pi t f) \mid^{1/2}} \int_{\mathbf{R}^N} dx\, e^{-\frac{1}{4}\, x\cdot(tf)^{-1}x} e^{i\,p\cdot x}$$

where f is a NxN matrix and $p \cdot x = \sum_{k=1}^N p_k x_k$, we want to represent

$$e^{-t\,j\cdot fj} \text{ as } \int dx\, G(t,x) e^{i\,j\cdot x} \text{ with } j\cdot fj = \sum_{kl} f_{kl} j_k j_l \,.$$

The operators $j_m := j(h_m)$ are fields smeared with real testfunctions h_m. As long as all j_m commute the above formula is true. We generalize this in the following way.

For fixed h_m we derive a PDE for $G(t,x)$ and try to solve it by the Ansatz $G(t,x) = C(t)\exp[-\frac{1}{4}\sum S(t)_{kl} x_k x_l]$

a) Let $\{h_k \in \mathcal{S}(\mathbf{R}), h_k \text{ real}\}$ be a partition of unity such that only neighbouring test-functions have overlapping supports.

For $j_k := j(h_k)$ we have $[j_k, j_{k+1}] = \frac{i}{2}$ and $[j_k, j_l] = 0$ if $| l - k | \geq 2$.

<u>Proof:</u> From $[j(x), j(y)] = -i\delta'(y-x)$ we see

$$\frac{1}{i}[j_k, j_{k+1}] = \int_{\mathbf{R}} h_k(x)h'_{k+1}(x)dx = \int_{\mathbf{R}} (h_k + h_{k+1})h'_{k+1}(x)dx - \underbrace{\int_{\mathbf{R}} h_{k+1}h'_{k+1}(x)dx}_{=0}$$

now h_k and h_{k+2} belong to the partition of unity defined above and because the supports of h_k and h_{k+2} do not overlap, there must be a point $a \in \operatorname{supp} h_{k+1}$ s.t. $\operatorname{supp} h_k \subseteq (-\infty, a)$, $h_{k+1}(a) = 1$, $\operatorname{supp} h_{k+2} \subseteq (a, \infty)$.

Therefore $(h_k + h_{k+1})h'_{k+1}$ equals h'_{k+1} if $x \leq a$ and $h_{k+1}h'_{k+1}$ if $x \geq a$ and we get

$$= \int_{-\infty}^{a} h'_{k+1}(x)dx + \int_{a}^{\infty} h_{k+1}(x)h'_{k+1}(x)dx = h_{k+1}(a) - \frac{1}{2}h^2_{k+1}(a) = \frac{1}{2}\,.$$

b) Define $j \cdot x = \sum_k j_k x_k, x_k \in \mathbf{R}$. We claim

$$j_k e^{i\,j\cdot x} = [\frac{1}{i}\partial_{x_k} - \frac{1}{4}(x_{k+1} - x_{k-1})]e^{i\,j\cdot x}\,.$$

<u>Proof:</u> The commutation relations $[j\cdot x, j_k] = [j_{k-1}x_{k-1} + j_{k+1}x_{k+1}, j_k] = \frac{i}{2}(x_{k-1} - x_{k+1})$ imply $[(j\cdot x)^l, j_k] = \frac{i}{2}\cdot(x_{k-1} - x_{k+1})\cdot l \cdot (jx)^{l-1}$ and therefore we have

$$\begin{aligned}
\frac{1}{i}\partial_{x_k}\, e^{i\,j\cdot x} &= \sum_{n\geq 1} \frac{i^{n-1}}{n!} \sum_{l=0}^{n-1} (j\cdot x)^l j_k (j\cdot x)^{n-1-l} \\
&= \sum_{n\geq 1} \frac{i^{n-1}}{n!} \sum_{l=0}^{n-1} ([(j\cdot x)^l, j_k] + j_k (j\cdot x)^l)(j\cdot x)^{n-1-l} \\
&= \sum_{n\geq 1} \frac{i^{n-1}}{n!} (\frac{i}{2}(x_{k-1} - x_{k+1})(j\cdot x)^{n-2} \sum_{l=1}^{n-1} l + n j_k (j\cdot x)^{n-1}) \\
&= \frac{1}{4}(x_{k+1} - x_{k-1})e^{i\,j\cdot x} + j_k e^{i\,j\cdot x}\,.
\end{aligned}$$

c) Let f be a real, symmetric NxN matrix. We want to derive a PDE for $G(t,x)$ such that

$$e^{-t\,j\cdot fj} = \int dx\, G(t,x) e^{i\,j\cdot x}\,.$$

Differentiating both sides with respect to t

$$\int dx\, (\partial_t G)(t,x) e^{i\,j\cdot x} = \partial_t e^{-t\,j\cdot fj} = -\int dx\, G(t,x) \sum f_{kl} j_k j_l e^{i\,j\cdot x}$$

$$= -\int dx\, G(t,x) \sum f_{kl} [\frac{1}{i}\partial_{x_k} - \frac{1}{4}(x_{k+1} - x_{k-1})][\frac{1}{i}\partial_{x_l} - \frac{1}{4}(x_{l+1} - x_{l-1})]e^{i\,j\cdot x}$$

by partial integration

$$= -\int dx\, e^{i\,j\cdot x} \sum f_{kl} [\frac{-1}{i}\partial_{x_l} - \frac{1}{4}(x_{l+1} - x_{l-1})][\frac{-1}{i}\partial_{x_k} - \frac{1}{4}(x_{k+1} - x_{k-1})]G(t,x)$$

$$\Rightarrow \partial_t G(t,x) = \sum f_{kl}[\frac{1}{i}\partial_{x_l} + \frac{1}{2}(\delta x)_l][\frac{1}{i}\partial_{x_k} + \frac{1}{2}(\delta x)_k]G(t,x)$$

where δ is the antisymmetric matrix given by $(\delta x)_l = \frac{1}{2}(x_{l+1} - x_{l-1})$.

d) To solve the PDE for $G(t,x)$ we make the Ansatz $G(t,x) = C(t)exp[-\frac{1}{4}x \cdot S(t)x]$ with a symmetric matrix $S = S^T$. We get for the l.h.s.

$$\partial_t G = (\frac{\dot{C}}{C} - \frac{1}{4}x \cdot \dot{S}x)G$$

and for the r.h.s.

$$\begin{aligned}
&-\sum f_{kl}[\frac{1}{i}\partial_{x_l} + \frac{1}{2}(\delta x)_l][\frac{i}{2}(Sx)_k + \frac{1}{2}(\delta x)_k]G(t,x) \\
&= -\sum f_{kl}\{\frac{1}{2}S_{kl} + \frac{1}{2i}\delta_{kl} + [\frac{i}{2}(Sx)_l + \frac{1}{2}(\delta x)_l][\frac{i}{2}(Sx)_k + \frac{1}{2}(\delta x)_k]\}G(t,x) \\
&= -[\frac{1}{2}Tr(Sf - i\delta f) + \frac{1}{4}x \cdot (iS + \delta^T)f(iS + \delta)x]G(t,x)\,.
\end{aligned}$$

Equating both sides we get

$$\{\frac{\dot{C}}{C} + \frac{1}{2}Tr(Sf - i\delta f) - \frac{1}{4}x \cdot [\dot{S} - (iS + \delta^T)f(iS + \delta)]x\}G(t,x) = 0\,.$$

This must be true for all x and therefore

$$\text{(i)}\ \frac{\dot{C}}{C} + \frac{1}{2}Tr(Sf) = 0$$

$$\text{(ii)}\ \dot{S} = -SfS - \delta f\delta + i(Sf\delta - \delta fS)\,.$$

(Because of the antisymmetry $\delta^T = -\delta$ we have $Tr\delta f = 0$!) Multiplying the second eqn. by f we get finally

$$\text{(iii)}\ \dot{S}f = -(Sf)^2 - (\delta f)^2 + i[Sf, \delta f]\,.$$

Now $Sf = \delta f \cot(t\delta f)$ solves the equation $\dot{S}f = -(Sf)^2 - (\delta f)^2$ and commutes with δf . Therefore it is a solution of (iii) too.

The factor $C(t)$ is determined as follows

$$\begin{aligned}
\partial_t \ln C &= \frac{\dot{C}}{C} = -\frac{1}{2}Tr(Sf) = -\frac{1}{2}Tr[\delta f \cot(\delta f)] \\
&= -\frac{1}{2}Tr\partial_t \ln[4\pi\delta^{-1}\sin(t\delta f)]. \\
\Longrightarrow C(t) &= \exp(-\frac{1}{2}Tr \ln[4\pi\delta^{-1}\sin(t\delta f)]) \\
&= \mid det(4\pi\delta^{-1}\sin(t\delta f)) \mid^{-\frac{1}{2}} \\
&\approx \mid det(4\pi t f) \mid^{-\frac{1}{2}} \ \text{for small } t
\end{aligned}$$

is the solution for $t > 0$ up to a constant factor ! From the Taylor expansion

$$x \cot x = 1 - \frac{1}{3}x^2 - \frac{1}{45}x^4 - \cdots$$

we see that

$$\begin{aligned} S(t) &= \cot(t\delta f)\delta = (tf)^{-1} - \frac{1}{3}t\delta f\delta - \cdots \\ &\approx (tf)^{-1} \text{ for small } t \end{aligned}$$

and so

$$\begin{aligned} G(t,x) &= C(t)e^{-\frac{1}{4}x\cdot S(t)x} \\ &= \frac{1}{\mid det(4\pi\delta^{-1}\sin(t\delta f))\mid^{\frac{1}{2}}} e^{-\frac{1}{4}x\cdot\cot(t\delta f)\delta x} \\ &\approx \frac{1}{\mid det(4\pi t f)\mid^{\frac{1}{2}}} e^{-\frac{1}{4}x\cdot(tf)^{-1}x} \text{ for small } t \\ &\to \delta(x) \text{ as } t \text{ approaches } 0 \end{aligned}$$

which obviously is the right initial condition for $G(t,x)$.

e) Wick ordering :

$: j_k j_l := j_k j_l - (\Omega, j_k j_l \Omega)$ because $j(x)$ is a free field . Define $(\mathcal{W}_+)_{kl} = (\Omega, j_k j_l \Omega)$. f is a symmetric matrix, therefore only the combination $\mathcal{W}_{kl} = \frac{1}{2}[(\mathcal{W}_+)_{kl} + (\mathcal{W}_+)_{lk}]$ shows up in the exponent, and we get

$$\begin{aligned} e^{-t:j\cdot fj:} &= e^{-t\,j\cdot fj} e^{t\,Tr(f\mathcal{W})} \\ &= \exp\{-\frac{1}{2}Tr\{\ln[4\pi\delta^{-1}\sin(t\delta f)] - 2tf\mathcal{W}\}\} \int dx\, e^{-\frac{1}{4}x\cdot\cot(t\delta f)\delta x} e^{i\,j\cdot x} . \end{aligned}$$

<u>Remark:</u> This formula - even if quite formal - may be of interest by itself, for it gives a finite dimensional approximation of $e^{-:j^2:(f)}$ expressed as an integral over $e^{i\,j\cdot x}$.

f) Now we take the vacuum expectation value of $e^{-t:j\cdot fj:}$. Because $j(x)$ is a free field we have

$$(\Omega, e^{i\,j\cdot x}\Omega) = e^{-\frac{1}{2}\sum x_k x_l (\Omega, j_k j_l \Omega)} = e^{-\frac{1}{2}x\cdot\mathcal{W}x}$$

and therefore

$$\begin{aligned} (\Omega, e^{-t:j\cdot fj:}\Omega) &= \exp(-\frac{1}{2}Tr\{\ln[4\pi\delta^{-1}\sin(t\delta f)] - 2tf\mathcal{W}\}) \int dx\, e^{-\frac{1}{4}x\cdot[\cot(t\delta f)\delta + 2\mathcal{W}]x} \\ &= \exp(-\frac{1}{2}Tr\{\ln[\cos(tf\delta) + 2(f\delta)^{-1}\sin(tf\delta)f\mathcal{W}] - 2tf\mathcal{W}\}) . \end{aligned}$$

Finally we put $t = -\frac{i}{2}$ and we get as a finite dimensional approximation for the characteristic functional

$$\begin{aligned} E(f) &= (\Omega, e^{\frac{i}{2}:j\cdot fj:}\Omega) \\ &= \exp(-\frac{1}{2}Tr\{\ln[\cos(\frac{i}{2}f\delta) - \sin(\frac{i}{2}f\delta)(\frac{i}{2}f\delta)^{-1}if\mathcal{W}] + if\mathcal{W}\}) . \end{aligned}$$

3. Continuum limit

a) In the continuum limit we have the following correspondence

$$\begin{aligned} f_{kl} = f_{lk} \text{ real } &\longrightarrow f(x,y) = f(y,x) \text{ real} \\ -\frac{i}{2}\delta_{kl} &\longrightarrow \Delta(x,y) := -\frac{i}{2}\delta'(y-x) = \frac{1}{2}(\Omega, [j(x), j(y)]\Omega) \\ \mathcal{W}_{kl} &\longrightarrow \mathcal{W}(x,y) = \frac{1}{2}(\Omega, [j(x)j(y) + j(y)j(x)]\Omega) \end{aligned}$$

i.e. all matrices go over to integral operators - e.g.

$$\begin{aligned} \sum j_k f_{kl} j_l &\longrightarrow \int\int dxdy f(x,y) j(x) j(y) \\ Tr f\mathcal{W} &\longrightarrow \int\int dxdy f(x,y) \mathcal{W}(x,y) \end{aligned}$$

Therefore we have in the continuum limit

$$(\Omega, e^{i\frac{1}{2}:j\otimes j:(f)}\Omega) = \exp(-\frac{1}{2}Tr\{\ln[\cos(f\Delta) - (f\Delta)^{-1}\sin(f\Delta) if\mathcal{W}] + if\mathcal{W}\}).$$

b) Because of Wick ordering $: j \otimes j : (f) = \int\int dxdy f(x,y) : j(x)j(y) :$ we are allowed to use testfunctions of the form $f(x,y) = f(\frac{1}{2}(x+y))\delta(y-x)$ and we obtain in this way

$$\frac{1}{2} : j \otimes j : (f) = \frac{1}{2}\int dx f(x) :j^2: (x) \equiv \Theta(f)$$

Using this special testfunction we have constructed the characteristic functional $E(f)$ of the Wick square Θ

$$E(f) = (\Omega, e^{i\Theta(f)}\Omega) = \exp E^T(f)$$

with the truncated functional

$$\begin{aligned} E^T(f) &= \sum_{n=1}^{\infty} \frac{i^n}{n!}\mathcal{W}_n^T(f^{\otimes n}) \\ &= -\frac{1}{2}Tr\{\ln[\cos(f\Delta) - (f\Delta)^{-1}\sin(f\Delta) if\mathcal{W}] + if\mathcal{W}\}. \end{aligned}$$

c) Let us check our result by computing the first truncated n-point-functions by expanding the right hand side of $E^T(f)$ and ordering in powers of f

$$\begin{aligned} E^T(f) &= -\frac{1}{2!}\mathcal{W}_2^T(f\otimes f) - \frac{i}{3!}\mathcal{W}_3^T(f\otimes f\otimes f) + \frac{1}{4!}\mathcal{W}_4^T(f\otimes f\otimes f\otimes f) + \cdots \\ &= -\frac{1}{2!}Tr\frac{1}{2}[(f\mathcal{W})^2 - (f\Delta)^2] - \frac{i}{3!}Tr[(f\mathcal{W})^3 - (f\Delta)^2(f\mathcal{W})] \\ &+ \frac{1}{4!}Tr[3(f\mathcal{W})^4 - 4(f\Delta)^2(f\mathcal{W})^2 + (f\Delta)^4] + \cdots \end{aligned}$$

Unfortunately expressions like $Tr(f\mathcal{W})^2, Tr(f\Delta)^2$, and so on do not exist ! Only in certain combinations the infinities cancel ! Such a rearrangement can be done in the following way.

By definition we have

$$(\mathcal{W}+\Delta)(x,y) = (\Omega, j(x)j(y)\Omega) =: \mathcal{W}_+(x,y) \text{ resp.}$$
$$(\mathcal{W}-\Delta)(x,y) = (\Omega, j(y)j(x)\Omega) =: \mathcal{W}_-(x,y)\,.$$

Furthermore – because of the antisymmetry of $\delta'(y-x)$ – we put $0 = Tr(f\mathcal{W})^2 f\Delta = Tr(f\Delta)^3 = Tr(f\mathcal{W})^3 f\Delta = Trf\mathcal{W}(f\Delta)^3$.(These formulae are quite obvious if we replace $\delta'(y-x)$ by an antisymmetric testfunction δ'_ϵ which approaches δ' as ϵ goes to 0 .)

$$E^T(f) = -\frac{1}{2!}Tr\frac{1}{2}(f\mathcal{W}_+ f\mathcal{W}_-) - \frac{i}{3!}Tr(f\mathcal{W}_+ f\mathcal{W}_+ f\mathcal{W}_-)$$
$$+\frac{1}{4!}Tr[(f\mathcal{W}_+)^3 f\mathcal{W}_- + (f\mathcal{W}_+)^2(f\mathcal{W}_-)^2 + (f\mathcal{W}_+ f\mathcal{W}_-)^2] + \cdots$$

These are the usual truncated Wightman functions of $\Theta(f)$ written as traces !

4. Discussion

As remarked in the introduction it is very desirable to know as much as possible about positivity properties of the considered Wightman functional $\mathcal{W}_\Theta$. For example we know that $\mathcal{W}_\Theta$ is twice divisible because there exists a field $\Theta_{\frac{1}{2}}$ with central charge $\frac{1}{2}$ .These divisibility properties should show up in $E^T(f)$ too . As a first step in this direction we cast the expression for $E^T(f)$ in a form which resembles very much a Lévy-Khintchine representation for infinitely divisible characteristic functions .

Starting from $\mathcal{W} = \mathcal{W}_+ - \Delta$ we get in a first step

$$\cos(f\Delta) - i\sin(f\Delta)\Delta^{-1}\mathcal{W} = \cos(f\Delta) + i\sin(f\Delta) - i\sin(f\Delta)\Delta^{-1}\mathcal{W}_+$$
$$= e^{if\Delta} - i\sin(f\Delta)\Delta^{-1}\mathcal{W}_+$$

and therefore

$$\ln[\cos(f\Delta) - i\sin(f\Delta)\Delta^{-1}\mathcal{W}] + if(\mathcal{W}_+ - \Delta)$$
$$= \ln[e^{if\Delta} - i\sin(f\Delta)\Delta^{-1}\mathcal{W}_+] - \ln e^{if\Delta} + if\mathcal{W}_+$$
$$= \ln[1 + \frac{1}{2}(e^{-2if\Delta} - 1)\Delta^{-1}\mathcal{W}_+] + if\mathcal{W}_+\,.$$

In a second step we use the representation

$$-\ln(1+u) = \int_0^\infty d\xi \frac{e^{-\xi}}{\xi}(e^{-\xi u} - 1)$$

to obtain

$$E^T(f) = \int_0^\infty d\xi\, \frac{e^{-\xi}}{2\xi} Tr\{\exp[-\frac{\xi}{2}(e^{-2if\Delta} - 1)\Delta^{-1}\mathcal{W}_+] - 1 - i\xi f\mathcal{W}_+\}\,.$$

This looks very similar to the Lévy-Khintchine representation for infinitely divisible characteristic functions . But even if we fix f and restrict ourselves to the truncated characteristic function $e^T(t) := E^T(tf)$ we were not able to prove that $e^T(t)$ is conditionally positive definite.

Acknowledgement

I want to thank D. Buchholz very much for his constant encouragement; he contributed the main idea and many of the details.

REFERENCES

1. M. Lüscher and G. Mack, *The energy momentum tensor of a critical quantum field theory in 1+1 dimensions* , *Hamburg* unpublished (1976).
2. B. Schroer, *A Trip to Scaling Land*, Proc. V. Brasilian Symposium on Theoretical Physics, E. Ferreira ed., Livros tecnicos e scientificos, Rio de Janeiro (1974).
3. D. Friedan, Z. Qui and S. Shenker, *Conformal Invariance, Unitarity, and Critical Exponents in Two Dimensions*, Phys. Rev. Lett. **52** (1984), 1575 - 1578.
4. G. C. Hegerfeldt, *Prime field decomposition and infinitely divisible states on Borchers' tensor algebra*, Commun. math. Phys. **45** (1975), 137 - 151.
5. K. Baumann and G. C. Hegerfeldt, *Is the Wick square infinitely divisible?*, J. Math. Phys. **18** (1977), 171 -172.

INSTITUT FÜR THEORETISCHE PHYSIK UNIVERSITÄT GÖTTINGEN D-3400 GÖTTINGEN, GERMANY

Acknowledgement

REFERENCES

COVARIANCE SUB-ALGEBRAS CONNECTED WITH SYMMETRY GROUPS OF C*-ALGEBRAS

H.J. BORCHERS

ABSTRACT. Let $\{\mathcal{A}, G, \alpha\}$ be a C*-dynamical system with G a topological group. Let $\mathcal{A}_c^*$ be the subset of elements of $\mathcal{A}^*$ on which α'_g acts in a strongly continuous manner. It is shown that $\mathcal{A}_c^*$ is the pre-dual of a von Neumann-algebra $\mathcal{M}_c \subset \mathcal{A}^{**}$. The algebra $\mathcal{M}_c$ consists, except for some trivial reduction, of those elements of $\mathcal{A}^{**}$ which map, by right and left multiplication, the submodule $\mathcal{A}_c^*$ of $\mathcal{A}^*$ into itself. The algebra $\mathcal{M}_c$ is covariant under the automorphisms α_g .

Dedicated to John Klauder at the occasion of his 60th birthday

I. Introduction

In [4] we investigated W*-dynamical systems $\{\mathcal{M}, G, \alpha\}$, where $\mathcal{M}$ is a von Neumann-algebra, G a topological group, and α a homomorphism of G into $\mathrm{Aut}(\mathcal{M})$. But we did not assume that the action $g \to \alpha_g x$, $g \in G$, $x \in \mathcal{M}$ is weakly continuous. This situation is of particular interest if one has a C*-dynamical system $\{\mathcal{A}, G, \alpha\}$ and $\mathcal{M}$ is the enveloping von Neumann-algebra, $\mathcal{M} = \mathcal{A}^{**}$. One usually is interested in finding covariant representations $\{\pi, U(g), \mathcal{H}\}$ such that

(i) $U(g)$ is a continuous unitary representation of the group G and
(ii) $U(g)$ implements the automorphisms α_g i.e., we have

$$\pi(\alpha_g x) = U(g)\pi(x)U^*(g), \quad g \in G,\ x \in \mathcal{A}.$$

If one is dealing with a von Neumann-algebra $\mathcal{M}$, one can ask for normal covariant representations.
In order to answer this question we introduced the space $\mathcal{A}_c^*$ as follows:

$$\mathcal{A}_c^* := \{\phi \in \mathcal{A}^*; g \to \phi \circ \alpha_g \quad \text{is continuous at the identity of } G \text{ if } \mathcal{A}^* \text{ is furnished with the norm-topology}\}$$

and, in the case of von Neumann-algebras, the space $\mathcal{M}_{*,c}$ correspondingly. The existence of covariant representations is described by the following result (see [4]).

Theorem:
Let $\{\mathcal{A}, G, \alpha\}$ be a C^*-dynamical system and $\{\pi, \mathcal{H}\}$ a representation of $\mathcal{A}$. Then we can find a covariant representation $\{\hat{\pi}, U(g), \hat{\mathcal{H}}\}$ with

(i) $\hat{\pi}$ *and* π *are quasi-equivalent*

(ii) $U(g)$ *is a continuous unitary representation on* $\hat{\mathcal{H}}$ *implementing the automorphisms* α_g

iff

(α) *all* π*-normal states* $F(\pi)$ *belong to* $\mathcal{A}_c^*$

(β) $F(\pi)$ *is invariant under the transposed group-action.*

Allowing to pass to quasi-equivalent representations has two advantages, namely, it solves all multiplicity problems and, furthermore, it avoids the necessity to look at projective representations of the group.

An analogous result can also be formulated for von Neumann-algebras.
Since the theorem uses only properties of $\mathcal{A}^*$ or $\mathcal{M}_*$, respectively, it is sufficient to investigate the von Neumann-case, since $\mathcal{A}^*$ is the pre-dual of $\mathcal{A}^{**}$.

For this set we derived the following properties in [4]:

(1) $\mathcal{M}_{*,c}$ *is a norm-closed linear sub-space of* $\mathcal{M}_*$.

(2) $\mathcal{M}_{*,c}$ *is invariant under the action of the group:* $\alpha'_g\, M_{*,c} = \mathcal{M}_{*,c}\,,\ g \in G$.

(3) *With* $\phi \in \mathcal{M}_{*,c}$ *also* ϕ^* *and* $|\phi|$ *are contained in* $\mathcal{M}_{*,c}$. *This means in particular* $\mathcal{M}_{*,c}$ *is linearly generated by its positive elements.*

In this note we want to derive some further properties of the space $\mathcal{M}_{*,c}$. We will show that $\mathcal{M}_{*,c}$ is isomorphic to the pre-dual of a von Neumann-algebra $\mathcal{N}_c$ and that this algebra is equivalent to a sub-von Neumann algebra $\mathcal{M}_c \subset \mathcal{M}$. This algebra is also characterized as follows:

$\mathcal{M}_c$ *is the set of elements in* $\mathcal{M}$ *mapping the sub-module* $\mathcal{M}_{*,c}$ *of* $\mathcal{M}_*$ *into itself, divided by the annihilator of* $\mathcal{M}_{*,c}$.

The investigations are based on the Tomita-Takesaki-theory [12], [11] and in particular, on the theory of the natural cone introduced by Araki [1], [2] and Connes [5], [6]. Especially, we need the characterization of von Neumann-algebras in terms of self-dual cones which is due to A.Connes [6]. The extension of these results from faithful normal states to faithful normal weights can be found in the paper of U.Haagerup [7].

In the next section we investigate that part of the natural cone which is induced from $\mathcal{M}_{*,c}$. Using the characterization of von Neumann-algebras by properties of the natural cone we show in the last section that $\mathcal{M}_{*,c}$ is the pre-dual of a von Neumann-algebra. Moreover we show that this von Neumann-algebra is isomorphic to a sub-von Neumann-algebra $\mathcal{M}_c$ of $\mathcal{M}$.

II. $\mathcal{M}_{*,c}$ and the cone $\mathcal{H}_c^+$

We identify $\mathcal{M}$ with its standard representation on a Hilbert-space $\mathcal{H}$ given by a normal faithful weight on $\mathcal{M}$. Let J denote the modular involution and $\mathcal{H}^J$ the space of vectors real under J. By $\mathcal{H}^+$ we denote the natural cone given by that standard representation. $\mathcal{H}^+$ is a self-dual cone in $\mathcal{H}^J$ and there exists to every positive linear functional $\omega \in \mathcal{M}_*^+$ a unique vector $\xi_\omega \in \mathcal{H}^+$, such that $\omega(x) = (\xi_\omega, x\,\xi_\omega)$ for all $x \in \mathcal{M}$. The map $\mathcal{M}_*^+$ onto

$\mathcal{H}^+$, given by the last formula, is a topological homeomorphism between the two cones, wich preserves order.

By $\mathcal{H}_c^+$ we denote the set of $\xi \in \mathcal{H}^+$ such that $\omega_\xi \in \mathcal{M}_{*,c}$. In [4] we derived for $\mathcal{H}_c^+$ the following properties:

(1) *$\mathcal{H}_c^+$ is a closed sub-cone of $\mathcal{H}^+$.*

(2) *$\mathcal{H}_c^J = \mathcal{H}_c^+ - \mathcal{H}_c^+$ is a closed linear sub-space of $\mathcal{H}^J$.*

(3) *$\mathcal{H}_c^+$ is a self-dual cone in $\mathcal{H}_c^J$.*

(4) *Let p_c denote the projection onto $\mathcal{H}_c = \mathcal{H}_c^J + \imath\mathcal{H}_c^J$ then for every $\xi \in \mathcal{H}^+$ the vector $p_c\xi$ is contained in $\mathcal{H}_c^+$. Hence by definition of $\mathcal{H}_c^+$ we obtain $p_c\mathcal{H}^+ = \mathcal{H}_c^+$.*

We first introduce two algebras which we need later

II.1 Definition:

(1) *We define*

$$\mathcal{M}_c^0 = \{A \in \mathcal{M}\,;\, [A, p_c] = 0\}.$$

(2) *We put*

$$\mathcal{M}_m^0 = \{A \in \mathcal{M}\,;\, A\omega \in \mathcal{M}_{*,c}\ \ \omega A \in \mathcal{M}_{*,c}\ \ \forall \omega \in \mathcal{M}_{*,c}\}.$$

(3) *Let E_c be the smallest projection in $\mathcal{M}$ with $E_c p_c = p_c$.*

All these objects are invariant under α_g. First note that both sets are von Neumann algebras. The two algebras are not different. We have

II.2 Lemma:

(1) *The two algebras $\mathcal{M}_c^0$ and $\mathcal{M}_m^0$ coincide.*

(2) *Every element in $\mathcal{M}_c^0$ commutes with E_c.*

Proof:

(1) If $A \in \mathcal{M}_c^0$ then it commutes with p_c which implies that with $\xi \in \mathcal{H}_c$ also $A\xi$ belongs to $\mathcal{H}_c$, hence $\mathcal{M}_c^0 \subset \mathcal{M}_m^0$. Conversely let $U \in \mathcal{M}_m^0$ be unitary then $UJUJ\mathcal{H}_c^+ = \mathcal{H}_c^+$ which implies that $UJUJ$ commutes with p_c. Since p_c and J commute we obtain $[p_c, U] + J[p_c, U]J = 0$. Replacing U by $\imath U$ we see that U belongs to $\mathcal{M}_c^0$. Since the von Neumann algebra is generated by its unitaries, it follows that $\mathcal{M}_m^0 \subset \mathcal{M}_c^+$.

(2) $A \in \mathcal{M}_c^0$ implies that it maps the Hilbert space $\mathcal{H}_c$ into itself. Consequently it maps also *closure* $\{\mathcal{M}'\mathcal{H}_c\} = E_c\mathcal{H}$ into itself. □

II.3 Definition:

We set

$$\mathcal{M}_c = \mathcal{M}_c^0 E_c.$$

Again this is a sub-von Neumann algebra of $\mathcal{M}$. In the investigations of this section we will use only the properties (1)...(4) of $\mathcal{H}_c^+$, so that the results can be used for arbitrary sub-cones of $\mathcal{H}^+$ with these properties. For the following investigations we need some

II.4 Notations:

Let f be a face of $\mathcal{H}_c^+$, then we denote by f^c the complementary face in $\mathcal{H}_c^+$, it is:

$$f^c = \{\xi \in \mathcal{H}_c^+ ; \xi \perp f\}.$$

By $f^\perp$ we denote the complementary face of f in $\mathcal{H}^+$:

$$f^\perp = \{\xi \in \mathcal{H}^+ ; \xi \perp f\}.$$

We remark that the map $f \to f^c$ is only defined for faces of $\mathcal{H}_c^+$, while $f \to f^\perp$ is also defined for faces of $\mathcal{H}^+$.

For every face f of $\mathcal{H}_c^+$ we associate two faces of $\mathcal{H}^+$, namely, $F^+(f) = (f^c)^\perp$ and $F^-(f) = (f^\perp)^\perp$.

Note that $f^c, f^\perp, F^-(f)$, and $F^+(f)$ are closed faces. If possible we will denote the faces of $\mathcal{H}_c^+$ by small letters and those of $\mathcal{H}^+$ by capital letters.

The elementary properties of these faces are described in the following

II.5 Lemma:

For any face f of $\mathcal{H}_c^+$ we obtain:

(a) $f \subset F^-(f) \subset F^+(f)$.

(b) $\xi \in \mathcal{H}^+$ *then* $p_c\xi \in (f^c)^c$ *iff* $\xi \in F^+(f)$.

(c) $p_c F^-(f) = p_c F^+(f) = (f^c)^c$.

(d) $(F^+(f))^\perp = F^-(f^c)$ *and* $(F^-(f))^\perp = F^+(f^c)$.

(e) *The set* $\{\eta \in \mathcal{H}^+;$ such that there exists $\xi \in f$ with $\eta \leq \xi\}$ *is dense in* $F^-(f)$.

(f) f *is dense in* $(f^c)^c$.

Proof:

(a) Note first that the operations $f \to f^c$ and $f \to f^\perp$ reverse the order of inclusion. Moreover, the relation $f \subset (f^c)^c$ follows directly from the definition. From $f^c \subset f^\perp$ we conclude therefore, $f \subset (f^\perp)^\perp \subset (f^c)^\perp$, which is the first statement by the definition of F^- and F^+.

(b) We know that $\xi \in \mathcal{H}^+$ implies $p_c\xi \in \mathcal{H}_c^+$. Hence $p_c\xi \in (f^c)^c$ iff $(p_c\xi, \eta) = 0$ for all $\eta \in f^c \subset \mathcal{H}_c^+$. Consequently $(p_c\xi, \eta) = (\xi, \eta) = 0$ and therefore $\xi \in (f^c)^\perp = F^+(f)$ and vice versa.

(c) The last result implies $p_c F^+(f) = (f^c)^c$. On the other hand $f^c \subset f^\perp$ and $p_c f^\perp = f^c$, since $p_c f = f$. Therefore, $\eta \in (f^c)^c$ and $\xi \in f^\perp$ implies $(\eta, \xi) = (\eta, p_c\xi) = 0$, from which we conclude $\eta \in F^-(f)$. Hence $(f^c)^c \subset F^-(f) \subset F^+(f)$ which yields (c).

(d) We have by definition $F^+(f) = (f^c)^\perp$ and hence $(F^+(f))^\perp = \left((f^c)^\perp\right)^\perp = F^-(f^c)$. Inserting now f^c for f we obtain $(F^+(f^c))^\perp = F^-((f^c)^c)$. Now remark, if $\xi \in f^\perp$ then $(\xi, \eta) = 0$ for all $\eta \in f$ and hence $p_c\xi \in f^c$. But this implies for $\eta \in (f^c)^c$ the equation $(\xi, \eta) = (p_c\xi, \eta) = 0$ and hence $f^\perp = \left((f^c)^c\right)^\perp$ or $F^-(f) = F^-((f^c)^c)$. This leads to $F^-(f) = (F^+(f^c))^\perp$ or $(F^-(f))^\perp = (F^+(f^c))^{\perp,\perp} = (F^+(f^c))$ since $(F^+(f^c))$ is a closed face of $\mathcal{H}^+$. (See A.Connes, [6] Théorème 4.2).

(e) Let F be the face $\{\eta \in \mathcal{H}^+ ; \eta \leq \xi \, for some \, \xi \in f\}$. Then by the above mentioned result of A.Connes $\bar{F} = (F^{\perp})^{\perp}$. Let $\eta \in f^{\perp}$, then $(\eta, \xi) = 0$ for all $\xi \in f$ and hence for all $\xi \in F$, since the scalar product preserves order. This shows $f^{\perp} \subset F^{\perp}$. On the other hand by definition of F we have $f \subset F$ and hence $F^{\perp} = f^{\perp}$. But this yields $\bar{F} = (F^{\perp})^{\perp} = (f^{\perp})^{\perp} = F^{-}(f)$.
(f) By the proof of (c) we know $(f^c)^c \subset F^{-}(f)$. Hence by (e) we find for a given $\xi \in (f^c)^c$ and $\epsilon > 0$ elements $\eta \in F^{-}(f)$ and $\zeta \in f$ such that $\eta \leq \zeta$ and $\|\xi - \eta\| < \epsilon$. Now $p_c\eta \leq p_c\xi = \zeta$ and hence $p_c\eta \in f$. On the other hand $\|\xi - p_c\eta\| = \|p_c(\xi - \eta)\| \leq \|\xi - \eta\| < \epsilon$. This shows that f is dense in $(f^c)^c$. □

In order to formulate the next results we need some more

II.6 Notations:

(a) *Let F be a face in $\mathcal{H}^+$ then $\sum_0^3 (\imath)^n F$ is a linear sub-space of $\mathcal{H}$. The projection onto its closure will be denoted by P_F.*

(b) *If we perform the same construction in $\mathcal{H}_c$ with a face f then we will denote the corresponding projection by p_f.*

II.7 Remark:

If F is a face of $\mathcal{H}^+$ (the natural cone of a von Neumann-algebra $\mathcal{M}$) then P_F is of the form $P_F = EJEJ$, where E is a projection in $\mathcal{M}$. In particular, $P_F\mathcal{H}$ is the standard representation-space for the algebra $\mathcal{M}_E$. In particular, $\bar{F}$ is the natural cone of $\mathcal{M}_E$. This implies that $\sum_0^3 (\imath)^n \bar{F}$ is closed. (For details see A.Connes [6] Théorème 4.2 and Lemme 4.4). At the moment we do not know whether or not $\sum_0^3 (\imath)^n \bar{f}$ is a closed sub-space of $\mathcal{H}_c$. That it is, indeed, closed is a consequence of the following

II.8 Lemma:

For any face f of $\mathcal{H}_c^+$ follows:

(a) $[P_{F^+(f)}, p_c] = [P_{F^-(f)}, p_c] = 0\,.$
(b) $p_f = p_c P_{F^-(f)} = p_c P_{F^+(f)}\,.$
(c) $[P_{F^+(f)}, P_{F^-(f)}] = 0\,.$
(d) $P_{F^-(f)} = P_{F^+(f)} P_{F^-(\mathcal{H}_c^+)}\,.$

Proof:
(a) From Lemma II.5 we know $p_c P_{F^+(f)} \mathcal{H}^+ = \bar{f} \subset P_{F^+(f)}\mathcal{H}^+$. Hence we obtain $p_c P_{F^+(f)}\mathcal{H} = p_c \sum (\imath)^n F^+(f) = \sum (\imath)^n \bar{f} \subset P_{F^+(f)}\mathcal{H}$. But this implies $p_c P_{F^+(f)}\mathcal{H} = P_{F^+(f)} p_c P_{F^+(f)}\mathcal{H}$ and hence $P_{F^+(f)} p_c P_{F^+(f)} = p_c P_{F^+(f)}$, which shows that the two projections commute. The same argument holds for $P_{F^-(f)}$.
(b) Due to the commuting of p_c and $P_{F^-(f)}$ the product is a projection. Since $p_c F^-(f)\mathcal{H}^+ = \bar{f}$ we learn that $p_c P_{F^-(f)}\mathcal{H} = \sum (\imath)^n \bar{f}$ is closed and equal to $p_f\mathcal{H}$. The same argument is again true for $P_{F^+(f)}$.
(c) First note $f^c \subset \mathcal{H}_c^+$, and hence $F^+(f) \supset \mathcal{H}_c^{+,\perp}$. Let now $\xi \in \mathcal{H}^+$ then by Remark II.7 one has $P_{F^-(\mathcal{H}_c^+)}\xi \in F^-(\mathcal{H}_c^+)$ and $P_{F^+(f)} P_{F^-(\mathcal{H}_c^+)}\xi \in F^+(f)$. Take $\eta \in (\mathcal{H}_c^+)^{\perp}$ then both equations lead to $(\eta, P_{F^+(f)} P_{F^-(\mathcal{H}_c^+)}\xi) = (\eta, P_{F^-(\mathcal{H}_c^+)}\xi) = 0$, hence we have $P_{F^+(f)} P_{F^-(\mathcal{H}_c^+)}\xi \in F^-(\mathcal{H}_\xi^+)$ for every $\xi \in \mathcal{H}^+$. Since such vectors generate $\mathcal{H}$ the equation

$P_{F^-(\mathcal{H}_c^+)}P_{F^-(\mathcal{H}_c^+)} = P_{F^-(\mathcal{H}_c^+)}$ follows, which is equivalent to (c).
(d) Since for $\xi \in \mathcal{H}^+$ the vector $P_{F^-(\mathcal{H}_c^+)}\xi$ belongs to $F^-(\mathcal{H}_c^+)$ there exist for every $\epsilon > 0$ vectors $\eta \in F^-(\mathcal{H}_c^+)$ and $\zeta \in \mathcal{H}_c^+$ with $\eta \leq \zeta$ and $\|P_{F^-(\mathcal{H}_c^+)}\xi - \eta\| \leq \epsilon$. So we obtain $\|P_{F^+(f)}P_{F^-(\mathcal{H}_c^+)}\xi - P_{F^+(f)}\eta\| \leq \epsilon$ and $P_{F^+(f)}\eta \leq P_{F^+(f)}\zeta \in \bar{f}$. This shows $P_{F^+(f)}P_{F^-(\mathcal{H}_c^+)}\mathcal{H}^+ \subset F^-(f)$. Since the inverse inclusion $F^-(f) \subset F^-(\mathcal{H}_c^+)$ holds both sides coincide. □

As a consequence of the last lemma we obtain:

II.9 Corollary:

$\mathcal{H}_c^+$ is a homogeneous cone in the sense of A.Connes ([6] Definition 5.1)

Proof:
If f is any face of $\mathcal{H}_c^+$ then we have to verify the equation

$$\exp\{t(p_f - p_{f^c})\}\mathcal{H}_c^+ = \mathcal{H}_c^+$$

for every $t \in \mathbf{R}$. Since $\mathcal{H}^+$ is a homogeneous cone it follows for every face F of $\mathcal{H}^+$

$$\exp\{t(P_F - P_{F^\perp})\}\mathcal{H}^+ = \mathcal{H}^+.$$

Choosing $F = F^+(f)$ then by Lemma II.2 $(F^+(f))^\perp = F^-(f^c)$. Multiplying the above equation by p_c we obtain

$$\begin{aligned}\mathcal{H}_c^+ &= p_c\mathcal{H}^+ = p_c \exp\{t(P_{F^+(f)} - P_{F^-(f^c)})\}\mathcal{H}^+ \\ &= \exp\{t(p_f - p_{f^c})\}p_c\mathcal{H}^+ = \exp\{t(p_f - p_{f^c})\}\mathcal{H}_c^+.\end{aligned}$$

This gives the desired result.

III. Orientation of $\mathcal{H}_c^+$

For the investigation of this section we introduce (see A. Connes [6]) the sets

III.1 Definition:

We denote by
(a) $\mathcal{D}(\mathcal{H}^+) = \{\delta \in \mathcal{L}(\mathcal{H});\ e^{t\delta}\mathcal{H}^+ \subset \mathcal{H}^+ \quad \forall t \in \mathbf{R}\}$
(b) $\mathcal{D}(\mathcal{H}_c^+) = \{\delta \in \mathcal{L}(\mathcal{H}_c);\ e^{t\delta}\mathcal{H}_c \subset \mathcal{H}_c \quad \forall t \in \mathbf{R}\}$
where $\mathcal{L}(\mathcal{H})$ denotes the set of all bounded linear operators on $\mathcal{H}$.

For facially homogeneous self-dual cones the following characterization of $\mathcal{D}(\mathcal{H}^+)$ has been given by A.Connes:

(*a*) *Let $\mathcal{H}$ be a facially homogeneous self-dual cone and $\delta \in \mathcal{L}(\mathcal{H})$ then $\delta \in \mathcal{D}(\mathcal{H}^+)$ iff*
 (i) $J\delta = \delta J$ and
 (ii) *For $\xi, \eta \in \mathcal{H}^+$ with $(\xi, \eta) = 0$ it follows $(\delta\xi, \eta) = 0$.*
(*b*) *If $\mathcal{H}^+$ is the natural cone of $\mathcal{M}$ then $\delta \in \mathcal{D}(\mathcal{H}^+)$ iff $\delta = x + JxJ$ for a suitable $x \in \mathcal{M}$.*

We start the investigation from Remark II.7 showing that to every face $f \subset \mathcal{H}_c^+$ is associated a minimal face $F^-(f)$ and hence a projection $E(f) \in \mathcal{M}$ with $P(F^-(f)) = E(f)JE(f)J$. We first want to investigate these special projections:

III.2 Lemma:

Let f be a closed face of $\mathcal{H}_c^+$ and let $F^-(f)$ be the minimal closed face of $\mathcal{H}^+$ containing f. Let $E(f) \in \mathcal{M}$ be the unique projection such that $P(F^-(f)) = E(f)JE(f)J$ holds then $E(f) \in \mathcal{M}_c$.

Proof:

We know from Lemma II.8 that $P(F^-(f)) = E(f)JE(f)J$ commutes with p_c. Then δ_{F^-} commutes with p_c. On the other hand, we know that δ_{F^-} is of the form $\delta_{F^-} = 1/2(E(f) + JE(f)J)$.

$e^{t/2E(f)}Je^{t/2E(f)}J$ maps $\mathcal{H}_c^+$ onto $\mathcal{H}_c^+$, consequently for $\omega \in \mathcal{M}_{*,c}^+$ we have

$$e^{t/2E(f)}\omega e^{t/2E(f)}Je^{tE(f)}J \in \mathcal{M}_{*,c}^+ .$$

The left-hand side of this expression has some analyticity property namely the first factors are entire analytic in t and the last is entire antianalytic. Therefore the inclusion holds also for complex t. But this can only be true if $e^{t/2E(f)}\omega e^{t/2E(f)} \in \mathcal{M}_{*,c}^+$. Hence $e^{t/2E(f)} \in \mathcal{M}_m^0$ and by differentiation $E(f) \in \mathcal{M}_m^0$. But since $E(f)$ is smaller than E_c we obtain $E(f) \in \mathcal{M}_c$. □

III.3 Definition:

We put

$$\mathcal{D}_c(\mathcal{H}^+) = \{\delta \in \mathcal{D}(\mathcal{H}^+) \text{ with } \delta = A + JAJ,\ A \in \mathcal{M}_c\}$$

Using the last result we obtain:

III.4 Proposition:

(1) *For every $\delta \in \mathcal{D}(\mathcal{H}^+)$ one has $\pi_0(\delta) \in \mathcal{D}(\mathcal{H}_c^+)$ with*

$$\pi_0(\delta) = p_c \delta p_c$$

where p_c denotes the projection onto $\mathcal{H}_c$.

(2) *If we denote by π_c the restriction of π_0 to $\mathcal{D}_c(\mathcal{H}^+)$ then π_c defines a bijection between $\mathcal{D}_c(\mathcal{H}^+)$ and $\mathcal{D}(\mathcal{H}_c^+)$.*

Proof:

(1) Since p_c and δ both commute with J, also $p_c\delta p_c$ commutes with J. Assume next $\xi, \eta \in \mathcal{H}_c^+$ with $(\xi, \eta) = 0$ then follows from $(\xi, p_c\delta p_c\eta) = (\xi, \delta\eta) = 0$ that $p_c\delta p_c$ belongs to $\mathcal{D}(\mathcal{H}_c^+)$.

(2) The spaces $\mathcal{D}(\mathcal{H}^+)$ and $\mathcal{D}(\mathcal{H}_c^+)$ are both invariant under involution and weakly closed. Therefore, one can pass to the self-adjoint and positive part. The sets $\mathcal{D}_{sa}(\mathcal{H}^+)_1^+$ and $\mathcal{D}_{sa}(\mathcal{H}_c^+)_1^+$ are both weakly compact and convex. The extremal elements are of the form

$$\delta_F = \frac{1}{2}(1_{\mathcal{H}} + P(F) - P(F^{\perp}))$$

or

$$\delta_f = \frac{1}{2}(1_{\mathcal{H}_c} + p(f) - p(f^{\perp}))$$

respectively. Bellissard and Iochum [3] have shown that for every facially homogeneous self-dual cone $\mathcal{H}^+$ the elements in $\mathcal{D}_{sa}(\mathcal{H}^+)$ permit an integral representation

$$\delta = \int_{-\|\delta\|}^{+\|\delta\|} \lambda d\delta_{F(\lambda)}.$$

This holds in particular for our cone $\mathcal{H}_c^+$. Since we know that every δ_f is of the form $\pi_0(\delta_{F(f)})$ with $\delta_{F(f)} = \frac{1}{2}(1_{\mathcal{H}} + P(F^+(f)) - P(F^-(f^c)))$ we see that every self-adjoint element in $\mathcal{D}(\mathcal{H}_c^+)$ is of the form $\pi_0(\delta)$.

Moreover, we know by Lemma III.2

$$P(F^+(f)) = (E(f) + 1 - E_c) + J(E(F) + 1 - E_c)J$$
$$P(F^-(f^c)) = E(f^c) + JE(f^c)J$$

which implies $\delta_{F^-(f)} \in \mathcal{D}_c(\mathcal{H}^+)$. Consequently π_c is surjective.

Since π_c is the multiplication by the projector p_c one has $\|\pi_c\delta\| \leq \|\delta\|$. On the other hand, the carrier of p_c in $\mathcal{M}_c$ is 1. Therefore the reconstruction of δ from $\pi_c\delta$ by means of the integral representation shows that (for self-adjoint) δ and $\pi_c\delta$ have the same spectrum. This shows $\|\delta\| = \|\pi_c\delta\|$ and hence ker $\pi_c = 0$. Consequently π_c can be extended to all δ in $\mathcal{D}_c(\mathcal{H}^+)$ and defines an isomorphism. □

Combining all results we obtain:

III.5 Theorem:

(1) *The cone $\mathcal{H}_c^+$ is facial homogeneous and oriented and is, therefore, the natural cone of a von Neumann-algebra $\mathcal{N}_c$.*

(2) *The von Neumann-algebra $\mathcal{N}_c$ is isomorphic to the sub-von Neumann-algebra $\mathcal{M}_c \subset \mathcal{M}_{E_c}$ where*

  (α) *E_c is the smallest projection in $\mathcal{M}$ which is larger than the support projections of all states belonging to $\mathcal{M}_{*,c}$;*

  (β) *$\mathcal{M}_c$ is the set of operators in $\mathcal{M}_{E_c}$ which are right and left multipliers of $\mathcal{M}_{*,c}$;*

  (γ) *The automorphisms α_g are automorphisms of $\mathcal{M}_c$.*

(3) *$\mathcal{M}_{*,c}$ is the pre-dual of $\mathcal{M}_c$.*

Proof:

We know from Lemma II.9 that $\mathcal{H}_c^+$ is a homogeneous cone. From Proposition III.4 we know that we obtain $\mathcal{D}(\mathcal{H}_c^+)$ by multiplying all elements $A + JAJ, \quad A \in \mathcal{M}_c$ by P_c. Since this operator commutes with J and with $\mathcal{M}_c$ and since π_c is a bijection we see that $\mathcal{H}_c^+$ is oriented by the orientation of $\mathcal{M}_c$. The isomorphism property of π_c shows the second statement of the theorem. The statement (γ) is a consequence of the invariance of $\mathcal{M}_{*,c}$.

The third statement is due to the fact that $\mathcal{H}_c^+$ generates all functionals in $\mathcal{M}_{*,c}$ and that $\mathcal{H}_c^+$ is the natural cone of $\mathcal{M}_c$. □

III.6. Concluding Remarks:

Looking at a W^*-dynamical system $\{\mathcal{M}, G, \alpha\}$ with properties described before, we saw that the space $\mathcal{M}_{*,c}$ is predual of a von Neumann algebra $\mathcal{N}_c$ which is isomorphic to a subalgebra $\mathcal{M}_c \subset \mathcal{M}$. We have derived some more structure. In Proposition III.4 we have constructed a projection form $\mathcal{D}(\mathcal{H}^+)$ onto $\mathcal{D}(\mathcal{H}_c^+)$. Since the orientation on both cones are compatible with each other we have, indeed, a projection from $\mathcal{M}$ onto $\mathcal{N}_c \cong \mathcal{M}_c$. Such a projection is the same as a conditional expectation. The kernel of this projection is in general quite large. If we restrict to the algebra $\mathcal{M}_{E_c}$ then we obtain by definition of E_c a faithful normal conditional expectation $\mathcal{M}_{E_c} \to \mathcal{M}_c$.

In general the two algebras $\mathcal{M}_{E_c}$ and $\mathcal{M}_c$ do not coincide. The deviation of the smaller algebra can be characterized by the index of V. Jones [9] or its generalization by Kosaki [10]. Whether this index has some interpretation in terms of $(\mathcal{M}_{E_c})_*$, considered as orbit-space of the group G, is not known to me.

References

1. H. Araki, *Some Properties of Modular Conjugation Operator of a von Neumann Algebra and a Radon-Nikodym Theorem with a Chain Rule*, Pac. J. Math. **50** (1974), 309-354.
2. H. Araki, *Positive Cones for von Neumann Algebras In: Operator Algebras and Applications*, Proc. of Symposia **38** part 2, A.M.S. Providence (1982).
3. J. Bellissard and B. Iochum, *Spectral theory for facially homogeneous symmetric selfdual cones*, Math. Scand. **45** (1979)), 118-126.
4. H. J. Borchers, *C*-Algebras and Automorphism Groups*, Commun. Math. Phys. **88** (1983), 95-103.
5. A. Connes, *Sur le théorème de Radon Nikodym pour les poids normaux fidèles semifinis*, Bull. Sci. Math 2ème Sér. **97** (1973), 253-258.
6. A.Connes, *Caractérisation des algèbres de von Neumann comme espaces vectoriels ordonnés*, Ann. Inst. Fourier **26** (1974), 121-155.
7. U. Haagerup, *The standard form of von Neumann algebras*, Math. Scand. **37** (1975), 271-283.
8. B. Iochum, *Cones Autopolaires et Algèbres de Jordan*, Thèse Université de Provence (1982).
9. V.F.R. Jones, *Index for subfactors*, Invent. Math. **71** (1983), 1-25.
10. H. Kosaki, *Extension of Jones' theory of index to arbitrary factors*, J. Funct. Anal. **66** (1986), 123-140.
11. M.Takesaki, *Tomita's Theory of Modular Hilbert Algebras and its Applications*, Lecture Notes in Math. Vol.**128**, (1970) Springer-Verlag, Heidelberg.
12. M.Tomita, *Standard Forms of von Neumann Algebras*, Fifth Functional Analysis Symposium of Math. Soc. of Japan, Sendai (1967).

Institut für Theoretische Physik, Universität Göttingen, D 3400 Göttingen, GERMANY

In [illegible]

References

[illegible]

[illegible] Comm. [illegible] Math. [illegible]

AFFINE COHERENT STATES AND WAVELETS

INGRID DAUBECHIES

ABSTRACT. Wavelets have many roots in older theories. One of these roots is given by the affine coherent states of Aslaksen and Klauder. This paper presents a short historic review of wavelets and their links to physics, mathematics and engineering.

Contribution in honor of the 60th birthday of John R. Klauder

In the last five or ten years "wavelets" have attracted increasing attention; at present they are considered a "hot" topic. Why this sudden success? Part of it can only be explained by forces more sociological than mathematical. Recent history has seen catastrophe theory, fractals and neural networks (to name only a few fields) capture the imagination of non-specialists, to the point where some of the new discoveries would be touted as miracle cures for almost any mathematical or engineering problem. Wavelets seem to have entered this sociological process.

Fortunately, this "sociological factor" accounts for only part of the popularity of wavelets. A far more serious reason for their appeal is that they are the result of a synthesis of ideas originating in many different fields, over the last twenty or thirty years. The term "wavelets" or "wavelet transform" stands for a mathematical tool that embodies ideas already present in pure harmonic analysis (Calderón's formula), physics (renormalization group theory, affine coherent states), electrical engineering (subband coding, constant Q-filters) and image processing (multiscale representations). The new synthesis, in the form of "wavelet theory" can readily be applied to all the above fields (and more), leading in every case to an elegant framework or a new point of view conducive to new ideas, and in some instances to significant new developments.

Because wavelets have roots in many different disciplines, "waveleteers" can come from many different backgrounds. Not all roads lead to wavelets, but many do! In my own case, the road started with affine coherent states, first introduced by Aslaksen and Klauder (1968, 1969). In this paper, I would like to give a short review of wavelets and show how they are related to the original affine coherent state constructions.

1. Affine coherent states

A family of affine coherent states in $L^2(\mathbb{R})$ is obtained by letting the $ax + b$ group act on a suitable "fiduciary vector". More precisely, choose $\psi \in L^2(\mathbb{R})$ such that

$$C_\psi = \int d\xi \ |\xi|^{-1} \ |\hat{\psi}(\xi)|^2 < \infty \quad , \tag{1}$$

where $\hat{\psi}$ denotes the Fourier transform of ψ, and define the affine coherent states $\psi^{a,b}$, for $a \in \mathbb{R}\backslash\{0\}$, $b \in \mathbb{R}$, by

$$\psi^{a,b}(x) = |a|^{-1/2} \ \psi\left(\frac{x-b}{a}\right) .$$

One checks easily that, for any $f, g \in L^2(\mathbb{R})$,

$$\int_{-\infty}^{\infty} da \ a^{-2} \int_{\mathbb{R}} db \ \langle g, \psi^{a,b}\rangle\langle\psi^{a,b}, f\rangle = 2\pi C_\psi \langle g, f\rangle \quad ,$$

which can also be rewritten as

$$\int_{-\infty}^{\infty} da \ a^{-2} \int_{\mathbb{R}} db \ \psi^{a,b}\langle\psi^{a,b}, f\rangle = 2\pi \ C_\psi f \tag{2}$$

or

$$\frac{1}{2\pi C_\psi} \int_{-\infty}^{\infty} da \ a^{-2} \int_{\mathbb{R}} db \ |\psi^{a,b}\rangle\langle\psi^{a,b}| = \mathrm{Id} \quad , \tag{3}$$

where both (2) and (3) have to be understood in a weak sense. The whole construction can be generalized to n dimensions. These affine coherent states are thus seen to be a special case of "coherent states", as described e.g. in the primer by John Klauder in Klauder and Skagerstam (1985); (2) or (3) is the corresponding "resolution of the identity". They were first introduced in Aslaksen and Klauder (1968, 1969); later, Paul (1985) showed that with a special choice for ψ they play the same role for the hydrogen atom Hamiltonian as the canonical coherent states (generated by the action of the Weyl-Heisenberg group on a Gaussian fiduciary vector) for the harmonic oscillator Hamiltonian.

2. Wavelets: the continuous transform

The term "wavelets" was coined by A. Grossmann and J. Morlet in the early 80 s. Originally, J. Morlet had proposed wavelets as a tool for signal analysis, in particular for the analysis of seismic data. A standard technique is the short-time or windowed Fourier transform, which consists in multiplying the signal $f(t')$ with a window function $g(t'-t)$ (if g is itself centered around 0, then this is centered around t), and computing the Fourier transform of this product,

$$(T^{\mathrm{win}} f)(t, \omega) = \int dt' \ f(t') \ g(t'-t) \ e^{-i\omega t'} . \tag{4}$$

The resulting function $T^{\text{win}} f$ depends on two variables, and it endeavors to give a representation of the frequency content *locally in time* (unlike the Fourier transform, from which the time-aspect is not easy to read off). This is important for so-called nonstationary signals, of which typical examples are sound recordings of speech or music. (Music notation itself is a time-frequency representation!) The windowed Fourier transform (4) can also be viewed as the result of taking inner products of f with the family of functions $g^{\omega,t}(t') = e^{i\omega t'} g(t'-t)$ which, incidentally, is a family of coherent states associated with the Weyl-Heisenberg group. These functions $g^{\omega,t}$ all consist of the same envelope function g, translated to the appropriate time t, and "filled in" with oscillations of frequency ω, as given by $e^{i\omega t'}$.

Many nonstationary signals have the feature that their high frequency features are much "shorter-lived" than lower frequency components. This is the case, for instance, for a music recording: the "attack" of the notes played on most instruments contains much higher frequencies than the main body of the note, and is short lived compared to the characteristic time period of the ground harmonic of the note. The windowed Fourier transform described above uses the same window width for all frequencies, and does not exploit the differences in longevity of the different frequency components in such signals. This is what led J. Morlet to propose a different kind of transform, which came to be named the wavelet transform. His idea was to take a reasonable g, use it as an envelope for some (fixed) number of oscillations, and to generate a new family from this one oscillating function by translating it in time and *dilating* it; compressing led to short-lived, high frequency functions, stretching to longer-lived, low frequency functions. A. Grossmann recognized a family of coherent states associated with the $ax + b$-group in this construction. Typically, J. Morlet would choose g Gaussian, e.g. $g(x) = e^{-x^2/2}$, and $\omega = 4\pi$, leading to a fiduciary vector $\psi(x) = e^{4\pi i x} e^{-x^2/2}$ for the affine coherent state family. Note that $\hat{\psi}(\xi) = \sqrt{2\pi}\ \exp\left[-\frac{1}{2}(\xi + 4\pi)^2\right]$, which violates (1); this can be corrected by defining $\psi(x) = e^{-x^2/2}\left(e^{4\pi i x} - e^{-8\pi^2}\right)$; in practice this correction term can be safely omitted, because it is so small. Using the coherent state machinery, Grossmann and Morlet, together with various collaborators, then studied this continuous wavelet transform (reconstruction, interpolation formulas, etc.) and proposed several applications. (See e.g. Grossmann and Morlet (1984), Grossmann, Morlet and Paul (1985,1986), Grossmann, Kronland-Martinet and Morlet (1989).) All these start from the wavelet transform

$$(T^{\text{wav}} f)(a,t) = \int dt'\ \frac{1}{\sqrt{|a|}}\ \psi\left(\frac{t'-t}{a}\right) f(t') \quad ;$$

one possible reconstruction formula is then (as given by (2))

$$f(s) = C \int\int \frac{da\, dt}{a^2}\ (T^{\text{wav}} f)(a,t)\ \frac{1}{\sqrt{|a|}}\ \psi\left(\frac{s-t}{a}\right). \tag{5}$$

3. Wavelets: discretely labelled families

In practice, one seldom if ever computes $(T^{\text{wav}} f)(a,t)$ for continuously varying labels a and t; instead one restricts to a discrete lattice of values. One very logical choice is to take

$$a = a_0^m \quad \text{with} \quad m \in \mathbb{Z} \quad \text{and} \quad a_0 > 1 \quad \text{fixed} \quad ;$$

since the width of the $\psi^{a,b}$ is proportional to a, the discretization of b should also be proportional to $a = a_0^m$, so that very narrow wavelets ($|m|$ large, $m < 0$) have a fine translation step, wide wavelets ($|m|$ large, $m > 0$) a large translation step, i.e.

$$b = nb_0 a_0^m \quad \text{with} \quad n \in \mathbb{Z} \quad \text{and} \quad b_0 > 0 \quad \text{fixed} \quad .$$

The corresponding family of wavelets is then

$$\psi_{m,n}(t) = a_0^{-m/2}\ \psi(a_0^{-m}t - nb_0) \quad .$$

Except for very special choices of ψ and a_0, b_0, there is no equivalent of the perfect reconstruction formulas (2) or (3) for such discretely labelled wavelet families. Even without an easy reconstruction formula, one can still study when the inner products $\langle \psi_{m,n}, f\rangle$ suffice to characterize $f \in L^2(\mathbb{R})$, and whether (and how) f can be reconstructed in a numerically stable way from these inner products. A. Grossmann found that the answer is given by using the concept "frame", proposed by Duffin and Schaeffer (1952) in a different context. Concretely, the $\psi_{m,n}$ constitute a *frame* if for some $A > 0$, $B < \infty$, and all $f \in L^2(\mathbb{R})$

$$A\|f\|^2 \leq \sum_{m,n} |\langle \psi_{m,n}, f\rangle|^2 \leq B\|f\|^2 \ . \tag{6}$$

Whenever this is satisfied, f can be computed by an algorithm with exponential convergence, the first order term of which looks like a discretized version of (5),

$$f = \frac{2}{A+B} \sum_{m,n} \langle \psi_{m,n}, f\rangle\ \psi_{m,n} + O\left(\frac{B}{A} - 1\right) \ . \tag{7}$$

If B and A are close, then the rest term in (7) can be neglected. For an extensive discussion, see Daubechies (1990).

There also exist special choices for ψ and a_0, b_0 for which $A = B$ in (6). For these tight frames, the rest term in (7) vanishes, and one has an exact reconstruction formula. Examples can be found in Daubechies, Grossmann and Meyer (1986); this collaboration was the result of a fortuitous meeting with Y. Meyer, who recognized wavelets as an old friend from harmonic analysis.

4. Harmonic analysis: Calderón's formula

Another root of wavelets is a formula in harmonic analysis, proposed in the early 60 s by Calderón, and used for the study of singular integral operators (see Calderón (1961)). It uses convolutions with scaled functions. Suppose, for simplicity, that ψ is a real, even function in $L^1(\mathbb{R}) \cap L^2(\mathbb{R})$, with $\int dx\ \psi(x) = 0$. Define, for $t > 0$,

$$\psi_t(x) = \frac{1}{t}\ \psi\left(\frac{x}{t}\right)$$

and

$$(Q_t f)(x) = (\psi_t * f)(x) = \int dy\ \frac{1}{t}\ \psi\left(\frac{x-y}{t}\right)\ f(y) \quad .$$

Then, for all $f \in L^2(\mathbb{R})$,

$$f = C \int_0^\infty \frac{dt}{t}\ Q_t Q_t f \quad , \tag{8}$$

where

$$C = \left[2\pi \int_0^\infty \frac{dt}{t}\ \hat{\psi}(t)^2\right]^{-1} \quad .$$

In many arguments where this formula is used, a partition of unity is inserted between the two Q_t operators in (8), and the integral over t is split into different contributions; the different terms resulting from these splittings can then be handled by different techniques. This very useful tool, known to the community of harmonic analysts, is in fact the resolution of identity (2) in another guise, as can easily be checked by writing out the convolutions in (8). Localization in phase space by means of dilation and translations had therefore a long tradition in this field, and it was exciting to see how it had surfaced completely independently in signal analysis. Building on this tradition, M. Frazier and B. Jawerth (1988) developed under the name "φ-transform" discrete expansions of the type described above in §3, simultaneously with but independently from Daubechies, Grossmann and Meyer (1986).

5. Orthonormal wavelet bases

The discrete version of the windowed Fourier transform (4) uses, instead of the $g^{\omega,t}(t') = e^{i\omega t'} g(t'-t)$, the discretely labelled $g_{m,n}(t') = e^{im\omega_0 t'} g(t'-nt_0)$, where m, n range over $\mathbb{Z}$, and $\omega_0, t_0 > 0$ are fixed. If g is chosen to be Gaussian, then another result of John Klauder's shows that this family spans $L^2(\mathbb{R})$ if and only if $\omega_0 t_0 \leq 2\pi$ (see Bargmann, Butera, Girardello and Klauder (1971), or also Perelomov (1971)). For $\omega_0 \cdot t_0 = 2\pi$, the family is still "overcomplete": one can remove one function (say, $g_{0,0}$), and the remainder still span all of $L^2(\mathbb{R})$ (Bacry, Grossmann and Zak (1975)). For $\omega_0 \cdot t_0 < 2\pi$, one can even remove an infinite number without losing any elements of the closed linear span. No choice of ω_0, t_0 therefore leads to an orthonormal basis for $L^2(\mathbb{R})$ if g is Gaussian. Balian (1981) and Low (1985) stated independently that this result can be generalized: if g is "nice" (to be specified below), then the $g_{m,n}$ *never* constitute an orthonormal basis. More precisely, if a family of $g_{m,n}$ constitutes an orthonormal basis ($g = \chi_{[0,1]}$, $\omega_0 = 2\pi$, $t_0 = 1$ is an obvious example), then either $\int dt\ t^2|g(t)|^2 = \infty$ or $\int d\xi\ \xi^2\ |\hat{g}(\xi)|^2 = \infty$.

After Daubechies, Grossmann and Meyer (1986), Y. Meyer set out to prove an analogous theorem, about non-existence of nice bases, for wavelets. Surprisingly (I for one was certainly surprised) he ended up with an extremely nice explicit construction in which ψ is a Schwartz function (more precisely, $\hat{\psi}$ is C^∞ and has compact support) and the $\psi_{j,k}(x) = 2^{-j/2}\ \psi(2^{-j}x - k)$; $j,k \in \mathbb{Z}$ are an orthonormal basis for $L^2(\mathbb{R})$. (See Meyer (1985), Meyer (1990).) It later turned out that J.-O. Stromberg (1982) had already constructed C^k functions ψ with exponential decay that led to orthonormal wavelet bases, but his discovery went unfortunately largely unnoticed at the time. Meyer's construction triggered other constructions, by Lemarié (1988) and Battle (1987). A different point of view, introduced by S. Mallat and borrowed from vision analysis, led to a framework, now called *multiresolution analysis* (see Mallat (1989), Meyer (1990)), in which all these bases find a natural interpretation. In this framework, $L^2(\mathbb{R})$ is seen as the "limit" of a hierarchy of approximation spaces V_j, with $\ldots V_2 \subset V_1 \subset V_0 \subset V_{-1} \subset V_{-2} \subset \ldots$, and the wavelets at scale 2^{-j} give the "detail" lost in going from V_{j-1} to V_j:

$$\text{Proj}_{V_{j-1}} f = \text{Proj}_{V_j} f + \sum_k \psi_{j,k}\langle\psi_{j,k}, f\rangle \ .$$

Moreover, this framework lends itself to a fast algorithm for decomposing functions into wavelet components, and reconstructing them again. This algorithm, which turns out to be a form of subband filtering as practised in electrical engineering, was the inspiration for my own construction of compactly supported wavelet bases (i.e. ψ is compactly supported – see Daubechies (1988)), which are now being used in applications ranging from signal processing to numerical analysis and the study of turbulence, to name a few. (See e.g. Ruskai et al. (1992).)

6. Concluding remarks

Wavelets have turned out to be a very versatile tool, originating as the synthesis of ideas from different fields, and leading to applications in many fields as well. In some applications they represent merely a new twist on existing techniques (such as e.g. subband coding), in others they lead to more novel developments. In all these applications they are however merely a tool, and, as always, a tool is no better than the craftsman who uses it. Despite their present popularity, wavelets will not do miracles all by themselves: to make them work for you, you must understand both the problem you apply them to and the tool itself ... Even then, they will not solve all the problems in the world, as they sometimes are now touted to do! Nevertheless, they are an exciting and promising development, to which one can arrive by many different roads. For me, all that started with affine coherent states.

Thank you, John!

References

1. E. W. Aslaksen and J. R. Klauder, *Unitary representations of the affine group*, J. Math. Phys. **9** (1968), 206–211; *Continuous representation theory using the affine group*, J. Math. Phys. **10** (1969), 2267-2275.
2. H. Bacry, A. Grossmann and J. Zak, *Proof of the completeness of lattice states in the kq-representation*, Phys. Rev. **B12** (1975), 1118–1120.
3. R. Balian, *Un principe d'incertitude fort en théorie du signal ou en mécanique quantique*, C.R. Acad. Sc. Paris (Série 2) **292** (1981), 1357-1362.
4. V. Bargmann, P. Butera, L. Girardello and J. R. Klauder, *On the completeness of coherent states*, Reports on Mod. Phys. **2** (1971), 221-228.
5. G. Battle, *A block spin construction of ondelettes. Part I: Lemarié functions*, Comm. Math. Phys. **110** (1987), 601–615.
6. A. P. Calderón, *Intermediate spaces and interpolation, the complex method*, Studia Math. **24** (1961), 113–190.
7. I. Daubechies, *Orthonormal bases of compactly supported wavelets*, Comm. Pure Appl. Math. **41** (1988), 909–996.
8. I. Daubechies, *The wavelet transform, time-frequency localization and signal analysis*, IEEE Trans. Inf. Th. **36** (1990), 961–1005.
9. I. Daubechies, A. Grossmann and Y. Meyer, *Painless nonorthogonal expansions*, J. Math. Phys. **27** (1986), 1271-1283.
10. R. J. Duffin and A. C. Schaeffer, *A class of nonharmonic Fourier series*, Trans. Am. Math. Soc. **72** (1952), 341–366.
11. M. Frazier and B. Jawerth, *The φ-transform and applications to distribution spaces*, Function Spaces and Application (M. Cwikel *et al.*, eds.), Lecture Notes in Mathematics 1302, Springer, Berlin, 1988, pp. 233–246; *A discrete transform and decompositions of distribution spaces (to appear)*, J. Func. Anal..
12. A. Grossmann and J. Morlet, *Decomposition of Hardy functions into square integrable wavelets of constant shape*, SIAM J. Math. Anal. **15** (1984), 723–736.
13. A. Grossmann, J. Morlet and T. Paul, *Transforms associated to square integrable group representations, I. General results*, J. Math. Phys. **27** (1985), 2473–2479.
14. A. Grossmann, J. Morlet and T. Paul, *Transforms associated to square integrable group representations, II. Examples*, Ann. Inst. H. Poincaré **45** (1986), 293–309.
15. A. Grossmann, R. Kronland-Martinet and J. Morlet, *Reading and understanding continuous wavelet transforms*, Wavelets (J. M. Combes, A. Grossmann and Ph. Tchamitchian, eds.), Springer, Berlin, 1989, pp. 2–20.
16. J. R. Klauder and B.-S. Skagerstam, *Coherent States*, World Scientific, Singapore, 1985.
17. P. G. Lemarié, *Une nouvelle base d'ondelettes de $L^2(\mathbb{R}^n)$*, J. de Math. Pures et Appl. **67** (1988), 227–236.
18. F. Low, *Complete sets of wave packets*, A passion for physics – Essays in honor of Geoffrey Chew, World Scientific, Singapore, 1985, pp. 17–22.
19. S. Mallat, *Multiresolution approximation and wavelets*, Trans. Am. Math. Soc. **315** (1989), 69-88.
20. Y. Meyer, *Principe d'incertitude, bases hilbertiennes et algèbres d'opérateurs*, Séminaire Bourbaki, nr. 662, 1985-1986.
21. Y. Meyer, *Ondelettes et opérateurs, I: Ondelettes, II: Opérateurs de Calderón-Zygmund, III: Opérateurs multilinéaires*, Hermann, Paris, 1990; *English translation*, Cambridge University Press, in preparation.
22. T. Paul, *Ondelettes et mécanique quantique, Ph.D. Thesis (1985)*, Université de Marseille (France); see also Paul and Seip (1991).

23. A. M. Perelomov, *On the completeness of a system of coherent states*, Teor. Mat. Fiz. **6** (1971), 213–224.
24. M. B. Ruskai, G. Beylkin, R. Coifman, I. Daubechies, S. Mallat, Y. Meyer and L. Raphael (eds.), *Wavelets: theory and application*, Jones and Bartlett, Boston, 1992.
25. J. O. Stromberg, *A modified Franklin system and higher order spline systems on* $\mathbb{R}^n$ *as unconditional bases for Hardy spaces*, Conf. in honor of A. Zygmund, Vol. II (W. Beckner *et al.*, eds.), Wadsworth math. series, 1982, pp. 475–493.

RUTGERS UNIVERSITY AND AT&T BELL LABORATORIES, USA

COHERENT–STATE PATH–INTEGRALS AND THEIR RELATIONS TO WAVELETS

B. De Facio

Abstract. A brief review of coherent states and of wavelets is given and then several relations among the two structures are indicated. It is pointed out that many aspects of wavelets in physics are variants of old ideas in the coherent states.

Dedicated to John R. Klauder, scientist and friend par excellence.

1. Introduction

The Feynman path integral [1,2] has provided deep physical insights and perplexing mathematical questions. For both of these reasons it has been near the center of John Klauder's research [3-8] for over thirty years. The author was first drawn into path integrals during a series of lectures by Klauder at the 1971 Boulder Symposium on Theoretical Physics, and functional integrals have been a personal tar baby ever since. Once grasped they are not easy to release nor do they reveal their secrets easily! Klauder's 1960 and 1963 papers gave the first *coherent states* [9] which have proved to be an essential structure for my approach to path integrals. The apt name coherent states is due to Glauber [10] who rediscovered them when he created the modern formulation of quantum optics. Kaiser [11,12] has given a phase–space formulation of relativistic quantum mechanics using the coherent states, and the author has given several different studies using path–integrals, including [13-16].

This short paper will contain a highly abridged review of coherent states, of wavelets and will show several connections between these two structures. There is little or nothing new being presented on coherent states and wavelets here, but the point that the two structures are closely related does **not** seem to have been properly appreciated. Actually many aspects of **wavelets** are included in Klauder's approach to the coherent states. Indeed Kaiser and Streater [17] have stated that Aslasken and Klauder [5] gave "the mother of all wavelets," and this author agrees. The two key properties are:

1) **continuity** of the coherent states in their label, say $z \in C^1$, and
2) the existence of a **reproducing kernel** for the coherent states as first shown by Klauder [3,4] and then studied by many others.

The continuity in the label z makes the coherent states an overcomplete family of states, **OFS** and the reproducing kernel guarantees a full inversion structure. A particularly useful introduction to the coherent states is given in [9] along with 69 reprints of papers from all areas of physics.

The point to this short note is to show that many of the fashionable **wavelet** methods under study today are coherent states and that their discretization corresponds to certain path integrals.

The organization of this paper is that the review of the coherent states is presented in Section 2, wavelets are reviewed in Section 3, some relations between coherent states and wavelets are given in Section 4, and the conclusions are presented in Section 5.

2. Coherent states

There are several approaches to the coherent states, see [3,4,9,10,11,17,18], and a hybrid of Klauder and Kaiser's approaches will be used here. A key ingredient is an underlying Lie group expressing the symmetry of the system.

The quantum theory of a one–degree of freedom Bose system with observables position Q and momentum P can be expressed at its deepest level by the Weyl form of the canonical relations

$$U(\alpha)V(\beta) = e^{-i\alpha\beta/\hbar}V(\beta)U(\alpha) \tag{1}$$

where $U(\alpha)$ and $V(\beta)$ are strongly continuous one parameter unitary operators with

$$U(\alpha) = e^{-i\alpha Q/\hbar}, \tag{2a}$$

and

$$V(\beta) = e^{-i\beta P/\hbar}. \tag{2b}$$

The uncertainty principle is built into these relations and the proof of the von Neumann uniqueness theorem follows naturally. This theorem states that all irreducible representations of the canonical commutation relation are unitarily equivalent providing that the number of degrees of freedom are finite. The Fock representation, F, of the physical Hilbert space, H , is generated by the action of the creation and destruction operators (a^+, a) on the Fock vacuum $|O_F\rangle$. The creation and destruction are non–self adjoint operators which satisfy the canonical commutation relations

$$[a, a] = [a^+, a^+] = 0, \tag{3a}$$

and

$$[a, a^+] = 1. \tag{3b}$$

The Fock vacuum $|O_F\rangle$ is a unique no particle vector in H such that

$$a\,|O_F\rangle = 0\,. \tag{4}$$

The Fock states $|n\rangle$, are eigenstates of the number operator, $N = a^+a$, with positive integer eigenvalues $n, n \in Z_+$.

For each complex variable $z \in C^1$, the coherent states can be generated by the unitary displacement operator

$$D(z) = e^{za^+ - \bar{z}a} \tag{5}$$

acting on the Fock vacuum to yield

$$|z\rangle = D(z)|O_F\rangle \tag{6a}$$

The Baker–Campbell–Hausdorff formula can be applied to (6a) to give

$$|z\rangle = e^{-|z|^2/2}\sum_{n=0}^{\infty}[\frac{z^n}{\sqrt{n!}}|n\rangle], \tag{6b}$$

in terms of an infinite coherent sum of number states. Equation (6b) is sometimes called a "Glauber state" and is clearly analytic in the label z. In general it only needs to be continuous. Inspection of (5) shows that the coherent states are in an exponential Hilbert space [6]. The completeness relation in Fock space is given by the outer product

$$\sum_n |n\rangle\langle n| = 1_H \tag{7}$$

where 1_H is the unit operator in the system Hilbert space. It is straightforward to show that

$$\frac{1}{\pi}\int |z\rangle\langle z|\, d^2z = 1_H, \tag{8}$$

see, e.g., [9]). The completeness properties hold for lattices with one point in the Planck cell of the phase space with area/(per degree of freedom) $2\pi\hbar = h$. Operators $B(a^+, a)$ which are polynomials in a^+ and a can be uniquely determined from their diagonal elements, thereby showing the overcomplete property of the coherent states.

The *affine* or *dilation* or *Ax + b group* [5,7] can be defined through the action of the operator D_{Ab} (A an invertible, real, $n \times n$ matrix of scale, $\vec{b} \in R^n$ is a translation and $\vec{x} \in R^n$) on $\vec{x}$

$$D_{Ab}\,\vec{x} = A^{-1}\,\vec{x} - b \tag{9a}$$

and on each $\psi \in L^2(R^n)$ as

$$(D_{Ab}\psi)(\vec{x}) = |A|^{-1/2}\psi(A^{-1}\,\vec{x} - \vec{b}), \tag{9b}$$

where the factor $|A|^{-1/2}$ in front of the right hand side is to preserve the L^2 norm. In order to use the coherent state structure given in this section it is useful to adapt the parameterization [5,7,21,22]

$$A = e^{\wedge} \tag{10}$$

or its inverse $\wedge = ln(A)$ which was used in [5] and [7]. Observe that if A is real and invertible the logarithm in the inverse relation to (10) is well behaved for all A. In a one–dimensional phase space, A reduces to $a > 0$, $a \in R^1$ and $b \in R^1$. It is useful to decompose D_{Ab} or $D_{\wedge b}$ into the semi–direct product of a scale change $S(a)$ or $S(\wedge)$ and a translation $T(\vec{b})$ with

$$T(\vec{0}) = S(\vec{0}) = 1$$

$$T(\vec{a})T(\vec{b}) = T(\vec{a}+\vec{b})$$
$$S(\lambda)S(\mu) = S(\lambda+\mu)$$
$$S(\lambda)T(\vec{a}) = T(e^{\lambda}\vec{a})S(\lambda). \tag{11}$$

Two physically important quantities [20] are the autocorrelation functions for the translated function; the convolution integral and the scale auto–correlation of a complex valued signal f are given by the

$$\langle f, T(a)f\rangle := \int_{-\infty}^{\infty} \frac{f^*(p)f(p-a)dp}{|p|} \tag{12}$$

and

$$\langle f, S(\lambda)f\rangle := \int_{-\infty}^{\infty} \frac{f^*(p)f(e^{\lambda}p)dp}{|p|}. \tag{13}$$

In optics, (12) represents the *contrast* of the signal and (13) is a scale distortion of the signal. In field theory and statistical mechanics these two equations represent two point functions, where (13) would describe a massless field.

Remarks:

(i) In the 1-d case $A \to a \in R^1_+(a>0)$ and $b \in R^1$ giving a phase–space $\Gamma = R^1_+ \otimes R^1$, $(a,b) \in \Gamma$. This was used for Fourier optics in [22].

(ii) A real cartesian d–dimensional generalization is readily obtained from $b \to \vec{b}$ and $a \to A$ *s.t.* $|A| \neq 0$. But this is not satisfactory in general because cylindrical, spherical and other curved surfaces are not smoothly represented since the rotations are not preserved in this framework. A recent idea by Kaiser [11,12] will be mentioned which allows arbitrary rotations to be adjoined to translations and scale changes, naturally.

3. Wavelets

A **wavelet** [21-30] is a "designer trig function" which acts as a basis in phase–space with improved localization and which is non–stationary in its variable. They are defined recursively. The reader will be reminded of the definition of a one–variable wavelet first in the continuous and then in the discrete cases.

Consider a one–variable, finite–energy signal $f(x) \in L^2(R^1)$ and let $\psi(x) \in L^2(R^1)$ be a real–valued function such that:

(i) the L^2-closure of the linear span of the dilations of ψ

$$\psi_{ab}(x) = (D_{ab}\psi)(x) = |a|^{-1/2}\psi(\frac{x-b}{a}) \tag{14}$$

span L^2 for $(a,b) \in \Gamma$. Of course (14) can be changed into exponential form by using (10);

(ii) the function ψ must be *admissible* in the sense that the constant

$$c_{\psi} := \int_0^{\infty} |\hat{\psi}(k)|^2 \frac{dk}{|k|} \tag{15}$$

satisfies the condition

$$0 < c_\psi < \infty. \tag{16}$$

Equation (16) implies that ψ has mean zero,

$$\hat{\psi}(0) = \int_{-\infty}^{\infty} \psi(x)dx = 0. \tag{17}$$

A function ψ which satisfies the conditions (i) and (ii) above is called a "mother wavelet" following Daubechies [23]. At each fixed $\vec{x}$, the dilates "live in" the phase–space

$$\Gamma = R^1_+ \otimes R^1, \tag{18}$$

i.e., each $(a, b) \in \Gamma$. The scale parameter a in Γ carries an interesting discrete multiresolution structure [30]. Consider the case $a = 2, b \in Z$ with $\Gamma = Z_+ \otimes Z$, Z is the set of all integers, so that the dilates of the mother wavelet are given by

$$\psi_{jk} = 2^{-j/2}\psi(2^{j/2}x - k). \tag{19}$$

The real space $V_o = V = L^2(R^1)$ can be decomposed s.t. $\forall j$

$$\cdots \supset V_j \supset V_{j+1} \supset \cdots \tag{20}$$

$$V_j = V_{j+1} \oplus W_{j+1} \tag{21a}$$

$$V_j \cap W_j = \{0\} \tag{21b}$$

with

$$V_j = S^j V. \tag{21c}$$

The spaces W_j contain a "father wavelet" ϕ whose dilates span W. Since ϕ is a wavelet the dilates of ϕ span $W_j, \forall j$, and it is admissible but has mean one

$$\hat{\phi}(0) = \int_{-\infty}^{\infty} \phi(x)dx = 1, \tag{22}$$

in contrast to the mother wavelet. The functions in each V_j have scale 2^j. To calculate the pair (ϕ, ψ) a finite set of masking coefficients $\{h(n)\}_{n=1}^N$ and $\{g(n)\}_{n=1}^N$ exist but are non–unique corresponding to every binomial process [23] which can correspond to the Fourier transforms of ψ and ϕ. Various ansätze can be chosen with increasing complexity, including

(i) the <u>box</u> function

$$b(x) = \begin{cases} 1, & |x| < 1/2 \\ 0, & |x| > 1/2 \end{cases} \tag{23a}$$

or (ii) the <u>hat</u> function

$$h(x) = \begin{cases} |x|, & |x| < 1/2 \\ 0, & |x| > 1/2 \end{cases} \tag{23b}$$

to start an iteration for ϕ and ψ. The masking coefficients for (i) and (ii) are

$$c_o = c_1 = 1$$

$$c_r = 0 \quad (r \geq 2), \tag{23c}$$

for the box function, and

$$c_0 = c_2 = 1/2$$

$$c_1 = 1$$

$$c_r = 0 \quad (r \geq 3), \tag{23d}$$

for the hat function. Then the father wavelet can be determined recursively from the iteration

$$\phi_j(x) = \sum_{k=1}^{M} c_k \phi_{j-1}(2x - k) \tag{24a}$$

using the coefficients above and $\phi_0(2x)$ as the ansatz. Similarly, the mother wavelet is given by

$$\psi_j(x) = \sum_{k=1}^{M} (-1)^k c_{M-k} \psi_{j-1}(2x - k), \tag{24b}$$

and has the stated properties. The *recursive* structure of these simple wavelets is clear from (24). Of course, other masking coefficients and other ansätze could be used. One knows that the level contours of the spherical functions carry the shapes of boundaries where data is imposed. One of the questions we are investigating at Missouri, is to try to determine the meaning of the shapes of different wavelets in function space.

The continuous wavelet transform can be written as

$$c_{ab} = c_{ab}(f, \psi) := \langle f, \psi_{ab} \rangle \tag{25a}$$

and the discrete wavelet series as

$$c_{jk} = c_{jk}(f, \psi) := \langle f, \psi_{jk} \rangle. \tag{25b}$$

The **direct problem** of wavelet analysis is that given a signal $f \in L^2(R^1)$, find a large enough subset of the wavelet transforms $\{c_{ab} \mid (a, b) \in \Gamma_A, \Gamma_A \subset \Gamma\}$ where Γ_A is a restriction of phase–space Γ to represent f. The **inverse problem** of wavelet analysis, the **reconstruction problem**, is that given a large enough subset of wavelet transforms $\{c_{ab} \mid (a, b) \in \Gamma_A, \Gamma_A \subset \Gamma\}$, find an approximation to f, f_A on Γ_A. For the discrete case, $c_{ab} \rightarrow c_{jk}$. For $f \in L^1 \cap L^2$ the **Grossmann–Morlet reconstruction formulas** [21] are

$$f(x) = \frac{1}{c_\psi} \int_0^\infty \frac{da}{|a|^2} \int_{-\infty}^{\infty} db \, c_{ab} \, \psi_{a,b}(x), \tag{26a}$$

in the transform case and

$$f(x) = \sum_{j,k} c_{jk}\psi_{jk}(x), \tag{26b}$$

in the series case. The reconstruction **exists** when ψ is admissible and $f \in L^1 \cap L^2$. **Uniqueness** requires additional conditions such as the Faddeev hypothesis that $f \in L^{1,1} = H^1$, where $H^1(R^S)$ is the Sobolev space. The recursive nature of wavelets is natural for construction in both the direct and inverse problems. Concerning noise, we quote Moses and Quesada, "We believe that periodicities in scale may be as important as periodicities in translation, especially when noise is present." Several Missouri Physics theses [31] have found this to be true with the quibble that it is the very **non–periodicity** of certain islands in phase space which is interesting and useful. The lack of periodicity in scale, in fact, gives wavelets their capability to analyze non–stationary time–series.

Kaiser [12,17] has given a generalization of one–dimensional signals $f(x)$ to higher s–dimensional signals, $s > 1$, which properly include rotations. In this analysis, it is satisfactory to require that f have the same decay at infinity as the Grossmann–Morlet hypothesis of $f \in L^1 \cap L^2$. Since atomic and nuclear potentials have rapidly converging partial wave expansions, Kaiser's approach is applicable and should prove useful for these scattering problems.

4. Relations between Coherent States and Wavelets

In Section 2 the group structure of the affine coherent states were indicated. However, an interesting feature which has long been known [5,7] was saved for this section. A standard Weyl–representation of phase–space has quantum states

$$|p,q\rangle \in H, \tag{27}$$

where (p,q) are the c-number labels of the observables P,Q. However, in the affine case $p > 0$ is required and sharp values of q cannot be realized because Q is not self–adjoint and, therefore, cannot be an observable [5]. Aslasken and Klauder showed that a new self–adjoint observable

$$B := \frac{1}{2}(PQ + QP) \tag{28a}$$

can be defined. In terms of the creation and destruction operators (28a) can be rewritten as

$$B = a^2 + (a^+)^2 \tag{28b}$$

which is suggestive of a two–photon coherent state [32,33]. For each positive real squeezing parameter, y, the Yuen squeeze operator [33] is given by

$$Y_y(y) := \exp\left[\frac{1}{2}y\left((a^+)^2 - a^2\right)\right], \tag{29}$$

which generates a two–photon squeezed state

$$|z,y\rangle = Y_y(y)D(z)|O_F\rangle. \tag{30}$$

To interpret this state express (a, a^+) in terms of the momentum and position observables (p, q) and then calculate that

$$Y(y)\begin{pmatrix} p \\ q \end{pmatrix} Y(y)^{-1} = \begin{pmatrix} e^y p \\ e^{-y} q \end{pmatrix}, \tag{31}$$

which clearly shows both the squeezing and the preservation of the uncertainty principle.

The group theory of the two–photon squeezed states was presented by Han, Kim and Noz [34]. In previous papers referenced therein, these authors studied the $(2 \oplus 1)$-dimensional Lorentz group by using the one–mode squeezed states in light–cone coordinates using the local isomorphism $ISp(2) \approx Sp(2)$. This is in the usual physicists notation, whereas mathematicians use $Sp(1)$. They have also reversed this argument to study the Lorentz transformation properties of extended hadrons, thereby really "having it both ways." The two–mode squeezed states have $Sp(4)$ as the symmetry of their linear canonical transformations which is locally isomorphic to the DeSitter or $(3 \oplus 2)$-dimensional Lorentz group. Observe that the $(4 \oplus 1)$-dimensional Lorentz group is also called a DeSitter group. The familiar $SU(2)$ and $SU(1,1)$ groups are subgroups of the DeSitter group and the correspondence of the two–mode states to one–mode states was obtained by integrating the other mode from the Wigner distribution function. The authors were able to study the statistical mechanics of one–mode radiation fields this way, and relativistic quark models of hadrons were studied also.

D'Ariano and De Facio [35] have formulated a direct and inverse wavelet for the density matrices of quantum optics for a class of problems. Let Θ be an observable, then

$$\Theta|\theta\rangle = \theta|\theta\rangle \tag{32}$$

and the eigenstates $|\theta\rangle$ form a resolution of the unit operator 1_H with respect to the measure $d\mu(\theta)$

$$\int d\mu(\theta)|\theta\rangle\langle\theta| = 1_H. \tag{33}$$

The class of physical systems must be restricted to the subset of systems for which the generating functional of moments of observables Θ in the state with density operator ρ

$$\langle e^{i\alpha\Theta}\rangle := Tr\left[e^{i\alpha\Theta}\rho\right], \tag{34}$$

exists, i.e., there are no long–time tails so that all moments exist and the series converges and where the density operator exists. It is a standard exercise to show that in this case

$$\langle e^{i\alpha\Theta}\rangle = \int d\mu(\theta)e^{i\alpha\theta}P(\rho;\theta), \tag{35}$$

where $P(\rho;\theta)$ is the probability distribution

$$P(\rho;\theta) = Tr\left[|\theta\rangle\langle\theta|\rho\right]. \tag{36}$$

Because of the trace, P is independent of the basis. Let χ be a window

$$\chi(\theta) = \begin{cases} 1, & \theta < \theta_M \\ 0, & \theta > \theta_M \end{cases} \tag{37}$$

for some fixed number θ_M and then the low-bandpass approximation to (35) is given by

$$\left\langle e^{i\alpha\Theta}\right\rangle_\gamma = \int d\mu(\theta) e^{i\alpha\theta}\gamma(\theta)P(\rho;\theta). \tag{38}$$

Let $\psi(x)$ be a wavelet as in Section 3 so that ψ_{ab} are the $(ab)^{th}$ dilate of ψ and the coefficient

$$c_\psi := \int_{-\infty}^{\infty} \frac{|\hat{\psi}(k)|^2}{|k|} dk \tag{39a}$$

satisfies

$$0 < c_\psi < \infty. \tag{39b}$$

The wavelet transform of $P(\rho;\theta)$ is now

$$\begin{aligned} c_{ab} &= c_{ab}(\rho, \psi) \\ &:= \frac{1}{|a|^{1/2}} \int d\mu(\theta)\psi\left(\frac{\theta - b}{a}\right) P(\rho, \theta) \end{aligned} \tag{40}$$

Since ψ is admissible, if ρ exists and is a Hilbert–Schmidt class operator the *inverse*

$$\rho = \frac{1}{c_\psi} \int_{0+}^{\infty} \frac{da}{|a|^2} \int_{-\infty}^{\infty} db c_{ab}\psi_{ab} \tag{41}$$

also exists but may not be unique. For uniqueness ρ and c_{ab} must have a *continuous* limit as $a \downarrow 0$.

The approximate Daubechies wavelets (24 a,b) are analytic at each iteration N so that an operator valued wavelet

$$\Psi^{(N)}(\theta)|\theta\rangle = \psi^{(N)}(\theta)|\theta\rangle \tag{42}$$

can be defined. The behavior of the limit as $N \to \infty$ is not obvious and needs additional investigation which is already in progress. If the operator limit on the left hand side of (42) exists, then a Heisenberg picture dilation can be written as

$$\begin{aligned} D^H_{ab}(\Theta) &= \frac{\Theta - b}{a} \\ &= e^{-a_1}(\Theta - b) \end{aligned} \tag{43}$$

where $a_1 = \ln(a)$.

It is straightforward to generate other wavelets using coherent states. Limitations of space (and time !) preclude showing how the affine-group, coherent state path-integrals can give discrete wavelet series.

5. Conclusions

Many published papers and books are pouring out on wavelets, or better, numerous versions of a few ideas are appearing by many authors[1]. For this reason it is desirable that the coherent states and path integrals not be "rediscovered" or ignored. Both structures have much more to teach physics and mathematics.

Acknowledgements

This work was supported in part by AFOSR grant 90-307. It is dedicated to Professor John R. Klauder both in thanks for all of the high quality research which he has completed and published and in anticipation of much more to come. Thor is thanked for 7 1/4 years of company during my own work. My graduate students, C.R. Thompson (now former), D.M. Patterson and A. Van Nevel are thanked for their excitement in theoretical physics and their long hours of excellent work. Professor G. Kaiser is thanked for reading the manuscript and making several useful suggestions.

References

1. R.P. Feynman, *Space-time approach to non-relativistic quantum mechanics*, Rev. Mod. Phys. **20** (1948), 367-387.
2. R.P. Feynman, *An operator calculus having applications in quantum electrodynamics*, Phys. Rev. **84** (1951), 108-128.
3. J.R. Klauder, *The action option and a Feynman quantization of spinor fields in terms of ordinary c-numbers*, Ann. Phys. (NY) **11** (1960), 123-168.
4. J.R. Klauder, *Continuous representation theory I. Postulates of continuous representation theory*, J. Math. Phys. **4** (1963), 1055-1058; *II. Generalized relation between quantum and classical dynamics*, ibid. **4** (1963), 1058-1073.
5. E.W. Aslasken and J.R. Klauder, *Unitary representations of the affine group*, J. Math Phys. **9** (1968), 208-211; *Continuous representation theory using the affine group*, ibid. **10** (1969), 2264-2275.
6. J.R. Klauder, *Exponential Hilbert space, Fock space revisited*, J. Math Phys. **11** (1970), 609-630.
7. J.R. Klauder, *Affine Fields and the operator formalism of augmented scalar fields*, J. Math Phys. **18** (1977), 1711-1719.
8. J.R. Klauder, *Quantization* is *geometry, after all*, Ann Phys (NY) **188** (1988), 120-141.
9. J.R. Klauder and B.S. Skagerstam, *Coherent States*, World Scientific, Singapore, 1985.
10. R.J. Glauber, *Photon correlations*, Phys. Rev. Lett. **10** (1963), 84-86; *The quantum theory of optical coherence*, Phys. Rev. **130** (1963), 2529-2539.
11. G. Kaiser, *Phase-space approach to relativistic quantum mechanics I.*, J. Math. Phys. **18** (1977), 952-959; *II.*, ibid. **19** (1978), 502-507; *III.*, ibid. **22** (1981), 705-714.
12. G. Kaiser, *Quantum Physics, Relativity and Complex Spacetime*, North Holland, Amsterdam, 1990.
13. B. DeFacio and C.L. Hammer, *Bilinear quantum field theories and the coherent states*, J. Math. Phys. **17** (1976), 267-281.
14. B. DeFacio and O. Brander, *Coherent-state path-integral formulation of inverse scattering theory*, Rend. del Circ. Mat. di Palermo (Series II) **17** (1987), 185-208.
15. B. DeFacio, G.W. Johnson and M. Lapidus, *Feynman's operational calculus as a generalized path integral*, in G.Kalliampur Festschrift (S.Mitri, ed.), Springer-Verlag, Berlin and New York, in press.
16. B. DeFacio, G.W. Johnson and M. Lapidus, *Feynman's operational calculus and quantum models* (in preparation).
17. G. Kaiser and R.F. Streater, *Windowed Radon transforms, analytic signals and the wave equation*, in Wavelets, a Tutorial Vol. II. (C.Chui, ed.), Academic Press, Boston, 1992.

---

[1] This comment emphatically does not apply to Jawerth, Daubechies or Kaiser !

18. A. Perelomov, *Generalized coherent-states and their applications*, Springer-Verlag, Berlin and New York, 1986.
19. E. Onofri, *A note on coherent state representations of Lie groups*, J. Math. Phys. **16** (1975), 1087-1089.
20. H.E. Moses and A.F. Quesada, *The power spectrum of the Mellin transform with application to scaling physical quantities*, J. Math. Phys. **15** (1974), 948-952.
21. A. Grossmann and J. Morlet, *Decomposition of Hardy functions into square-integrable wavelets of constant shape*, SIAM J. Math. Anal. **15** (1984), 723-736.
22. B. DeFacio, C.R. Thompson and G.V. Welland, *Wavelet version of Fourier Optics*, SPIE **1351** (1990), 21-37.
23. I. Daubechies, *Orthonormal basis of compactly supported wavelets*, Comm. Pure Appl. Math. **41** (1988), 909-996.
24. M. Frazier and B. Jawerth, *A discrete transform and the decomposition of distribution spaces*, J. Funct. Anal. **93** (1990), 34-170.
25. M. Frazier, B. Jawerth and G. Weiss, *Littlewood-Paley theory and the study of function spaces CBMS* **79**, SIAM, Philadelphia, 1991.
26. I. Daubechies, *Wavelets CMBS* **81**, SIAM, Philadelphia, 1992.
27. Y. Meyer, *Ondelettes et Opérateurs, Vol.I*, Herman, Paris, 1990; *Vol.II*, 1991.
28. C.K. Chui, *An introduction to wavelets*, Acad. Press, Boston, 1992.
29. G. Kaiser, *An algebraic theory of wavelets I. Operational calculus and complex structure*, SIAM J. Appl. Math. **23** (1992), 222-243.
30. S. Mallet, *Multiresolution approximations and wavelet orthonormal basis of* $L^2(R^1)$, Trans. Amer. Math. Soc. **315** (1989), 69-87.
31. C.R. Thompson, *Some applications of wavelets to physics*, PhD Dissertation, Missouri (1992); D.M. Patterson, *Applications of Wavelets to ultrasonic inverse scattering analysis and simulated annealing to nuclear magnetic resonance pulse shaping*, M.S. Thesis, Missouri (1992); A.Van Nevel, *The application of wavelets to nonlinear coherent structures in the bubble formation for a fluid driven by a pressure wave*, M.S. Thesis, Missouri (in progress).
32. D. Stoler, *Generalized coherent states*, Phys. Rev. D **4** (1971), 2309-2312.
33. H.P. Yuen, *Two-photon coherent states of the radiation field*, Phys. Rev. A **1** (1976), 2226-2243.
34. D. Han, Y.S. Kim and M.E. Noz, *Linear canonical transformations of coherent states in Wigner phase space. III. Two-mode states*, Phys. Rev. A **41** (1990), 6233-6244.
35. G.M. D'Ariano and B. DeFacio, *A quantum wavelet for quantum optics* (submitted).

Department of Physics & Astronomy, Missouri University, Columbia, MO 65211, USA

PARTICLE AND WAVE DESCRIPTIONS ARE NOT EQUIVALENT

HIROSHI EZAWA

ABSTRACT. It is demonstrated with a simple model of non-interacting Bose system in one-dimensional space that, contrary to prevailing expectations, the transformation mediating the wave- and the particle-pictures does not exist in the Fock space in the continuum limit.

Contribution in honor of the 60th anniversary of Professor John Klauder

1. Introduction

It is a great honour to be invited to take part in the celebration of Professor John R. Klauder's sixtieth birthday. With deep gratitude for the constant help he has extended to me for many years, I wish him many happy returns of the day. Since we often discussed the question of inequivalent representations of the canonical variables in quantum mechanics, it should be appropriate to take up the problem again at this occasion.

In quantum mechanics, the wave-particle duality is the most fundamental cornerstone of the theoretical framework, not to mention the established experimental fact demonstrated most beautifully by the process of single-particle build up of interference pattern in actual (not Gedanken) two-slit experiments with electrons [1] and with photons [2].

The duality is accomodated in quantum mechanics by the transformation theory. Namely, a quantum mechanical system can be pictured either as a wave field obeying an appropriate wave equation or as an assembly of particles each having position and momentum variables, and the two pictures, although excluding each other as the number-phase uncertainty relation attests, are said to be transformable into one-another. One and the same system can then be described equivalently in the wave- and the particle-pictures, and moreover the two pictures are complementary to each other.

In this article, we shall demonstrate non-existence of such a transformation by taking a simple model of non-interacting spinless Bose system. More precisely, we begin with a discretized lattice space to formulate an approximate wave picture for the Bose system and then construct a transformation operator to a particle picture, also approximate in the lattice space. The procedure so far is well-known. Then, we show that all the matrix elements of the transformation operator with respect to basis vectors of the state space of the particle picture tend to zero in the continuum limit where one lets the lattice spacing go to zero.

2. The Two Pictures and their Transformation

As a simple example, we consider the case of a free scalar field $\hat{\phi}(x)$ of mass μ in one-dimensional space. Let us first review some elementary facts, which will serve for fixing notations.

We begin with a lattice space of spacing ϵ in a finite "box" of length $L = 2N\epsilon$. The Hamiltonian is given by

$$\hat{\mathcal{H}} = \frac{\epsilon}{2} \sum_{s=-N}^{N} \left\{ \frac{1}{\epsilon^2} \hat{\pi}_s^2 + \left(\frac{\hat{\phi}_{s+1} - \hat{\phi}_s}{\epsilon} \right)^2 + \mu^2 \hat{\phi}_s^2 \right\} \tag{1.1}$$

in terms of the canonical variables $\left\{ \hat{\pi}_s, \hat{\phi}_s \right\}_I$, $I = \{-N, -N+1, \cdots, N\}$ of the *wave* picture with

$$\hat{\phi}_{N+1} = \hat{\phi}_{-N} \qquad \text{(periodic boundary condition)}, \tag{1.2}$$

satisfying the canonical commutation relations

$$\left[\hat{\pi}_s, \hat{\phi}_{s'} \right] = -i\, \delta_{s,s'} ,$$

$$[\hat{\pi}_s, \hat{\pi}_{s'}] = 0 , \qquad \left[\hat{\phi}_s, \hat{\phi}_{s'} \right] = 0 . \tag{1.3}$$

Then, the Heisenberg equation of motion for $\hat{\phi}(x)$ turns out to be a wave equation of the Klein-Gordon type.

To motivate switching to particle picture, we diagonalize the Hamiltonian (1.1) by the transformation to a new set of variables $\{\hat{p}_\nu, \hat{q}_\nu\}_I$ defined by

$$\begin{aligned} \hat{\phi}_s &= \alpha_{s,0} q_0 + \sum_{\nu=1}^{N} (\alpha_{s,\nu} \hat{q}_\nu + \alpha_{s,-\nu} \hat{q}_{-\nu}) \\ \hat{\pi}_s &= \alpha_{s,0} p_0 + \sum_{\nu=1}^{N} (\alpha_{s,\nu} \hat{p}_\nu + \alpha_{s,-\nu} \hat{p}_{-\nu}) \end{aligned} \tag{1.4}$$

where

$$\begin{aligned} \alpha_{s,0} &= \left(\frac{1}{2N+1} \right)^{1/2} , \\ \alpha_{s,\nu} &= \left(\frac{2}{2N+1} \right)^{1/2} \cos \frac{2\pi s \nu}{2N+1}, \\ \alpha_{s,-\nu} &= \left(\frac{2}{2N+1} \right)^{1/2} \sin \frac{2\pi s \nu}{2N+1}, \\ \nu &\in I_+ = \{1, \cdots, N\}. \end{aligned} \tag{1.5}$$

This is an orthogonal transformation,

$$\sum_{s=-N}^{N} \alpha_{s,\nu}\alpha_{s,\nu'} = \delta_{\nu,\nu'} , \tag{1.6}$$

and hence the canonical commutation relations (1.3) are preserved, viz., $[\hat{p}_\nu, \hat{q}_{\nu'}] = -i\delta_{\nu,\nu'}$, etc. The Hamiltonian becomes

$$\hat{\mathcal{H}} = \frac{\epsilon}{2} \sum_{\nu=-N}^{N} \left\{ \frac{1}{\epsilon^2}\hat{p}_\nu^2 + \omega_\nu^2 \hat{q}_\nu^2 \right\} , \tag{1.7}$$

where

$$\omega_\nu = \sqrt{\frac{4}{\epsilon^2}\sin^2\frac{\nu\pi}{2N+1} + \mu^2} \tag{1.8}$$

For the following calculations, it is convenient to use the Schrödinger representation for our canonical variables,

$$\hat{q}_\nu = x_\nu \cdot \quad , \quad \hat{p}_\nu = -i\frac{\partial}{\partial x_\nu} \qquad (\nu \in I), \tag{1.9}$$

each of which acts on an appropriate domain in the Hilbert space $\mathbb{H}$ of square-integrable functions, $\psi(x_{-N}, \cdots, x_N)$ on $\mathbb{R}^{2N+1}$. Then, the eigenstates and the corresponding eigenvalues of the Hamiltonian (1.7) are given respectively by

$$u_{n_{-N},\cdots,n_N}(x_{-N},\cdots,x_N) = \left[\prod_{\nu=-N}^{N} \frac{1}{\sqrt{n_\nu!}}(a_\nu^\dagger)^{n_\nu}\right] \Omega_N \tag{1.10}$$

and

$$E_{n_{-N},\cdots,n_N} = \sum_{\nu=-N}^{N} \left(n_\nu + \frac{1}{2}\right)\omega_\nu , \tag{1.11}$$

with

$$n_\nu = 0, 1, 2, \cdots \qquad (\nu \in I),$$

where

$$\Omega_N = \left\{\prod_{\nu=-N}^{N} \left(\frac{\epsilon\,\omega_\nu}{\pi}\right)^{1/4}\right\} \exp\left[-\frac{1}{2}\epsilon\,\omega_\nu x_\nu^2\right] \tag{1.12}$$

is the ground state and

$$a_\nu^\dagger = \sqrt{\frac{\epsilon\,\omega_\nu}{2}}\,\hat{q}_\nu + i\sqrt{\frac{1}{2\epsilon\,\omega_\nu}}\,\hat{p}_\nu . \tag{1.13}$$

The physical interpretation becomes clear in the continuum limit where $\epsilon \to 0, N \to \infty$ with $L = 2N\epsilon$ kept fixed. In this limit, (1.8) becomes

$$\omega_\nu \to \sqrt{k_\nu^2 + \mu^2} \tag{1.14}$$

with

$$k_\nu = \frac{2\nu\pi}{L},$$

so that ω_ν can be interpreted as the relativistic energy of a particle of mass μ having momentum k_ν. Similarly, it can be seen that the state (1.10) is the simultaneous eigenstate of the total momentum of the system with the eigenvalue,

$$P_{n_{-N},\cdots,n_N} = \sum_{\nu=-N}^{N} n_\nu k_\nu . \tag{1.15}$$

Then, in view of the forms of the energy and momentum eigenvalues, (1.11) and (1.15), the state (1.10) is interpreted as having n_ν particles in each momentum state k_ν, $\nu \in I$. Therefore, the set of the variables $\{p_\nu, q_\nu\}_I$ provides a *particle* picture, and the transformation (1.4) mediates the wave picture and the particle picture, both approximate for the finite lattice spacing ϵ.

3. Inequivalence in the Continuum Limit

We wish to show that the transformation mediating the wave- and the particle-pictures ceases to exist in the continuum limit, $\epsilon \to 0$, $N \to \infty$ with the size $L = 2N\epsilon$ of the "box" kept fixed.

The transformation (1.4) can be written as

$$\hat{q}_\nu = e^{i\hat{\Lambda}} \hat{\phi}_\nu e^{-i\hat{\Lambda}} \quad , \quad \hat{p}_\nu = e^{i\hat{\Lambda}} \hat{\pi}_\nu e^{-i\hat{\Lambda}} \ , \tag{2.1}$$

where

$$\hat{\Lambda} = \sum_{r,r'=-N}^{N} \Lambda_{r,r'} \hat{\phi}_r \hat{\pi}_{r'} \tag{2.2}$$

with the matrix $\Lambda = (\Lambda_{r,r'})$ satisfying

$$\left(e^{\Lambda}\right)_{s,\nu} = \alpha_{s,\nu} . \tag{2.3}$$

In fact, the Hadamard formula

$$e^{i\Lambda} \hat{\phi}_\nu e^{-i\Lambda} = \hat{\phi}_\nu + \left[i\hat{\Lambda}, \hat{\phi}_\nu\right] + \frac{1}{2!}\left[i\Lambda, \left[i\Lambda, \hat{\phi}_\nu\right]\right] + \cdots$$

leads to

$$e^{i\hat{\Lambda}} \hat{\phi}_\nu e^{-i\hat{\Lambda}} = \sum_{s=-N}^{N} \left(e^{\Lambda}\right)_{s,\nu} \hat{\phi}_s = \hat{q}_\nu \tag{2.4}$$

and similarly for $\hat{p}_\nu$. Since the matrix $\alpha = (\alpha_{\mu,\nu})$ is orthogonal, Λ is real, antisymmetric and in particular its diagonal elements are all zero, so that the ordering of the operators, $\hat{\phi}_r$ and $\hat{\pi}_{r'}$, in $\hat{\Lambda}$ in (2.2) is irrelevant. Consequently, $\hat{\Lambda}$ is self-adjoint and $e^{-i\hat{\Lambda}}$ is unitary.

Let us now see how the transformation $e^{-i\hat{\Lambda}}$ acts on the state space $\mathbb{H}$ by examining the rotation it induces on the system of basis vectors, which we take to be the complete orthonormal system of the eigenvectors (1.10) of $\hat{\mathcal{H}}$. Namely, we look at the expansion coefficients in

$$e^{-i\hat{\Lambda}}\,u_{n_{-N},\cdots,n_N} = \sum_{n'_{-N},\cdots,n'_N} u_{n'_{-N},\cdots,n'_N}\langle u_{n'_{-N},\cdots,n'_N}, e^{-i\hat{\Lambda}}u_{n_{-N},\cdots,n_N}\rangle. \tag{2.5}$$

We begin with one particular coefficient:

$$\langle \Omega_N, e^{-i\hat{\Lambda}}\Omega_N\rangle =$$
$$\int_{\mathbb{R}^{2N+1}}\left\{\prod_\nu\left(\frac{\epsilon\omega_\nu}{\pi}\right)^{1/4}\exp\left[-\frac{\epsilon\omega_\nu}{2}x_\nu^2\right]\right\}\cdot e^{-i\hat{\Lambda}}\left\{\prod_{\nu'}\left(\frac{\epsilon\omega_{\nu'}}{\pi}\right)^{1/4}\exp\left[-\frac{\epsilon\omega_{\nu'}}{2}x_{\nu'}^2\right]\right\}\prod_{\nu''}dx_{\nu''}. \tag{2.6}$$

But, with the Schrödinger representation $\{\xi_s, -i\partial/\partial\xi_s\}$ for $\left\{\hat{\phi}_s, \hat{\pi}_s\right\}$ one has

$$e^{i\hat{\Lambda}}\xi_s = \sum_r \left(e^\Lambda\right)_{r,s}\xi_r = x_s\,,$$

so that by (1.4)

$$e^{-i\hat{\Lambda}}x_s = \xi_s = \sum_\nu \alpha_{s,\nu}x_\nu = \sum_\nu \left(e^\Lambda\right)_{s,\nu}x_\nu$$

and hence

$$\langle \Omega_N, e^{-i\hat{\Lambda}}\Omega_N\rangle = \int_{\mathbb{R}^{2N+1}}\exp\left[-\frac{\epsilon}{2}\sum_{s,s'}x_s\left(\omega+e^{-\Lambda}\omega e^{\Lambda}\right)_{s,s'}x_{s'}\right]\prod_\nu\left(\frac{\epsilon\omega_\nu}{\pi}\right)^{1/2}dx_\nu\,, \tag{2.7}$$

where ω is the diagonal matrix having $\omega_s > 0$ on the diagonal. Thus,

$$\langle \Omega_N, e^{-i\hat{\Lambda}}\Omega_N\rangle = 2^{(2N+1)/2}\left[\det(1+K)\right]^{-1/2}, \tag{2.8}$$

with

$$K = \omega^{-1/2}e^{-\Lambda}\omega e^{\Lambda}\omega^{-1/2}\,, \tag{2.9}$$

which is a real symmetric matrix. Let its eigenvalues be $\{\kappa_s\}_I$. Then,

$$\det K = \prod_{s=-N}^{N}\kappa_s = 1. \tag{2.10}$$

Further,

$$\det(1+K) = \prod_{s=-N}^{N}(1+\kappa_s)$$

can be estimated as

$$\begin{aligned}[\det(1+K)]^2 &= 4^{2N+1}\prod_{s=-N}^{N}\left\{1+\frac{(\kappa_s-1)^2}{4\kappa_s}\right\} \\ &\geq 4^{2N+1}\left\{1+\frac{1}{4}\sum_{s=-N}^{N}\left(\kappa_s-2+\frac{1}{\kappa_s}\right)\right\}.\end{aligned} \tag{2.11}$$

Therefore, (2.8) is bounded by

$$|\langle\Omega_N, e^{-i\hat{\Lambda}}\Omega_N\rangle| \leq \left\{1+\frac{1}{4}\sum_{s=-N}^{N}\left(\kappa_s-2+\frac{1}{\kappa_s}\right)\right\}^{-1/4}. \tag{2.12}$$

Now, we wish to show that the right-hand side vanishes in the continuum limit $\epsilon \to 0, N \to \infty$ with $L = 2N\epsilon$ kept fixed. Since e^{Λ} is an orthogonal matrix, (2.9) gives

$$\sum_{s=-N}^{N}\kappa_s = \operatorname{tr} K = \sum_{r,s=-N}^{N}\frac{\omega_s}{\omega_r}\alpha_{r,s}^2\,.$$

But, $\omega_r = \omega_{-r}$ and

$$\alpha_{s,r}^2 + \alpha_{s,-r}^2 = \frac{2}{2N+1} \qquad \text{for} \qquad r \in I_+\,,$$

by (1.5), so that

$$\sum_{s=-N}^{N}\kappa_s = \frac{1}{2N+1}\left(\sum_{s=-N}^{N}\omega_s\right)\left(\sum_{r=-N}^{N}\frac{1}{\omega_r}\right).$$

We have

$$\begin{aligned}\sum_{s=-N}^{N}\omega_s &= \frac{4}{\epsilon}\sum_{s=1}^{N}\sin\frac{s\pi}{2N+1} \\ &= \frac{4}{\epsilon}\left(\cos\frac{\pi}{2(2N+1)}\right)\Big/\left(\sin\frac{\pi}{2(2N+1)}\right), \\ \sum_{s=-N}^{N}\frac{1}{\omega_s} &\geq (2N+1)\,\epsilon\int_{\frac{\pi}{2N+1}}^{\frac{N\pi}{2N+1}}\frac{d\theta}{\sqrt{\theta^2+\left(\frac{\epsilon\mu}{2}\right)^2}} \\ &= (2N+1)\,\epsilon\,\log\frac{4N\pi}{\mu L} + O(\epsilon)\,,\end{aligned}$$

the latter holding asymptotically as $N \to \infty$. Hence,

$$\sum_{s=-N}^{N}\kappa_s \geq \frac{8(2N+1)}{\pi}\log\frac{4N\pi}{\mu L} \qquad (N \to \infty). \tag{2.13}$$

In (2.12), further

$$\sum_{s=-N}^{N} (-2) = -2(2N+1),$$

$$\sum_{s=-N}^{N} \frac{1}{\kappa_s} > (2N+1)^2 \left[\sum_{s=-N}^{N} \kappa_s \right]^{-1} \sim \frac{(2N+1)\pi}{8} \left[\log \frac{4N\pi}{\mu L} \right]^{-1}, \qquad (N \to \infty)$$

are both dominated by (2.13) when N is large.

Thus, we conclude:

$$|\langle \Omega_N, e^{-i\hat{\Lambda}} \Omega_N \rangle| \leq \left[\frac{2(2N+1)}{\pi} \log \frac{4N\pi}{\mu L} \right]^{-1/4}, \qquad (N \to \infty) \tag{2.14}$$

the right-hand side vanishing in the continuum limit $N \to \infty$ and hence

$$\lim_{N \to \infty} \langle \Omega_N, e^{-i\hat{\Lambda}} \Omega_N \rangle = 0\,. \tag{2.15}$$

The same conclusion can be obtained for the other coefficients in the expansion (2.5), irrespective of which basis vector we take for the left-hand-side provided it stays in the Fock space, $\sum_\nu n_\nu < \infty$. This implies that the transformation mediating the wave- and the particle-pictures we have been considering ceases to exist in the Fock space in the continuum limit. The particle- and the wave-pictures are thus inequivalent to each other.

References

1. A. Tonomura, J. Endo, T. Matsuda, T. Kawasaki and H. Ezawa, Am. J. Phys. **57** (1989), 117.
2. Y. Tsuchiya, E. Inuzuka, T. Kurono and M. Hosoda, Advances in Electronics and Electron Physics (P. Hawkes, ed., Academic Press, New York) **64A** (1982), 21.

Department of Physics, Gakushuin University, Tokyo, JAPAN

BROWNIAN MOTION AND ITS CONDITIONAL DESCENDANTS

PIOTR GARBACZEWSKI

Contribution in honor of the 60th birthday of John R. Klauder

It happened before [1] that I have concluded my publication with a special dedication to John R. Klauder. Then, the reason was John's PhD thesis [2] and the questions (perhaps outdated in the eyes of the band-wagon jumpers, albeit still retaining their full vitality [3]): *(i) What are the uses of the classical (c-number,non-Grassmann) spinor fields, especially nonlinear ones, what are they for at all ? (ii) What are, if any, the classical partners for Fermi models and fields in particular ?* The present dedication, even if not as conspicuously motivated as the previous one by John's research, nevertheless pertains to investigations pursued by John through the years and devoted to the analysis of random noise. Sometimes, re-reading old papers and re-analysing old, frequently forgotten ideas might prove more rewarding than racing the fashions. Following this attitude, let us take as the departure point Schrödinger's original suggestion [4] of the existence of a special class of random processes, which have their origin in the Einstein-Smoluchowski theory of the Brownian motion and its Wiener's codification. The original analysis due to Schrödinger of the probabilistic significance of the heat equation and of its time adjoint in parallel, remained unnoticed by the physics community, and since then forgotten. It reappeared however in the mathematical literature as an inspiration to generalise the concept of Markovian diffusions to the case of Bernstein stochastic processes. But, it stayed without consequences for a deeper understanding of the possible physical phenomena which might underly the corresponding abstract formalism. Schrödinger's objective was to initiate investigations of possible links between quantum theory and the theory of Brownian motion, an attempt which culminated later in the so-called Nelson's stochastic mechanics [8] and its encompassing formalism [7] in which the issue of the Brownian implementation of quantum dynamics is placed in the framework of Markov-Bernstein diffusions. Schrödinger's discussion of the analogy between wave mechanics and random phenomena of classical statistical physics starts with recalling an obscurity present in the notion of probability (Born's postulate) adopted in quantum theory. For the purposes of the probabilistic interpretation, it seems that one should decide in advance, whether one is considering a probablity *after* one knows what has happened, or rather a probability of what *is to* happen. To conform with the classical notion of probability (event, i.e. sample space is needed to have the axiomatic definition of the probability space) the most natural way is to look at a classical probabilistic system, which structurally is as close as possible to the wave (Schrödinger) equation of quantum mechanics. In case of the free (V=0) propagation, the heat equation with its time adjoint well fit the purpose:

$$\begin{array}{lcl} i\partial_t\psi = -D\triangle\psi & & \partial_t\theta_* = \ \ D\triangle\theta_* \\ i\partial_t\overline{\psi} = \ \ D\triangle\overline{\psi} & t\rightarrow it & \partial_t\theta \ = -D\triangle\theta \end{array} \tag{1}$$

where the familiar imaginary time transformation is indicated as a pedestrian recipe to pass from quantum theory to statistical physics. Here $\psi, \overline{\psi}$ are complex, while θ_*, θ are real functions, and the diffusion constant D is left unspecified ($D = \hbar/2m$ gives rise to the Schrödinger equation in its standard form). Let us consider the transition probablity density (heat kernel, the Brownian law of random displacements) $h(x,s,y,t)$ for the Brownian motion $(Y_t, t_1 \leq t \leq t_2)$ on the interval $[t_1, t_2]$ i.e.

$$h(x,s,y,t) = [4\pi D(t-s)]^{-1/2}\exp[-\frac{(y-x)^2}{4D(t-s)}] \tag{2}$$

where Y_t takes values in R^1 (Brownian motion in one spatial dimension). If we prescribe the initial particle distribution $\rho_1(x)$ for the random variable Y_t, then all intermediate distributions of $(Y_t, t_1 \leq t \leq t_2)$ are determined in terms of $\rho_1(x)$ and $h(x,s,y,t)$, including the terminal one as well. We have indeed:

$$\rho(x,t) = \int h(z,t_1,x,t)\rho_1(z)\,dz \tag{3}$$

where $t_1 \leq t$. Starting from the classical Brownian law of random displacements we can ask the following question: *Assuming that a test particle originates from x_1 at t_1, and terminates its route in (not far from) x_2 at t_2, what is the probability to find it between x and $x + \triangle x$ at the intermediate time t with $t_1 \leq t \leq t_2$?* The pertinent intermediate probability distribution is given by the conditional transition probability density formula :

$$\rho(x,t) = P(x_1,t_1;x,t;x_2,t_2) = \frac{h(x_1,t_1,x,t)\,h(x,t,x_2,t_2)}{h(x_1,t_1,x_2,t_2)} \tag{4}$$

It is then obvious that the formula for $\rho(x,t)$ can always be rewritten as a product of solutions $\theta(x,t), \theta_*(x,t)$ of the heat equation and its time adjoint: $\rho(x,t) = (\theta\theta_*)(x,t)$ provided $t_1 \leq t \leq t_2$. Let us now define $\rho(x,t_0) = \rho_0(x)$ and $\rho(x,T) = \rho_T(x)$ to be the initial and final probability distributions *determined* by the Bernstein transition density P, with $t_1 < t_0 < T < t_2$. Although we know the general Brownian transition mechanism (the law of random displacements), h , the conditioning present in the Bernstein (in fact Brownian, since V=0) bridge construction allows us to formulate a new probabilistic problem. We are now revisiting the original question due to Schrödinger [5,6]: *What is the most likely way for the particles to evolve as t goes from t_0 to T, once we have prescribed in advance both the initial $\rho_0(x)$ and final (terminal) $\rho_T(x)$ probability densities for the process, given the prior transition mechanism ?* The answer is given by deriving from the original (prior) process $(Y_t, t_1 \leq t \leq t_2)$, the new one $(X_t, t_1 < t_0 \leq t \leq T < t_2)$ which is now known as the Markov-Bernstein process [7]. The above discussion can be rephrased in more phenomenological terms. Suppose that the observer is measuring a coordinate x of the event (particle entering the observation area A, the measurement accuracy does not matter) at time t_0 viewed as the initial time instant in the repeatable series of one particle x experiments. Accumulating the data one arrives at the empirical distribution which asymptotically is found to approximate a probability distribution $\rho_0(x)$. It is then taken to characterise the "state of the system" at time t_0. Assume also that the observer is collecting the coordinate data of these repeatable events (entering particles) in the detection area B at a later time T and let them approximate the terminal probability distribution $\rho_T(x)$. If $\rho_T(x)$ is far from what it should be according to the law of large numbers (i.e. when $\rho_T(x)$ is much different from $\int h(x,t_0,y,T)\rho_0(y)dy$, with h given before), then we arrive at the *core* of the original Schrödinger's discussion: ***What are the intermediate probability distributions $\rho(x,t)$ and what is the particular transition mechanism responsible for the probabilistic evolution from $\rho_0(x)$ to $\rho_T(x)$, if no external forces are affecting the particle, except for Brownian agitation (Brownian noise) ?*** Let us proceed more generally [6] than the previous discussion would suggest, and assume to have given a pair of diffusion equations in duality for real functions θ, θ_* :

$$\begin{aligned} \partial_t\theta_* &= \ \ D\triangle\theta_* - V\theta_*/2mD \\ \partial_t\theta \ &= -D\triangle\theta \ + V\theta \ /2mD \end{aligned} \tag{5}$$

D is a diffusion constant, m the mass of a particle subject to diffusion. The potential V is assumed to be continuous and bounded from below, which implies the existence of the strictly positive semigroup kernel generated by the operator $H = -2mD^2\triangle + V$. Let $h = h(x,s,y,t)$, $s \le t$ be the fundamental solution of the first diffusion equation. Then, the initially chosen function $\theta_*(x,-T/2)$, $0 \le T$ is propagated forward $\theta_*(x,t) = \int \theta_*(z,-T/2)h(z,-T/2,x,t)dz$, $-T/2 \le t$ while the terminal choice of $\theta(x,T/2)$, $0 \le T$ allows one to reproduce the past data $\theta(x,t)$, $t \le T/2$ through the backward propagation $\theta(x,t) = \int h(x,t,y,T/2)\theta(y,T/2)dy$. By virtue of the semigroup property of the kernel h we have also:

$$\begin{aligned} \theta_*(x,t) &= \int \theta_*(z,s)h(z,s,x,t)\,dz \\ \theta(x,s) &= \int h(x,s,z,t)\theta(z,t)\,dz \end{aligned} \tag{6}$$

where $(-T/2 \le s < t \le T/2)$; hence a solution of the dual system with the prescribed boundary data at $\pm T/2$ might be given, such that $\int \theta_*(x,t)\theta(x,t)\,dx = 1$ holds true for all times t in $[-T/2, T/2]$. We have here determined the Markov-Bernstein process [7], which allows one to propagate (hence both predict the future and reproduce the past, given the present) the probability distribution:

$$\rho(x,t) = \theta(x,t)\theta_*(x,t) \tag{7}$$

respectively forward and backward in time. Statistical predictions about the future can be accomplished by means of the forward transition probability density:

$$p(x,s,y,t) = h(x,s,y,t)\,\frac{\theta(y,t)}{\theta(x,s)} \tag{8}$$

while the past can be reproduced statistically by means of the backward density

$$p_*(x,s,y,t) = h(x,s,y,t)\,\frac{\theta_*(x,s)}{\theta_*(y,t)} \tag{9}$$

for the diffusion with fixed boundary probability distributions $\rho(x,-T/2)$ and $\rho(x,T/2)$.

With p and p_* in hand, we can straightforwardly evaluate the conditional expectation values, which are necessary to establish the mean forward and backward derivatives in time for functions of the random variable $X(t)$. The backward $(D_-X)(t) = b_*(x,t)$ and forward $(D_+X)(t) = b(x,t)$ drifts of the Markovian diffusions in the above read thus $b(x,t) = 2D\nabla\theta/\theta$ and $b_*(x,t) = -2D\nabla\theta_*/\theta_*$ so that the continuity equation follows:

$$\partial_t\rho = -\nabla(\rho v) = D\triangle\rho - \mathrm{div}(\rho b) = -D\triangle\rho - \mathrm{div}(\rho b_*) \quad \text{with} \quad v = (b+b_*)/2 \tag{10}$$

If we define $\theta = \exp(R+S)$, $\theta_* = \exp(R-S)$ with R and S being real functions, then there holds:

$$v = 2D\nabla S \qquad u = (b-b_*)/2 = 2D\nabla R \tag{11}$$

and the continuity equation can be rewritten as follows (with its gradient form, due to Nelson [8], included) :

$$(1/2D)\partial_t R = (-1/2)\triangle S - (\nabla R)(\nabla S) \;\rightarrow\; \partial_t u = -D\triangle v - \nabla(uv) \tag{12}$$

If the continuity equation holds true, then the necessary consequence of the dual system of diffusion equations is the (generalised) Hamilton-Jacobi equation:

$$V = 2mD\{\partial_t S + D(\nabla S)^2 + D[(\nabla R)^2 + \triangle R]\} \tag{13}$$

which can furthermore be rewritten as :

$$2mD^2[(\nabla R)^2 + \triangle R] = 2mD^2\,\frac{(\triangle\rho^{[1/2]})}{\rho^{1/2}} = Q \tag{14}$$

Except for the sign inversion, Q has the familiar functional form of the de Broglie-Bohm-Vigier "quantum potential" [5,6,9,10]. We can here argue in reverse, and recover the dual system of diffusion equations, given the continuity and the generalised Hamilton-Jacobi equations. By evaluating the forward and backward time derivatives of $b(x,t)$ and $b_*(x,t)$ we can verify that the gradient form of the Hamilton-Jacobi equation reads:

$$\partial_t v = 2D\triangle u + (1/2)\nabla u^2 + (1/2)\nabla v^2 + (1/m)\nabla V \tag{15}$$

which in turn implies the validity of the so-called Nelson-Newton law [6] with the sign inverted potential (conventionally interpreted as the signal that we are in the Euclidean framework, see however [9,10]):

$$(m/2)(D_+D_+ + D_-D_-)X(t) = \nabla V \tag{16}$$

This stochastic acceleration (in the conditional mean) formula was primarily rejected by Nelson [8] as the physically relevant characteristic of the Markovian diffusion. Notice that the above equation can be written in the form :

$$(\partial_t + v\nabla)v = (-1/m)\nabla(Q - V) = (1/m)\nabla(V - Q) \tag{17}$$

reminiscent of the momentum balance equation in the kinetic theory of gases and liquids, which should in principle apply to all conceivable osmotic diffusions.

To illustrate the previous discussion, let us come back to the previous ($V = 0$) considerations. We can ask for a probabilistic interpolation between the coinciding boundary distributions:

$$\rho(x, t_0) = \rho(x, T) = [\frac{\alpha}{2\pi D(\alpha^2 - \beta^2)}]^{1/2} \exp[-\frac{\alpha x^2}{2D(\alpha^2 - \beta^2)}] \tag{18}$$

in the time interval $[t_0, T]$. No physicist would expect such an evolution while having a traditional picture of the Brownian motion in memory. However it immediately comes out from the Bernstein (actually Brownian) bridge construction. Indeed, let us set $t_0 = -\beta = -T$, $t_1 = -\alpha = -t_2$, $0 < \beta < \alpha$ and choose $x_1 = 0$ as a source of particles. If we confine attention to these Brownian particles only, which after time 2α from their emission are bound to be back at (or at least not far away from) the initial location $x_1 = 0 = x_2$ then the Bernstein transition density P reduces to :

$$\begin{aligned} \rho(x,t) &= [\frac{\alpha}{2\pi D(\alpha^2 - t^2)}]^{1/2} \exp[-\frac{\alpha x^2}{2D(\alpha^2 - t^2)}] \\ &= (8\pi D\alpha)^{1/2} h(0, -\alpha, x, t) h(x, t, 0, \alpha) \end{aligned} \tag{19}$$

where the transition densities:

$$\begin{aligned} h(0, -\alpha, x, t) &= [4\pi D(t + \alpha)]^{-1/2} \exp[-\frac{x^2}{4D(t + \alpha)}] \\ h(x, t, 0, \alpha) &= [4\pi D(\alpha - t)]^{-1/2} \exp[-\frac{x^2}{4D(\alpha - t)}] \end{aligned} \tag{20}$$

solve the systems of heat equations in duality ($V = 0$), in the time interval $[-\alpha, \alpha]$. Notice that the pertinent boundary data are induced by the Bernstein (Brownian) bridge itself. By previous arguments we have given here the transition probability densities characterising Markovian diffusions, hence the drifts of the processes can be evaluated as follows:

$$\begin{aligned} b(x,t)\Delta t &\simeq \int p(x, t, y, t + \Delta t) y dy \; - \; x \\ b_*(x,t)\Delta t &\simeq x \; - \int p_*(z, t - \Delta t, x, t) z dz \end{aligned} \tag{21}$$

where Δt is a small time increment, b is the forward drift of the diffusion (mean velocity of outgoing –from x at t – particles) while b_* is its backward drift (mean velocity of incoming – to x at t – particles) $t_0 \leq t \leq T$. The drifts stand for substitutes of time derivatives, non-existent in the naive sense for Wiener paths. Here $(D_+X)(t) = b(x,t)$ is the left time derivative in the conditional mean, while $(D_-X)(t) = b_*(x,t)$ is the right one. Up to irrelevant multiplicative constants we can set :

$$\begin{aligned} \theta_*(x,t) &\sim h(0,-\alpha,x,t) \\ \theta(x,t) &\sim h(x,t,0,\alpha) \end{aligned} \qquad (-\beta \leq t \leq \beta < \alpha) \tag{22}$$

and consequently there holds:

$$\begin{aligned} b(x,t) &= 2D\frac{\nabla\theta}{\theta} = -\frac{x}{\alpha - t} = -\frac{x(\alpha+t)}{\alpha^2 - t^2} = u + v \\ b_*(x,t) &= -2D\frac{\nabla\theta_*}{\theta_*} = \frac{x}{\alpha+t} = \frac{x(\alpha - t)}{\alpha^2 - t^2} = -u + v \end{aligned} \tag{23}$$

We have here a natural decomposition into two terms, one of which (namely v) is odd while the other (namely u) is even with respect to the time reversal. In addition, we can easily recover the relations:

$$\begin{aligned} &u = 2D\nabla R && v = 2D\nabla S \\ &R = -\frac{\alpha x^2}{4D(\alpha^2 - t^2)} && S = -\frac{t x^2}{4D(\alpha^2 - t^2)} \\ &\theta_* \sim \exp(R - S) && \theta \sim \exp(R + S) \\ &\rho(x,t) = (\theta\theta_*)(x,t) \sim \exp 2R \quad \rightarrow && u = D\frac{\nabla\rho}{\rho} = 2D\nabla R \end{aligned} \tag{24}$$

Notice the validity of the momentum balance equation (acceleration formula) $(D_+D_+ + D_-D_-)X(t) = 0$.

Acknowledgement: I would like to express my warm thanks to Professors R. Kerner and J.P. Vigier for hospitality in Paris, to Professor Kerner again for the emergency file processing tutorial, and to Professor G.G. Emch for the time amnesty which made preparation of this contribution possible.

REFERENCES

1. P. Garbaczewski, J.Phys. **A 20** (1987), 1277.
2. J.R. Klauder, Ann.Phys.(NY) **11** (1960), 123.
3. P. Garbaczewski, *Classical and Quantum Field Theory of Exactly Soluble Nonlinear Systems*, World Scientific, Singapore, 1985.
4. E. Schrödinger, Ann. Inst. H.Poincaré **2** (1932), 269.
5. P. Garbaczewski, J.P. Vigier, *Brownian Motion and its Descendants According to Schrödinger*, submitted.
6. P. Garbaczewski, *Physical Significance of the Nelson-Newton Laws*, submitted.

7. J.C. Zambrini, J.Math.Phys. **27** (1986), 2397.
8. E. Nelson, *Quantum Fluctuations*, Princeton Univ.Press, Princeton, 1985.
9. P. Garbaczewski, Phys.Lett. **A 162** (1992), 129.
10. P. Garbaczewski, Phys.Lett. **A 164** (1992), 6.

Laboratoire de Physique Théorique, Université Paris VI, F-75231 Paris, FRANCE ; and Institute of Theoretical Physics, University of Wrocław, PL-50205 Wrocław POLAND *(permanent address)*

ON COMPLEX LANGEVIN

H. GAUSTERER

ABSTRACT. Some problems and peculiarities of complex Langevin are discussed. Special attention is given to the equivalence of the $2n$-variable Fokker-Planck equation and the n-variable pseudo Fokker-Planck equation with a complex drift term.

Contribution in honor of the 60th birthday of John R. Klauder

1. Introduction

Standard stochastic techniques like the Metropolis et al. algorithm [1] for solving multidimensional integrals have proven to be very successful. These techniques are normally called Monte Carlo (MC) integration or simulation. The above mentioned algorithm is just one of these approaches. Let us assume that we have a typical canonical ensemble for some variables $x_i \in \mathcal{X}$ with

$$dF(\{x\}) = f(\{x\}) \prod_i d\mu(x_i) \sim e^{-S(\{x\})} \prod_i d\mu(x_i),$$

where $S(\{x\}) \in \mathcal{R}$ is the Hamiltonian of a classical statistical mechanics system or the action of a lattice field theory. The idea is to generate a Markov sequence $\{\{X_n\} : n \geq 1\}$ with a distribution density $h(\{x_n\})$. This Markov sequence converges in distribution to $\{X\}$ with distribution density $f(\{x\})$. This is

$$h(\{x_n\}) \to f(\{x\}) \quad , \quad \text{for } n \to \infty.$$

For instance this can be achieved by constructing a proper transition density (standard MC) or solving numerically a Langevin equation. If S is a real valued function then a suitable implementation of these procedures will converge (see e.g. [2,3,4,5,6] or any other suitable textbook).

So, if $S(\{x\})$ is real valued then there is a probabilistic interpretation and probabilistic (stochastic) methods can be applied. Unfortunately there are several systems which are described by complex actions like quantum spins or fermions represented by a continuous action or the problem of lattice QCD at finite chemical potential. If $S(\{x\})$ is complex valued then there is no straightforward probabilistic interpretation and the above techniques can not be applied straightforwardly. For complex actions the Metropolis et al.

approach fails to apply since this would lead to complex or at least negative transition densities which are not defined. J.R. Klauder proposed in [2] that such systems could be simulated based on the Langevin method. The idea behind complex Langevin is that for a Langevin equation there is formally no restriction to real drift terms and thus one could try a numerical simulation [2]. The problem with this approach is that the mechanism when and why one gets the correct answer is somewhat obscured. Applied to complex action problems this technique has worked well for most of the attempts and has proven to be a powerful tool to solve such problems [7].

2. Complex Langevin: A Real Process in $2n$ Dimensions

To elucidate the problem we will restrict the discussion to systems with one degree of freedom $x \in \mathcal{R}$ which will be described by a complex action $S : \mathcal{R} \to \mathcal{C}$ with $\mathrm{Re}[S(x)]$ bounded from below.

$$\langle g(x) \rangle = \frac{1}{N} \int_{\mathcal{R}} g(x) e^{-S(x)} dx,$$

$$N = \int_{\mathcal{R}} e^{-S(y)} dy,$$

provided that for the partition function N we have $0 < N < \infty$. $g(x)$ is such that $g(x) \exp(-S(x)) \in L^1(\mathcal{R}, dx)$ and $g(z)$ is analytic and locally bounded in $\mathcal{C}$. A typical example would be polynomial. For S we have that $S(z)$ is at least continuously differentiable twice with respect to z[1], thus $S(z)$ is analytic. For that action one can construct formally a Langevin equation

$$dZ(\tau) = F(Z(\tau))d\tau + dW(\tau), \tag{2.1}$$

with the driving force

$$F(z) = -\frac{1}{2}\frac{dS(z)}{dz}.$$

$W(\tau)$ is a standard Wiener process with zero mean and covariance

$$< W(\tau_1)W(\tau_2) >= \min(\tau_1, \tau_2).$$

This is actually a two–dimensional process of the form

$$dX(\tau) = G(X(\tau), Y(\tau))d\tau + dW(\tau),$$

$$dY(\tau) = H(X(\tau), Y(\tau))d\tau$$

with $S(z) = u(x,y) + iv(x,y)$

$$G(x,y) = -\frac{1}{2}\frac{\partial u(x,y)}{\partial x} \quad , \qquad H(x,y) = \frac{1}{2}\frac{\partial u(x,y)}{\partial y}.$$

This two–dimensional process described by the above Langevin equation has a well defined distribution density $f(x,y,t)$. The first crucial question is whether this process converges to

[1] If $S(z)$ obeys this condition, then the Lipschitz condition holds for equation (2.1).

a strongly stationary process. The second crucial question is, if the process has converged, whether for the stationary density $\hat{f}(x,y)$ the following holds:

$$E(g(X+iY)) = \int_{\mathcal{R}^2} g(x+iy)\hat{f}(x,y)dxdy = \frac{1}{N}\int_{\mathcal{R}} g(x)e^{-S(x)}dx. \tag{2.2}$$

First, we will discuss whether the above process defined by the Langevin equation can converge to a stationary density. Now, for any diffusion process with a twice continuously differentiable distribution density with respect to x, y and once with respect to τ, there exists a Fokker-Planck (F-P) equation (Kolmogorov forward equation) [5]. Then for the above Langevin equation we have

$$\frac{\partial f(x,y,\tau)}{\partial \tau} = T_{FP} f(x,y,\tau) \tag{2.3}$$

with

$$T_{FP} = -G_x(x,y) - G(x,y)\frac{\partial}{\partial x} - H_y(x,y) - H(x,y)\frac{\partial}{\partial y} + \frac{1}{2}\frac{\partial^2}{\partial x^2}.$$

The question which arises now is whether this F-P equation has a stationary solution which is a stationary density. This is

$$T_{FP}\hat{f}(x,y) = 0.$$

Since $\hat{f}(x,y) \geq 0$ we can define a $\Phi(x,y)$ by

$$\hat{f}(x,y) = \frac{1}{N}e^{-\Phi(x,y)},$$

$$N = \int_{\mathcal{R}^2} e^{-\Phi(x,y)}dxdy.$$

This F-P equation unfortunately obeys neither the so–called potential condition nor the detailed balance condition. This is identical to the fact that the drift term cannot be written as a gradient of some 'potential'. Therefore there is no straightforward method to construct Φ. Further, we do not know whether there is a unique stationary density without investigating the drift term in more detail. This is one of the major differences to the real action case where the drift term is the gradient of the action. In that case Φ is known explicitly. For the moment we will assume that at least one $\hat{f}$ exists.

Using this stationary density one has

$$T_{FP} = \frac{1}{2}e^{-\Phi/2}(-T)e^{\Phi/2}$$

with $T = T_S + T_A$, where T_S is a Schrödinger type operator.

$$T_S = -\frac{\partial^2}{\partial x^2} + V(x,y)$$

with the potential

$$V(x,y) = \frac{1}{4}\Phi_x^2(x,y) - \frac{1}{2}\Phi_{xx}(x,y).$$

For T_A one gets

$$T_A = 2e^{\Phi/2}\left[\frac{\partial}{\partial x}A_1(x,y) + \frac{\partial}{\partial y}A_2(x,y)\right]e^{-\Phi/2}$$

with

$$A_1(x,y) = G(x,y) + \Phi_x(x,y), A_2(x,y) = H(x,y).$$

We have $V(x,y)$ bounded from below and $T_S \geq 0$. Since $\hat{f}(x,y)$ is a stationary distribution arising from the F-P equation, T_S is essentially self-adjoint on $C_0^\infty(\mathcal{R}^2)$ [8,9]. If the infimum of the spectrum $\sigma(T_S)$ is an eigenvalue then this will be a simple eigenvalue. T_A is also densely defined on $C_0^\infty(\mathcal{R}^2)$. Unfortunately $T = T_S + T_A$ is not a normal operator, since in general $[T_S, T_A] \neq 0$. We can therefore not assume that T has a spectral representation. It might, however, be the generator of a contracting semigroup.

First, we note that $\exp[-\Phi(x,y)/2]$ is an eigenstate with zero eigenvalue for T_S and T_A. This eigenvalue is not degenerate for T_S (see above) which is sufficient in our case.

$$T_S e^{-\Phi/2} = T_A e^{-\Phi/2} = 0.$$

$T = T_S + T_A$ is densely defined on $C_0^\infty(\mathcal{R}^2)$. Further we have

$$\mathrm{Re}(T\phi,\phi) \geq 0 \text{ for all } \phi \in D(T),$$

thus the operator is closable and the above equation holds also for the closure [8]. Since the same holds for the adjoint operator the assumptions for the Hille-Yosida theorem hold for T. Therefore T generates a contracting semigroup [8,10]. T_{FP} generates a contracting semigroup in $L^2(\mathcal{R}^2, 1/N\exp(-\Phi)dx^2)$. If T_{FP} has a simple zero eigenvalue, then we can conclude for the pointwise convergence [8,10]:

$$\lim_{\tau\to\infty} e^{L_{FP}\tau} f(x,y,\tau) = \hat{f}(x,y), \text{ almost everywhere.}$$

If the zero eigenvalue of T_{FP} is M-fold degenerate then there are M ergodic classes. In this case we have

$$\lim_{\tau\to\infty} e^{L_{FP}\tau} f(x,y,\tau) = \sum_{i=1}^{M} c_i \hat{f}_i(x,y), \text{ almost everywhere.}$$

This scenario will exist if the drift term is not locally bounded which we have excluded, or allows for run-away solutions of the deterministic problem [4,6]. If T_{FP} has a unique stationary density $\hat{f}(x,y)$ then the process $(X(\tau), Y(\tau))$ defined by equation (2.1) converges to a unique strongly stationary process with the distribution density $\hat{f}(x,y)$.

To check whether equation (2.2) holds we make the following ansatz.

$$E(g(X(\tau) + iY(\tau)) = \int_{\mathcal{R}^2} g(x+iy)f(x,y,\tau)dxdy = \int_{\mathcal{R}} g(x)h(x,\tau)dx, \qquad (2.4)$$

for all τ. $h(x,\tau)$ is twice continuously differentiable with respect to x and once with respect to τ. If we define now the process $(X(\tau), Y(\tau))$ by the Langevin equation (2.1) then we obtain for $f(x,y,\tau)$ the F-P equation (2.3) and for $h(x,\tau)$ a pseudo Fokker-Planck equation with a complex drift term. We have

$$\frac{\partial h(x,\tau)}{\partial \tau} = \frac{1}{2}\frac{\partial}{\partial x}\left[\frac{\partial S(x)}{\partial x} + \frac{\partial}{\partial x}\right] h(x,\tau) = \tilde{T}_{FP} h(x,\tau). \tag{2.5}$$

One immediately sees that $\tilde{T}_{FP}$ has a non-degenerate zero eigenvalue with the eigenfunction ('stationary density')

$$\hat{h}(x) \sim e^{-S(x)}.$$

The properties of this differential operator have been discussed in [11]. If one can show that this operator generates a contracting semigroup (see [11]) then one has

$$\lim_{\tau \to \infty} h(x,\tau) = \hat{h}(x), \text{ almost everywhere.}$$

Let us assume that T_{FP} is of the form that, if it has a stationary density, then this stationary density will be unique. Now, since equation (2.4) holds for all τ for certain solutions of equations (2.3) and (2.5) and since the operators $\exp(T_{FP}\tau)$ and $\exp(\tilde{T}_{FP}\tau)$ project, in this case, onto unique states for $\tau \to \infty$, we can conclude the following: If $\tilde{T}_{FP}$ is the generator of a contracting semigroup with a stationary solution as above, then T_{FP} has a unique stationary density $\hat{f}(x,y)$ and the equality in equation (2.2) holds. In general it will be very complicated or almost impossible to treat T_{FP} explicitly and to find for instance the stationary density. Already the stationary density for the very simple problem of $S(x) = \alpha x^2$ with $\alpha \in (0,\infty) \times i\mathcal{R}$ is surprisingly complicated (Haymaker and Peng, or Ambjorn and Yang in [7]). If we are in the situation where there is no unique stationary density, then there are solutions of T_{FP} for which (2.4) holds for all τ. But one can not be sure that if the process has converged[2] that the answer one gets from the simulation is the desired one. The scenario which will most likely occur in this situation is that the convergence will be so slow that within a realistic simulation time no convergence will be observed or that one experiences pseudo divergences, which are excursions of the process so that one obtains a floating point overflow. The conclusion is that if the process has converged 'nicely' one can be 'almost' sure that one got the correct answer.

The above discussion can be generalized to the n-dimensional case ($S : \mathcal{R}^n \to \mathcal{C}$) without further complications. It would be very interesting to find a general classification scheme for actions which are suitable for a complex Langevin simulation.

3. Outlook

Complex Langevin has proven to be a very powerful tool to approach problems which are defined by complex actions or Hamiltonians. The method is still, when applied to some realistic problem, speculative, since there is no guarantee for a positive outcome of

[2]It is extremely unlikely that a process can numerically converge under the assumption that the deterministic problem allows for run-away solutions.

the numerical experiment. This is to say that for most of the complex actions one does not even know whether $\tilde{T}_{FP}$ has the right properties apart from the complications with T_{FP}. This may sound a little bit desperate but it should be mentioned again that many of the problems tackled so far have behaved well.

Acknowledgement

I wish to thank J.R Klauder for his interest in my scientific career as well as his friendship. It is a pleasure to collaborate with him. I would like to take this opportunity to wish him a lot of exciting and interesting years to come.

References

1. N. Metropolis et al., J. Chem. Phys **21** (1953), 1087.
2. J.R. Klauder, *in: Recent Developments in High Energy Physics*, (H. Mitter and C.B. Lang, eds.), Springer, Wien, New York, 1983.
3. G.R. Grimmett and D.R. Stirzaker, *Probability and Random Processes*, Claredon Press, Oxford, 1982; W. Feller, *An Introduction to Probability Theory and its Application, Vol. 1*, John Wiley & Sons, New York, London, 1968; W. Feller, *An Introduction to Probability Theory and its Application, Vol. 2*, John Wiley & Sons, New York, London, 1971.
4. Y.V. Prohorov and Y.A. Rozanov, *Probability Theory*, Springer, Berlin, 1969.
5. L. Arnold, *Stochastic Differential Equations*, Wiley, New York, 1974.
6. H. Risken, *The Fokker-Planck Equation*, Springer, Berlin, 1984; C.W. Gardiner, *Handbook of Stochastic Methods*, Springer, Berlin, 1983.
7. H. Gausterer and J.R. Klauder, Phys. Lett. B **164** (1985), 127; H. Gausterer and J.R. Klauder, Phys. Rev. D **33** (1986), 3678; H. Gausterer and E.M. Ilgenfritz, Phys. Lett. B **193** (1987), 476; H. Gausterer and J.R. Klauder, Phys. Rev. Lett. **56** (1986), 306; H. Gausterer and J.R. Klauder, Phys. Lett. B **177** (1986), 391; N. Bilic, H. Gausterer and S. Sanielevici, Phys. Lett. B **198** (1987), 235; N. Bilic, H. Gausterer and S. Sanielevici, Phys. Rev. D **37** (1988), 3684; H. Gausterer, S. Sanielevici, M.F.L. Golterman and D.N. Petcher, Nucl. Phys.B (Proc. Suppl) **20** (1991), 581; J. Ambjorn, M. Flensburg and C. Peterson, Phys. Lett. B **159** (1985), 335; J. Ambjorn and S.K. Yang, Nucl. Phys. B **275** (1986), 18; F. Karsch and H.W. Wyld, Phys. Rev. Lett **55** (1985), 2242; J. Flower, S.W. Otto and S. Callahan, Phys. Rev. D **34** (1986), 598; R.W. Haymaker and J. Wosiek, Phys. Rev. D **37** (1988), 969; R.W. Haymaker and Y. Peng, Phys. Rev. D **41** (1990), 1269; H. Okamoto, K. Okano, L. Schülke and S. Tanaka, Nucl. Phys. B **324** (1989), 684; K. Okano, L. Schülke and B. Zheng, Phys. Lett. B **258** (1991), 421; H. Gausterer and C.B. Lang, Phys. Rev. D **36** (1987), 1229; J.R. Klauder and S. Lee, UFL preprint (1991).
8. M. Reed and B. Simon, *Methods of Modern Mathematical Physics II*, Academic Press, New York, 1975; M. Reed and B. Simon, *Methods of Modern Mathematical Physics IV*, Academic Press, New York, 1978.
9. J. Weidmann, *Lineare Operatoren in Hilberträumen*, B.G. Teubner, Stuttgart, 1976.
10. K. Yosida, *Functional Analysis*, Springer, Berlin, 1968.
11. J.R. Klauder and W.P. Petersen, J. Stat. Phys. **39** (1985), 53.

Institut für Theoretische Physik, Universität Graz, A-8010 Graz, AUSTRIA

PATH INTEGRALS AND NETWORK QUANTUM NUMBERS

R. GILMORE AND MARK JEFFERY

Contribution in honor of the 60th birthday of John R. Klauder

1. Introduction

The first successful quantum mechanical path integral calculation was done by deBroglie [1]. He showed how a non-interference, or single-valuedness, condition led directly to integer quantum numbers and the Bohr quantization condition [2] for the circular orbits of the hydrogen atom (Fig. 1a).

A comparable path integral argument can be used to describe the Aharonov-Bohm effect (Fig. 1b) [3]. A charged particle enters an annular region at port 1 and proceeds to travel to port 2 either through the upper or the lower region of the annulus. If the wavefunction entering port 1 is $\psi(1)$ and the amplitudes at this port for traversing the upper and lower paths are $\psi_+(1)$ and $\psi_-(1)$, the amplitude at port 2 is $\psi(2) = \psi_+(2) + \psi_-(2)$, where $\psi_+(2) = e^{ikl_+}\psi_+(1)$ and $\psi_-(2) = e^{ikl_-}\psi_-(1)$. Here $k = p/\hbar$ is the wave number of the particle of momentum p and l_+, l_- are the lengths of the paths from port 1 to port 2 along the upper and lower parts of the annulus. When the annulus surrounds a nonzero magnetic field, the amplitudes $\psi_\pm(2)$ are changed by the magnetic action of the particle along the path traversed

$$\begin{aligned} \psi_\pm(2) \rightarrow & \quad \psi_\pm(2)e^{i\phi_\pm} \\ \phi_\pm = & \quad 2\pi\frac{q}{hc}\int_1^2 \mathbf{A}\cdot d\mathbf{s} \end{aligned} \tag{1}$$

where the line integral is carried out in the upper or lower part of the annulus. The output intensity at port 2 exhibits magnetic field dependence [4]

$$\begin{aligned} |\psi(2)|^2 &= |\psi_+(2)e^{i\phi_+} + \psi_-(2)e^{i\phi_-}|^2 \\ &= |\psi_+(2) + \psi_-(2)e^{2\pi i\Phi/\Phi_0}|^2 \end{aligned} \tag{2}$$

where $\Phi = \oint \mathbf{A}\cdot d\mathbf{s}$ is the flux through the annulus and Φ_0 is the quantum of flux associated with a particle of charge $q : \Phi_0 = hc/q$.

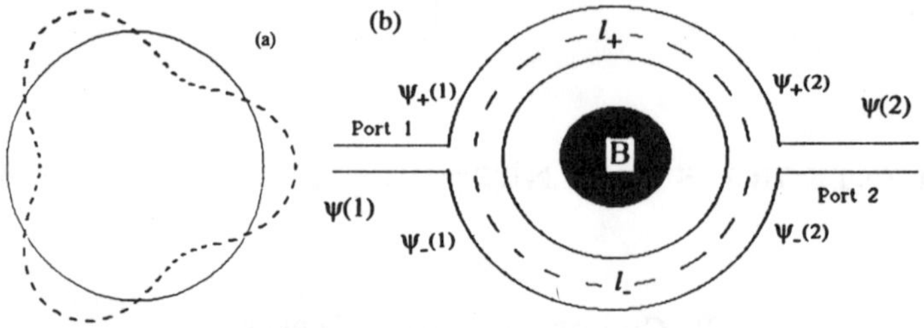

FIGURE 1: (A) DEBROGLIE ARGUED THAT AN ELECTRON COULD OCCUPY A CIRCULAR ORBIT AROUND A PROTON IF THE CIRCUMFERENCE OF THE ORBIT WAS AN INTEGER NUMBER OF WAVELENGTHS $\lambda = h/p$. OTHER ELECTRON STATES WOULD NOT BE ALLOWED BECAUSE OF DESTRUCTIVE INTERFERENCE. (B) AN ELECTRON TRAVERSING AN ANNULUS WILL EXHIBIT MAGNETIC INTERFERENCE PATTERNS IF THE ANNULUS ENCLOSES A MAGNETIC FIELD.

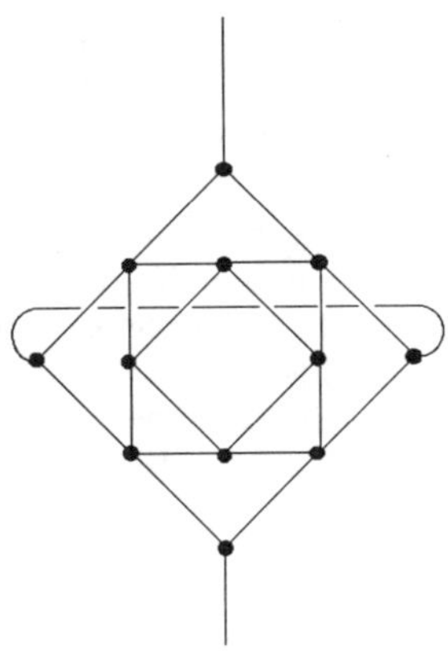

FIGURE 2: WHEN AN ELECTRON CAN TRAVERSE MANY DIFFERENT PATHS BETWEEN TWO POINTS, FINDING AN ANALOG OF DEBROGLIE'S INTERFERENCE, OR SINGLE VALUEDNESS, CONDITION IS NOT SIMPLE.

The two path integral calculations illustrated in Fig. 1 are relatively simple. In both instances it is sufficient to compute a phase change generated by a single passage around a closed loop. In cases where there are many distinct paths connecting the starting point to the end point (cf. Fig. 2) the computation of a path integral is not quite so intuitive.

In this work we will summarize the procedures for computing path integrals on quantum networks. These have one–dimensional multiply connected geometries. The computations will be carried out for arbitrary magnetic fields. The principal results are the following:

• In a network with n_l independent loops it is possible to construct up to n_l independent single valued wavefunctions;

• Each single valued wave function can support a supercurrent flow. The allowed supercurrent flows for each wave function are quantized in integer steps;

• For a fixed set of single valued wave functions the network energy is quantized by the integers which label the supercurrent flow;

• The minimum energy, over all quantized supercurrent flows, is the network Meissner effect.

2. Quantum Networks

The networks which we consider consist of one–dimensional wires connected together at various nodes. Two such networks are shown in Fig. 3. The network shown in Fig. 3a consists of six equal length wires joined three at a time at four nodes with tetrahedral symmetry. The network shown in Fig. 3b consists of ten wires with four nodes and tetrahedral symmetry. This consists of the tetrahedral network shown in Fig. 3a together with four additional wires which act as links to the external world. Since quantum networks with external links have been studied extensively [5,6,7], we will concentrate here on closed networks without external links.

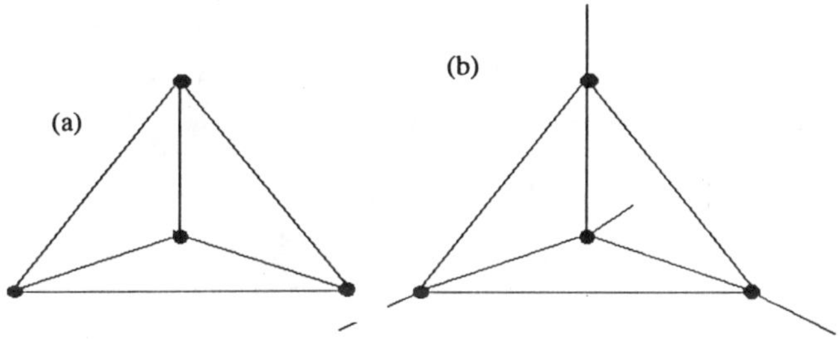

FIGURE 3: (A) SIX WIRES ARE CONNECTED WITH TETRAHEDRAL SYMMETRY AT FOUR NODES. (B) THE SIX WIRE NETWORK OF (A) IS CONNECTED TO FOUR EXTERNAL LEADS.

It is clear from Fig. 2 that the difficulty in constructing single–valued wavefunctions on quantum networks arises from the scattering properties of particles at the network nodes. It is therefore extremely useful to approach this problem directly by introducing quantities which describe the scattering properties of the individual components of a network.

These quantities are the S-(scattering) matrices for the nodes and links. The scattering matrix for an arbitrary network element (Fig. 4) is constructed as follows. An input i_1 to a passive linear system at port 1 produces an output at ports $1, 2, ..., n$ $(o_1, o_2, ..., o_n)$. These quantities are linearly related: $o_j = S_{j1} i_1$. The matrix element S_{j1} is therefore the ratio $o_j / i_1 = S_{j1}$. By linearity and symmetry, this is true for all ports. The S-matrix defines a linear input-output relationship

$$\begin{aligned} [o] &= S[i] \\ o_s &= S_{sr} i_r \end{aligned} \tag{3}$$

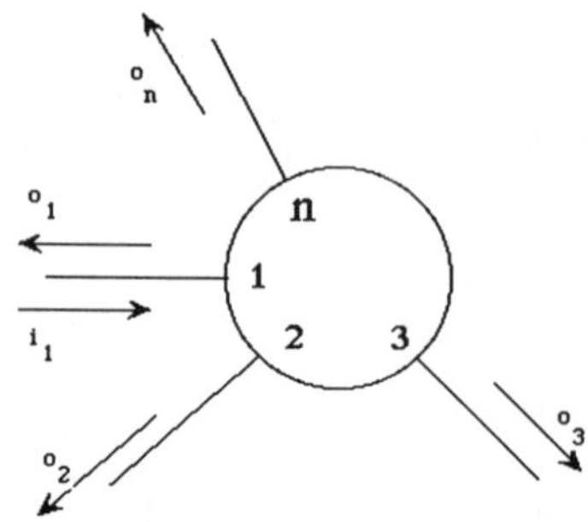

FIGURE 4: AN ARBITRARY COMPONENT OF A NETWORK HAS n EXTERNAL LEADS, OR INPUT PORTS. THE OUTPUT AMPLITUDES o_s ARE PROPORTIONAL TO THE INPUT AMPLITUDES FOR A PASSIVE LINEAR ELEMENT. THE PROPORTIONALITY CONSTANT IS AN ELEMENT OF THE S-MATRIX: $S_{sr} = \partial o_s / \partial i_r$.

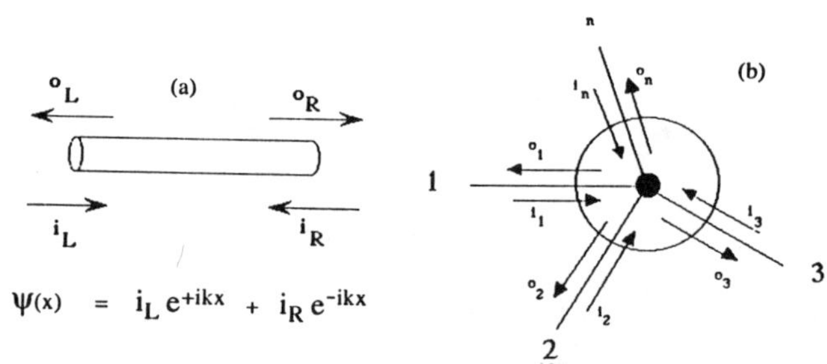

FIGURE 5: (A) THE WAVE FUNCTION ALONG A WIRE IS COMPLETELY DETERMINED BY SPECIFYING ANY TWO OF THE FOUR AMPLITUDES $i_L, o_L; i_R, o_R$. (B) THE INPUT AMPLITUDES AT A NODE ARE THE OUTPUT AMPLITUDES FROM THE WIRES CONNECTED AT THAT NODE. SIMILARLY, THE NODE OUTPUT AMPLITUDES ARE THE WIRE INPUT AMPLITUDES. THE NODE S- MATRIX Γ RELATES WIRE OUTPUT AMPLITUDES TO WIRE INPUT AMPLITUDES.

The wavefunctions along a wire (Fig. 5a) can be expressed in terms of the input and output amplitudes at the left hand end,

$$\psi(x) = i_L e^{ikx} + o_L e^{-ikx} \tag{4}$$

in terms of the amplitudes at the right hand end,

$$\psi(x) = i_R e^{-ikx} + o_R e^{ikx} \tag{5}$$

in terms of the input amplitudes i_L, i_R, or the output amplitudes o_L, o_R. Since the Schrödinger equation is second order, only two of these four amplitudes are independent.

The linear relations among them are simply summarized by the wire S-matrix

$$\begin{bmatrix} o_L \\ o_R \end{bmatrix} = \begin{bmatrix} S_{11} & S_{12} \\ S_{21} & S_{22} \end{bmatrix} \begin{bmatrix} i_L \\ i_R \end{bmatrix} \tag{6}$$

In any network, the inputs to the node are the outputs from the wires. The S-matrix for a node (Fig. 5b) provides a linear relation between the output amplitudes for the wires joined at that node and the wire input amplitudes

$$[i] = \Gamma[o] \tag{7}$$

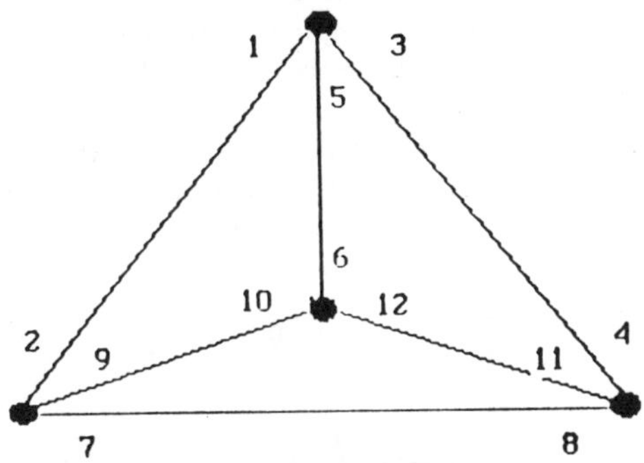

FIGURE 6: THE 12 PORTS FOR THE SIX WIRES IN THE TETRAHEDRAL NETWORK SHOWN IN FIG. 3A MAY BE NUMBERED AS SHOWN. AN INPUT AND OUTPUT AMPLITUDE IS DEFINED AT EACH PORT. THE INPUT AMPLITUDES AT THE TWO ENDS OF EACH WIRE DEFINE THE WAVE FUNCTION ALONG THAT WIRE.

The network shown in Fig. 6 consists of six wires and four nodes. Each wire is represented by a 2×2 S-matrix, each node by a 3×3 S-matrix. The input amplitudes $i_1, ..., i_{12}$ and output amplitudes $o_1, ..., o_{12}$ are related by the wire S-matrix

$$[o] = S[i] \tag{8}$$

This S-matrix is a direct sum of the six 2×2 wire S-matrices. Similarly, the inputs are related to the outputs by the node S-matrices

$$[i] = \Gamma[o] \tag{9}$$

Here Γ is a direct sum of the four 3×3 node S-matrices in a permuted basis [order $(1,3,5), (2,7,9), (4,8,12), (6,10,11)$].

The node and wire S-matrices have the following properties:

1. They are unitary. In the absence of magnetic fields they are time reversal invariant. That is, they are also symmetric;

2. In the presence of a magnetic field each S-matrix element is multiplied by a phase factor

$$S_{sr}(\mathbf{0}) \to S_{sr}(\mathbf{B}) = S_{sr}(\mathbf{0})e^{i2\pi\phi_{sr}} \tag{10}$$

$$\phi_{sr} = \frac{q}{hc}\int_r^s \mathbf{A}\cdot\mathbf{ds} = \Phi/\Phi_0 \tag{11}$$

The phase change induced by the magnetic field is proportional to the magnetic action $\Phi = \int_r^s \mathbf{A}\cdot\mathbf{ds}$ between ports r and s [4,6,7]. In particular:

- diagonal matrix elements are independent of $\mathbf{B}$;
- if ports r and s are very close the magnetic action between them can be neglected. We will assume that the nodes are sufficiently small so that the magnetic field effects at the nodes may be neglected;

3. For superconducting wires, the diagonal matrix elements of the wire S-matrices vanish, by an argument due to Laudauer [8]. Thus, for a network constructed of superconducting wires and operated below the transition temperature T_c, the wire S-matrices have the form

$$S(\mathbf{B}) = e^{i\psi}\begin{bmatrix} 0 & e^{i\alpha} \\ e^{-i\alpha} & 0 \end{bmatrix} \tag{12}$$

$$\alpha = 2\pi \int_r^s \mathbf{A}\cdot\mathbf{ds}/\Phi_0 \tag{13}$$

4. The current flow into any wire at the 'left hand' end is proportional to $|i_L|^2 - |o_L|^2$. By the unitarity of wire S-matrices, this is $|o_R|^2 - |i_R|^2$. Further, if the wire is superconducting, then by (12), $|o_L|^2 = |i_R|^2$ and the current flow into the left hand end of the wire is proportional to $|i_L|^2 - |i_R|^2$.

3. Network Matrix and Single Valuedness

The linear relation between the input amplitudes and the output amplitudes provided by the wire S-matrices and the linear relation between the wire output amplitudes and input amplitudes provided by the node S-matrices Γ can be combined

$$\begin{aligned} [o] &= S(\mathbf{B})[i] \\ [i] &= \Gamma[o] \end{aligned} \tag{14}$$

$$[i] = \Gamma S(\mathbf{B})[i] \tag{15}$$

The matrix $\Gamma S(\mathbf{B})$ is called the *network matrix.*

In general, the network matrix is not the identity matrix. The network matrix equation always has the trivial solution $[i] = 0$, which is uninteresting. There is a nontrivial solution only if the network matrix has an eigenvalue $+1$. In this case it can be shown that the corresponding wave function is single valued. We do not show this here; rather, we indicate in Fig. 7 how this can be done.

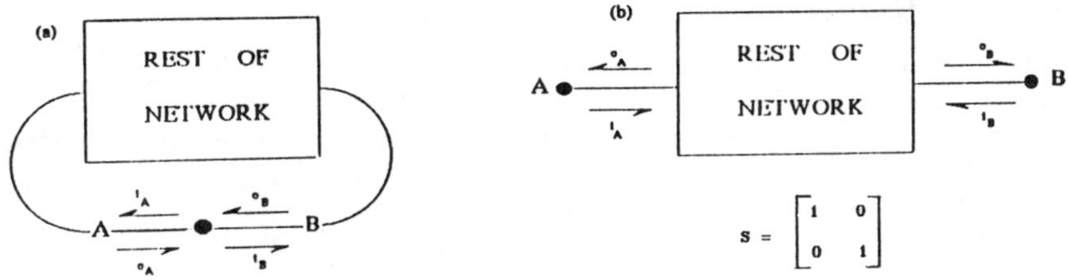

FIGURE 7: (A) AN EXTRA NODE IS PLACED ON ANY WIRE IN THE NETWORK. AN EIGENVECTOR OF THIS AUGMENTED NETWORK WHICH SATISFIES THE NETWORK QUANTIZATION CONDITION IS FOUND. (B) THE WIRE IS CUT AT THE NODE, PRODUCING A NETWORK WITH TWO EXTERNAL LEADS. THE S- MATRIX FOR THIS NETWORK AND EIGENSTATE IS AS SHOWN, SHOWING THAT $o_A = i_B$, $o_B = i_A$. THERE IS AN OVERALL PHASE CHANGE OF $2\pi m$ AS THE PARTICLE PASSES THROUGH THE NETWORK ALONG ALL POSSIBLE PATHS.

Choose any wire in the network, and place a node in it, as shown in Fig. 7a. Choose an eigenvector $[i]$ of the network matrix $\Gamma S(\mathbf{B})[i] = [i]$ with eigenvalue $+1$ for this augmented network. Now cut the wire at the node (Fig. 7b). The resulting network has two output leads (cf., Fig. 2). The amplitudes i_A, i_B, o_A, o_B of this eigenstate at the cut node are related to each other in this new network by the 2×2 S-matrix shown. This means that a particle can enter the network at A, traverse all possible paths through it, and on returning to B will arrive at the node without any phase change. This is the exact network analog of the deBroglie argument shown in Fig. 1a. This can only take place in eigenstates of the network matrix with eigenvalue $+1$. For eigenvectors with eigenvalues not equal to $+1$ there is a phase change on passage of the particle through all possible paths back to its starting point. This phase change is responsible for destructive interference of the wave function, exactly as in deBroglie's treatment of the hydrogen atom.

The network quantization condition is the condition that one (or more) eigenvalues of the network matrix $\Gamma S(\mathbf{B})$ have value $+1$. When the network quantization condition is satisfied, the following results are true:

1. The wave function constructed from the components of the eigenvector $[i]$ with eigenvalue $+1$, following the prescription given in (5) for each individual wire, is everywhere single valued;

2. When two or more eigenvalues of $\Gamma S(\mathbf{B})$ are $+1$, the corresponding wave functions span the linear vector space of single valued wave functions on the network;

3. The maximal number of independent single valued wave functions on a network is n_l, the number of independent loops in the network.

We emphasize that the network quantization condition is a nonlinear condition defined on the matrices which describe a linear theory. This implies that we can use this theory to compute more than probabilities.

4. Network Reduction

At this point it is useful to simplify our arguments by replacing the original network by a simpler but equivalent network. This is done as follows. The number of independent loops in a network, n_l, is related to the number of wires, n_w, the number of nodes, n_n, and the number of connected components, n_c, by Euler's relation

$$n_l - n_w + n_n = n_c \tag{16}$$

It is possible to reduce the number of nodes and wires so that only one node remains in each connected network component. We show in Fig. 8 the tetrahedral network and an equivalent reduced network obtained by eliminating the three wires in the bottom plane.

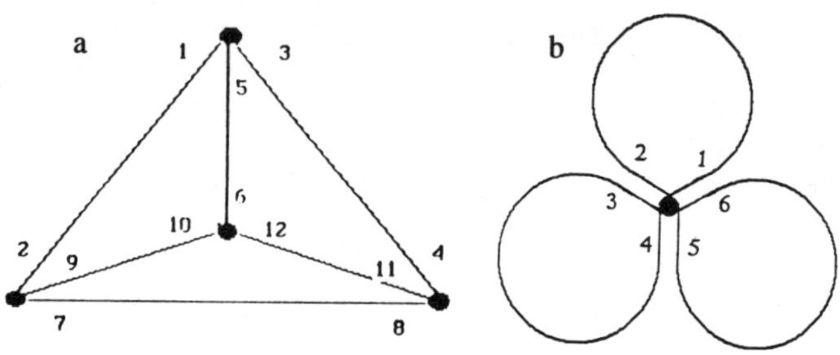

FIGURE 8: THE TETRAHEDRAL NETWORK CAN BE REDUCED TO A MINIMAL EQUIVALENT NETWORK WITH JUST ONE NODE AND HAVING ONE WIRE SURROUNDING EACH INDEPENDENT LOOP.

The elimination proceeds as follows. For the tetrahedral network, the integers $(n_l, n_w, n_n, n_c) = (3, 6, 4, 1)$. An equivalent network has $(n_l, n_w, n_n, n_c) = (3, 6-3, 4-3, 1)$, where $3 = n_w - (n_n - n_c)$. We retain three wires [with ports $(1,2), (3,4), (5,6)$] to surround the three independent loops in the reduced network. The remaining three wires are eliminated by partitioning the input and output amplitudes into those which remain $(R : 1-6)$ and those which are eliminated $(E : 7-12)$. The matrices S and Γ are similarly partitioned. Then we can eliminate the amplitudes o_E, i_E from the linear relations

$$\begin{bmatrix} i_R \\ i_E \end{bmatrix} = \begin{bmatrix} \Gamma_{RR} & \Gamma_{RE} \\ \Gamma_{ER} & \Gamma_{EE} \end{bmatrix} \begin{bmatrix} o_R \\ o_E \end{bmatrix} \tag{17}$$

$$\begin{bmatrix} o_R \\ o_E \end{bmatrix} = \begin{bmatrix} S_{RR} & 0 \\ 0 & S_{EE} \end{bmatrix} \begin{bmatrix} i_R \\ i_E \end{bmatrix} \tag{18}$$

to obtain [7]

$$[o_R] = S_{RR}[i_R]$$

$$\begin{aligned} [i_R] &= \left[\Gamma_{RR} + \Gamma_{RE} \left(S_{EE}^{-1} - \Gamma_{EE}\right)^{-1} \Gamma_{ER}\right] [o_R] \\ &= \Gamma' [o_R] \end{aligned} \tag{19}$$

For convenience, we choose a gauge in which the magnetic action vanishes between all nodes which are eliminated. In this gauge, the reduced node matrix is independent of magnetic field, and the S-matrices for the remaining wires possess a phase factor, the magnetic action, which is the flux through the area which that wire encloses.

The network matrix for the reduced network is

$$\Gamma' \times \begin{bmatrix} S_1(\alpha) & 0 & 0 & \cdots \\ 0 & S_2(\alpha) & 0 & \cdots \\ 0 & 0 & S_3(\alpha) & \cdots \\ \vdots & \vdots & \vdots & \ddots \end{bmatrix} \tag{20}$$

$$S_i(\alpha) = e^{i\psi} \begin{bmatrix} 0 & e^{i\alpha_i} \\ e^{-i\alpha_i} & 0 \end{bmatrix} \tag{21}$$

$$\alpha_i = 2\pi\Phi_i/\Phi_0 \tag{22}$$

The network matrix for the reduced network has $2n_l$ eigenvalues $\lambda_1(\alpha), \lambda_2(\alpha), ..., \lambda_{2n_l}(\alpha)$. Each is a function of the n_l fluxes $\alpha = (\alpha_1, \alpha_2, ..., \alpha_{n_l})$. By varying the fluxes α_i one or more of the eigenvalues λ_j may assume the value $+1$. By standard transversality arguments [9], no more than n_l eigenvectors of the network matrix $\Gamma' S(\alpha)$ may simultaneously satisfy the network quantization condition.

5. Flux Quantization

We first observe that the network matrix for the reduced network, $\Gamma' S_{RR}(\alpha)$, is invariant under the symmetry $\alpha_i \to \alpha_i + 2\pi m_i$, with m_i an integer.

Next, we search over the cube $-\pi < \alpha_i \leq +\pi$ in R^{n_l} for values α^o which saturate the network quantization condition. That is, for such values there is a maximum number of eigenvalues $\lambda = +1$, and this number is n_l. For the network consisting of one node and one loop shown in Fig. 9a, with node and wire matrices

$$\Gamma = e^{i\psi} \begin{bmatrix} i\,sin\theta\, e^{i\xi} & cos\theta \\ cos\theta & i\,sin\theta\, e^{-i\xi} \end{bmatrix} \qquad S(\mathbf{B}) = \begin{bmatrix} 0 & e^{i\alpha} \\ e^{-i\alpha} & 0 \end{bmatrix} \tag{23}$$

the network quantization condition reduces to $cos\psi = cos\alpha\, cos\theta$. This condition has two solutions when $|cos\psi/cos\theta| \leq 1$. When $\psi - \alpha = 2\pi m$ the eigenvector of $\lambda_1 = +1$ is $(i_1, i_2) = (1, 0)$, describing a counterclockwise flowing current. For $\psi + \alpha = 2\pi m$ the eigenvector of $\lambda_2 = +1$ is $(0, 1)$, describing a clockwise flowing current.

For the reduced network consisting of the two loops shown in Fig. 9b, there are at most four distinct values of $\alpha = (\alpha_1^o, \alpha_2^o)$ in the square $(-\pi, \pi] \times (-\pi, \pi]$ which describe currents which flow clockwise or counterclockwise in either of the two loops.

The solutions of the nonlinear network quantization conditions move about, coalesce and bifurcate in the usual way as the matrix elements in the node scattering matrix are varied [9].

The single valued wave functions are unchanged under the symmetry $\alpha^o \to \alpha^o + 2\pi\mathbf{m}$.

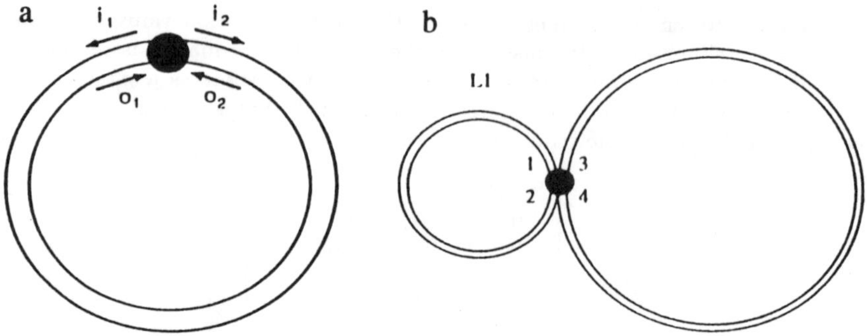

FIGURE 9: (A) A NETWORK CONSISTS OF A SINGLE LOOP AND A SINGLE NODE, WITH THE S-MATRICES SHOWN IN EQS. (23). (B) A NETWORK CONSISTS OF A SINGLE NODE AND TWO WIRES ENCIRCLING THE INDEPENDENT LOOPS.

6. Current Quantization

The current in each wire can be determined from the components of the eigenvectors $[i]$ which satisfy the network quantization condition. If the input amplitudes to the rth wire are i_{2r-1} and i_{2r}, the current in the rth wire, due to charge carriers described by the single valued wave function ψ_γ, is

$$j_{\gamma,r} \sim |i_{2r-1}|^2_\gamma - |i_{2r}|^2_\gamma \tag{24}$$

The proportionality can be made an equality by introducing a single scaling amplitude for each eigenstate, J_γ, so that

$$j_{\gamma,r} = J_\gamma \left(|i_{2r-1}|^2_\gamma - |i_{2r}|^2_\gamma\right) \tag{25}$$

The current in the rth wire is then the sum of the currents due to each of the eigenstates which satisfy the network quantization condition

$$J_r = \sum_\gamma J_\gamma \left(|i_{2r-1}|^2_\gamma - |i_{2r}|^2_\gamma\right) \tag{26}$$

These currents depend on the magnetic field present through the magnetic actions α_r, and are in turn responsible for generating these magnetic fields.

The magnetic flux through any loop is the sum of two contributions. These are the flux due to an externally applied magnetic field, $\mathbf{B}_{ext}$, which may be considered as uniform over the scale of a typical mesoscopic network, and the magnetic fields generated by the shielding currents flowing in the network:

$$(\alpha + 2\pi m)_r \Phi_0 = \mathbf{B}_{ext} \cdot \mathbf{A}_r + \Lambda_{rs} J_s \tag{27}$$

Here $\mathbf{A}_r$ is the vector representing the directed area of loop r and $\mathbf{B}_{ext} \cdot \mathbf{A}_r$ is the flux through loop r due to the external magnetic field. The current J_s in loop s establishes a magnetic field which flows through each of the loops. The magnetic flux through loop r, generated by current J_s in loop s, is determined by the mutual inductance matrix, Λ_{rs}. The areas $\mathbf{A}_r$ and mutual inductance matrix for the reduced network can be computed from the geometry of the original network by standard methods [7,10].

The current flow in each loop, J_s, can be computed by inverting (27)

$$J_s = [(\alpha + 2\pi m)_r \Phi_0 - \mathbf{B}_{ext} \cdot \mathbf{A}_r] \, \Lambda_{rs}^{-1} \tag{28}$$

This result shows that even when the single valued wave functions over the network are fixed, the current flow in the network is quantized. The current state of the network is indexed by the quantum numbers $\mathbf{m} = (m_1, m_2, ..., m_{n_l})$. These integers specify the flux state through each independent loop. The flux itself is not quantized, $[\Phi_r = (\alpha^o + 2\pi m)_r \Phi_0]$, but *differences* in flux states are given by integer multiples of $2\pi\Phi_0$.

7. Network Energy and Meissner Effect

The network energy is a quadratic function of the supercurrent flow: $E = \frac{1}{2} J_i \Lambda_{ij} J_j$. Since the supercurrents in each loop are linear functions of the external magnetic field and the network quantum numbers, the network energy is a quadratic function of $\mathbf{B}_{ext}$ and $\mathbf{m}$

$$\begin{aligned} &E(\mathbf{B}_{ext}, \mathbf{m}) = \\ &\frac{1}{2} [(\alpha + 2\pi m)_r \Phi_0 - \mathbf{B}_{ext} \cdot \mathbf{A}_r] \, \Lambda_{rs}^{-1} \, [(\alpha + 2\pi m)_s \Phi_0 - \mathbf{B}_{ext} \cdot \mathbf{A}_s] \end{aligned} \tag{29}$$

The network quantum numbers $\mathbf{m}$ describe a specific supercurrent flow in the network. As the external magnetic field is varied, the supercurrents readjust themselves so as to maintain a fixed flux through each of the loops in the network. The network energy exhibits a quadratic dependence on the external magnetic field, in any supercurrent flow state ($\mathbf{m}$). The network energy for the two loop network shown in Fig. 9b is shown in Fig. 10.

The ordering of the energy levels changes as a function of the external magnetic field. The state which minimizes the network energy also minimizes (in the mean square sense) the magnetic field within the network. The ground state energy is

$$E_g(\mathbf{B}_{ext}) = Min \; E(\mathbf{B}_{ext}, \mathbf{m}) \tag{30}$$

If we choose a sequence of magnetic quantum numbers $\mathbf{m}_1, \mathbf{m}_2, ...$ with increaseing energy level spectrum $E_g = E(\mathbf{B}_{ext}, \mathbf{m}_1) \leq E(\mathbf{B}_{ext}, \mathbf{m}_2) \leq E(\mathbf{B}_{ext}, \mathbf{m}_3) \leq ...$, then as we approach the ground state through this sequence $... \to \mathbf{m}_3 \to \mathbf{m}_2 \to \mathbf{m}_1$, the magnetic field is increasingly expelled from the network. If the network is very large so that there are many loops, the ground state energy will be approximately zero for any external magnetic field. That is, a large superconducting network will expel a magnetic field (Meissner effect). It is for this reason that we call the network energy minimization of (30) the network Meissner effect.

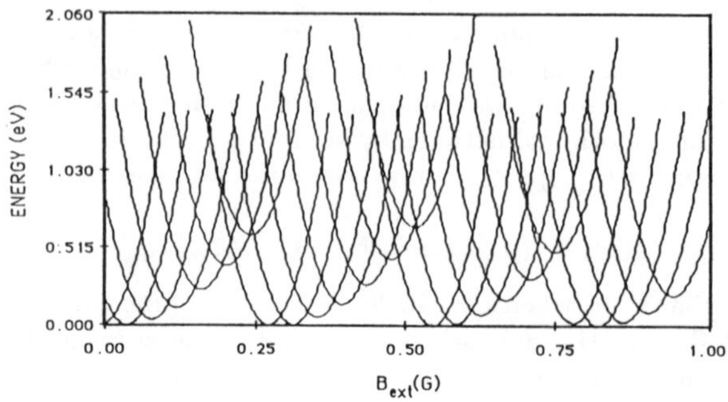

FIGURE 10: THE MAGNETIC ENERGY OF THE TWO LOOP NETWORK SHOWN IN FIG. 9B CONSISTS OF A SERIES OF PARABOLAS. EACH PARABOLA IS INDEXED BY THE NETWORK QUANTUM NUMBERS $\mathbf{m} = (m_1, m_2)$. AS THE EXTERNAL MAGNETIC FIELD IS CHANGED THE MINIMUM JUMPS FROM ONE BRANCH TO ANOTHER. THE MINIMUM ENERGY STATE IS THE STATE WHICH MINIMIZES THE MAGNETIC FIELD WITHIN THE NETWORK. THIS IS THE NETWORK MEISSNER EFFECT.

8. Summary and Discussion

Path integrals have been used since the beginning of quantum mechanics to determine properties of a quantum system. The allowed Bohr circular orbits of an electron in a hydrogen atom were determined by deBroglie using an argument which is classical in nature but which can also be regarded as the first quantum mechanical path integral argument (*cf* , Fig. 1a). The magnetic interference patterns in the Aharonov-Bohm effect are also determined by a path integral argument (*cf* , Fig. 1b). In fact, if one removes the external leads from the annulus in Fig. 1b and asks for the electron states which can circle the annulus without destructive self interference, one quickly returns to an argument identical in spirit, if different in detail, to deBroglie's earlier argument.

When many paths from an initial point to a final point are possible, then the sum over all classically allowed paths cannot generally be carried out in closed form. One is then reduced to various approximations, such as the stationary phase approximation. When this is carried to second order in $\hbar$ the Schrödinger equation is recovered; to fourth order, some of the 'relativistic' corrections to the Schrödinger equation are recovered.

There is one instance in which the sum over all possible paths can be carried out in closed form. This occurs for one–dimensional networks with an arbitrary multiply connected geometry. The difficulty with carrying out these sums is determining what happens to a particle when it encounters a network junction and has to decide which of several wires to be scattered into, and with which probability amplitudes. It is exactly this information which is summarized by the S-(scattering) matrices of the network components. In fact, the problem of describing the transmission of energy through a microwave network is identical to the problem of computing the sum of probability amplitudes for a particle traversing through all possible paths between two points in a network. We have previously solved this problem for networks with external leads. In that solution, the network is described by the S–matrices for the individual network components, the wires and the nodes. All internal amplitudes are eliminated by simple matrix procedures, leaving a linear relation between the input and the output amplitudes on the external leads. The network scattering matrix is expressed in terms of the inverse of the difference between two scattering matrices which, when expanded in a Taylor series, is equivalent to a sum over all possible paths through the network between external leads [7]. When the network has no external leads, the considerations are somewhat different. The difference is the same as that which occurs for the Aharonov-Bohm experiment when there are external leads (Fig. 1b) and magnetic interference effects are observed, and when there are no external leads and the only self-sustaining charge flows which can occur are determined by a single-valuedness argument (*cf*, Fig. 1a).

Our general results are as follows. For a network without leads, the properties are determined by the network matrix, $\Gamma S(\mathbf{B})$. Here $S(\mathbf{B})$ is the S-matrix for the wires, $\mathbf{B}$ is the magnetic field, and Γ is the scattering matrix for the nodes of the system. The matrix Γ also determines the topology of the network. A wave function on the network is everywhere single valued if and only if the associated input amplitudes at the two ends of each of the wires is an eigenvector of the network matrix with eigenvalue $+1$. A connected network with n_l independent loops can be transformed by standard network manipulations to an equivalent network with one node having n_l ports and n_l wires, one surrounding each independent loop. The equivalent reduced network is described by a $2n_l \times 2n_l$ network matrix, $\Gamma' S_{RR}(\alpha)$, where the n_l component vector $\alpha = (\alpha_1, \alpha_2, ..., \alpha_{n_l})$ describes the flux through each independent loop. The $2n_l$ eigenvalues $\lambda_1(\alpha), \lambda_2(\alpha), ..., \lambda_{2n_l}(\alpha)$ of the network matrix $\Gamma' S_{RR}(\alpha)$ depend on the n_l fluxes α_r, $1 \leq r \leq n_l$. As a result, up to n_l of the $2n_l$ eigenvectors can satisfy the network quantization condition. There may be as many as 2^{n_l} points α^o in the hypercube $(-\pi, +\pi]^{\otimes n_l} \subset R^{n_l}$ at which the network quantization condition is satisfied. Each such point describes a current flow in the "clockwise" or "counterclockwise" direction in each of the independent loops.

For any choice of α^o at which the network quantization condition is satisfied, the n_l single valued wave functions span the linear vector space of single valued wave functions. The wave functions themselves do not depend on the network quantum numbers. Rather, it is the coefficients of these wave functions which are quantized.

This comes about as follows. The fluxes through each independent loop, α^o, consist of the flux due to an external magnetic field and fluxes due to current flows in all of the loops: $\alpha\Phi = \mathbf{B}_{ext} \cdot \mathbf{A} + \Lambda J$. However, fluxes which are different by quantized flux units generate the same network matrix. Therefore, a set of single valued wave functions is associated with a quantized set of supercurrent flows by means of $(\alpha^o + 2\pi\mathbf{m})\Phi_0 = \mathbf{B}_{ext} \cdot \mathbf{A} + \Lambda J$. This is dual to the typical case in quantum mechanics where it is the wave functions that depend on quantum numbers while arbitrary superpositions of them are allowed.

The network energy is a quadratic function of both the external magnetic field and the network quantum numbers. The ground state energy is obtained, for fixed $\mathbf{B}_{ext}$, by minimizing over the network quantum numbers. This is the network Meissner effect. As the external magnetic field is varied, the supercurrent flows readjust themselves to maintain fixed fluxes through each loop in the network. At some point the network quantum number may change, resulting in a jump from a current state of higher energy to one of lower energy.

In the event the network contains weak links or weak nodes, the quantum numbers **m** no longer describe supercurrent eigenstates. Rather, they describe a basis of supercurrent flows all associated with one set of single valued wave functions. The set of current flows can be used as a basis in which to diagonalize an effective hamiltonian. Two asymptotic regimes have been recognized and measured for single loop networks. These are the flux mode regime, in which the flux **m** in each loop is almost a good quantum number, and the charge mode (Josephson) regime in which charge at each junction is almost a good quantum number [11]. Discussion of these two limiting cases, as well as all other intermediate cases, would take us too far away from the limits of this volume.

REFERENCES

1. L. deBroglie, *Investigations on quantum theory*, Annales de Physique **3** (1925), 22; reprinted in: G. Ludwig, Wave Mechanics, Pergamon, Oxford, 1968, pp. 73-93.
2. N. Bohr, *On the constitution of atoms and molecules*, Phil. Mag. **26** (1913), 1; reprinted in: D. ter Haar, The Old Quantum Theory, Pergamon, Oxford, 1967, pp. 132-159.
3. Y. Aharonov and D. Bohm, *Significance of electromagnetic potentials in the quantum theory*, Phys. Rev. **115** (1959), 485-491.
4. R. P. Feynman and A. R. Hibbs, *Quantum Mechanics and Path Integrals*, McGraw Hill, New York, 1965.
5. M. Jeffery, C. Green, S. Tyagi, and R. Gilmore, *Quantum network simulations of high-T_c ceramic superconductors*, Int. J. Mod. Phys. **B2** (1988), 1399-1405.
6. M. Jeffery, C. Green, S. Tyagi, and R. Gilmore, *Reproducible magnetic features of high-T_c superconductors in weak fields*, Phys. Rev. **B39** (1989), 9054-9060.
7. M. Jeffery and R. Gilmore, *Lattice dependent magnetic structure in quantum network simulations of high-T_c superconductors*, Phys. Rev. **B41** (1990), 2057-2072.
8. R. Landauer, *Electrical resistance of disordered one-dimensional lattices*, Phil. Mag. **21** (1970), 863-867.
9. R. Gilmore, *Catastrophe Theory for Scientists and Engineers*, John Wiley and Sons, New York, 1984.
10. D. M. Kerns and R. W. Beatty, *Basic Theory of Waveguide Junctions and Introductory Microwave Network Analysis*, Pergamon, Oxford, 1967.
11. T. D. Clark, *Macroscopic quantum objects*, in: Quantum Implications: Essays in Honor of David Bohm; B. J. Hiley and F. D. Peat, eds, Routledge and Kegan Paul, London, 1987, pp. 121-150.

DEPARTMENT OF PHYSICS AND ATMOSPHERIC SCIENCE, DREXEL UNIVERSITY, PHILADELPHIA, PA 19104 , USA

SYMMETRY AND SYMMETRY–BREAKING IN STATISTICAL PHYSICS

H. Grosse

Abstract. After a general introduction we discuss the role of solitons in one-dimensional polymers. Methods which allow to solve certain nonlinear evolution equations can be applied to the large class of ultralocal models. The solvability of vertex models can be traced back to solutions of the Yang-Baxter relation. Recently the notion of a symmetry has been enlarged. Models showing instead of a Lie-algebra symmetry a quantum group symmetry are discussed in the last section of the paper.

Contribution in honor of the 60th birthday of John R. Klauder
who discussed ultralocal models extensively, and
to whom we wish many happy recurrencies.

1. Introduction

The notion of a symmetry is extremely often used in theoretical physics. $\sigma\upsilon\nu$ $\mu\epsilon\tau\rho o s$ comes from the Greek and means "with measure" or regular. R. Feynman related it to objects with which we are doing something without having changed anything at the end. H. Weyl described it as the concordance with which parts are put together to the whole.

Left-right symmetry is realized by a few letters already. The monkey of Puritan gives another example. Two Z_2 symmetries for two orthogonal axes imply already a rotation symmetry. Cyclic groups and the dihedral groups are realized by snow flakes. It is easy to prove that crystals may have only Z_2, Z_3, Z_4 or Z_6 symmetry axes. Without translations 32 crystallographic groups occur in three dimensions, 10 in two dimensions. Adding translations, 230 groups for $d = 3$ and 17 for $d = 2$ occur. Escher's periodic drawings influenced studies in mathematics. But more recently (nonperiodic) tilings led to interesting new structures both from the physics point of view (quasicrystals) as well as from the mathematical point view (due to the relation to noncommutative geometry).

We may mention the platonic bodies too, which have the discrete nonabelian subgroups of the rotation group in three dimensions as their symmetry group. The big C_{60} molecules (fullerene) have a more complicated structure.

Mirror reflections give rise to handedness or chiralities. That parity is broken in nature by weak interactions is well-known. It led, unfortunately, also to broken friendship among the inventors. That chiral molecules occur in nature (left-handed aminoacid, right-handed DNA) is fascinating. Recently the hypothesis that parity violating neutral currents might be responsible for it has been put forward. Mathematically, superselection rules allow us to formulate sectors of a model such that no transitions occur.

Finally, we mention that no complete one-interval tone system exists. But symmetries occur in music pieces in various forms.

2. Solitons

That extremal solutions of a variational principle do not have the symmetry of the functional with which one starts with is a well-known phenomenon. The double well potential gives rise to "spontaneous" breakdown of the left-right symmetry or the continuous symmetry if one starts with the Mexican hat potential.

For special solutions of two-dimensional field theoretic models the expression soliton has been coined. In 1834 John Scott Russell observed waves going through each other and showing up after the interaction almost undisturbed but phase shifted. Korteweg and de Vries obtained their celebrated nonlinear partial differential equation shortly before the turn of the century. We might mention next the Fermi-Pasta-Ulam result. In a coupled system of oscillators the energy was not distributed in an equal amount among the particles involved. Gardner, Green, Kruskal, and Miura found the tricks to "solve" finally the KdV equation for $v(t,x)$:

$$v_t = 6vv_x - v_{xxx}. \tag{1}$$

There is a dispersive part and a nonlinear part. The first one would imply that wave packets disperse, the second part would give rise to the appearance of shock waves. The resulting inverse scattering transform method is easy to explain. One takes the initial values $v(0,x)$ as potential in a Schrödinger operator and goes over to the scattering data (reflection coefficient, eigenvalues and normalization constants). This nonlinear transformation leads to new variables which have a free time evolution. The inverse step back allows to recover the potential $v(t,x)$ at time t by solving a Gelfand-Levitan-Marchenko-type equation. If the reflection coefficient is identically zero, solutions of the last equation can be obtained explicitly. These solitons have stability properties and a class of initial values will develop asymptotically into a superposition of isolated solitons. It is remarkable that the symplectic structure for the field at time zero can be transformed into a Poisson algebra structure for the scattering data [1]. The classical Yang-Baxter equation turns out to be satisfied.

A simple model describing the interaction of electrons hopping along a chain of CH-molecules (polyacetylen) reveals the role of solitons. Alternating double and single bonds can be arranged in two different ways showing reflection symmetry. If u_n denotes the displacement of the n-th CH-molecule, $u_n = (-1)^n\varphi$ allows for alternating vibrations and introduces an order parameter φ. Distorsion costs elastic energy proportional φ^2, electronic energy proportional to $-\varphi^2 \ln|\varphi|$ is obtained from lowering states if a gap opens up. This Peierls' instability can be proven to occur in a simple Yukawa model coupling fermions to bosons (in two dimensions). The resulting double well potential allows for kinks interpolating between the two vacua. Charged solitons can move easily through the lattice and might be responsible for the increase of conductivity of polyacetylen under doping.

3. Phase Transitions

Although a large number of phase transitions occur in nature we shall discuss only models of (anti-) ferromagnetism respectively ferroelectricity. Assume that spins are assigned to lattice points and are allowed to point up or down. Ferromagnetic nearest neighbor interaction in a one-dimensional system will not give rise to an ordered phase at nonzero temperature. Such energy entropy arguments give intuitive insight but can be made rigorous too. Assume an ordered phase is stable at $T \neq 0$. We may flip half the spins by changing only two bindings. Thus δU is independent of the length N of the chain, δS is proportional to $\ln N$ and $\delta F = \delta U - T\delta S$ is negative contradicting the principle of minimal free energy for isothermal equilibrium states. This argument does not apply to higher dimensions. Peierls estimated the occurrence of islands of oriented spins and showed that phase transitions occur and give rise to nonanalyticity of the magnetization [2].

If we allow the classical spins to vary on a sphere or if we treat quantum spins with continuous symmetry, the phase structure changes drastically. In two dimensions the Mermin-Wagner-Hohenberg theorem tells us that the spontaneous magnetization will be zero. A continuous symmetry will not be spontaneously broken in two dimensions. For more dimensions spontaneous magnetization will occur due to the appearance of thick Bloch walls. If there are no long range forces present the Goldstone theorem will imply the occurrence of spin-wave excitations with zero frequency. In case of long range forces the Higgs mechanism will imply the appearance of massive particles in superconductors of type II.

A number of problems are still open for quantum spin models. In the continuum limit, the spectrum might show no gap, it might have a gap with a nondegenerate ground state or a degenerate ground state might occur. Examples occur in models which we discuss next.

4. Vertex Models – Quantum Spin Models

A simple model for ice can be taken as an illustration. Oxygen atoms should be surrounded by exactly two hydrogen atoms. Altogether sixteen vertex configurations can be described. The ice rule singles out six possibilities. For the symmetric six vertex model equal weights are given to reflection symmetric configurations. The resulting phase diagram depends on two parameters a/c and b/c; the parametrization $a = \sin(\gamma - \lambda)$, $b = \sin\lambda$, $c = \sin\gamma$ is convenient. The partition function is given by summation over all possible configurations

$$Z_{N,M} = \sum_{\mathrm{conf}} \prod_{n,m} \omega_{\mathrm{conf}}(n,m) = \mathrm{Tr}\ {T_N}^M \tag{2}$$

and can be written as a trace of a $2^N \times 2^N$ transfer matrix T_N.

A vertex can be described as a 4×4 matrix

$$L^{\beta_1\beta_2}_{\alpha_1\alpha_2} = \begin{pmatrix} a & 0 & 0 & 0 \\ 0 & b & c & 0 \\ 0 & c & b & 0 \\ 0 & 0 & 0 & a \end{pmatrix} \qquad \text{in} \qquad \mathbb{C}^2 \otimes \mathbb{C}^2; \tag{3}$$

the monodromy matrix acts in $\mathbb{C}^2 \otimes (\bigotimes_{j=1}^N \mathbb{C}^2)$ and describes the interaction from one row to the next:

$$M_N = L_1 \cdot L_2 \cdot \ldots \cdot L_N, \qquad \mathrm{tr}_0 M_N = T_N, \tag{4}$$

where periodic boundary conditions imply the trace over the auxiliary $\mathbb{C}^2$ space with index zero.

One-dimensional quantum spin models are naturally connected to two-dimensional vertex models [3]. We consider T_N as a function of λ and develop $\ln T_N(\lambda)$ into a power series. $T_N(0)$ becomes the lattice momentum shifting all spins by one lattice point.

$$\sin\gamma^{-1}\left(\frac{d}{d\lambda}\ln T_N(\lambda)\right)\Big|_{\lambda=0} = H_{XXZ} = \frac{1}{2}\sum_{j=1}^{N}(\sigma_j^x\sigma_{j+1}^x + \sigma_j^y\sigma_{j+1}^y + \Delta(\sigma_j^z\sigma_{j+1}^z + 1)) \tag{5}$$

where $\sigma_j^{x,y,z} = I \otimes \ldots \otimes I \otimes \sigma^{x,y,z} \otimes I \otimes \ldots \otimes I$ (with $\sigma^{x,y,z}$ at j-th place) form a basis for all $2^N \times 2^N$ matrices, and $\Delta = -\cos\gamma$. H_{XXZ} has a rotation symmetry for $\Delta = 1$ and becomes the Heisenberg model. A new type of symmetry still exists even for $\Delta \neq 1$ (besides the abelian rotation symmetry in the x, y plane). We shall return to this point later. Higher logarithmic derivatives of $T_N(\lambda)$ imply further interactions among next to nearest neighbors, etc.

We shall not discuss more complicated models like the eight-vertex model, the solid – on – solid models or the interaction around a face models. They enable one to go off criticality in the region $|a^2 + b^2 - c^2| < 2|ab|$ of the six-vertex model.

5. Quantum Integrability

A remarkable kind of symmetry allows one to "solve" a certain class of models. An infinite number of conserved quantities is implied from the fact that the transfer matrices to different spectral parameters commute:

$$[T_N(\lambda), T_N(\mu)] = 0. \tag{6}$$

This relation was used by Yang and Baxter and follows from the relation which is named after them. There exists a matrix $R(\lambda) = PL(\lambda)$ acting in $\mathbb{C}^2 \otimes \mathbb{C}^2$, such that

$$R_{12}(\lambda-\mu)L_{13}(\lambda)L_{23}(\mu) = L_{13}(\mu)L_{23}(\lambda)R_{12}(\lambda-\mu) \tag{7}$$

where indices denote the action in $\mathbb{C}^2 \otimes \mathbb{C}^2 \otimes (\mathbb{C}^2)^N$. Expressed in terms of the R-matrices themselves, (7) becomes

$$R_{23}(\lambda-\mu)R_{12}(\lambda)R_{23}(\mu) = R_{12}(\mu)R_{23}(\lambda)R_{12}(\lambda-\mu). \tag{8}$$

There exist rational, trigonometric, elliptic and hyperelliptic solutions of (8). Generalizations to the graded case and to $\mathbb{C}^n$ are straightforward. Multiparameter solutions exist too. (7) and (8) are identical to the assumption of factorization for S-matrices in two-dimensional models. The parameter becomes the rapidity. After suitable redefinition of the matrix L, the limit $\lambda \to i\infty$ leads to a nontrivial solution of the braid relation:

$$R_{12}R_{23}R_{12} = R_{23}R_{12}R_{23}. \tag{9}$$

A homomorphism $b_i \to \mathrm{End}\,(\bigotimes^N \mathbb{C}^2)$ with

$$b_i \to I \otimes \ldots \otimes I \otimes R_{ii+1} \otimes I \otimes \ldots \otimes I \tag{10}$$

yields a representation of the braid group B_N generated by $\{b_1, \ldots, b_{N-1}, b_1^{-1}, \ldots, b_{N-1}^{-1}\}$ with

$$b_i b_i^{-1} = I \qquad b_i b_j = b_j b_i \quad \text{if} \quad |i-j| \geq 2, \qquad b_i b_{i+1} b_i = b_{i+1} b_i b_i. \tag{11}$$

From the gauged six-vertex model one obtains the solution of (9)

$$R = \begin{pmatrix} 1 & 0 & 0 & 0 \\ 0 & 0 & t & 0 \\ 0 & t & 1-t^2 & 0 \\ 0 & 0 & 0 & 1 \end{pmatrix}, \qquad t = e^{i\gamma}, \qquad (R-1)(R+t^2) = 0. \tag{12}$$

R has only two different eigenvalues and therefore solves a quadratic relation (12). We may map the two eigenvalues to zero and one $b_i \to \sigma_i$: $b_i = -(1+t^2)\sigma_i + 1$ and obtain the Hecke algebra

$$\sigma_i^2 = \sigma_i, \qquad \sigma_i \sigma_{i+1} \sigma_i - \left(\frac{t}{1+t^2}\right)^2 \sigma_i = \sigma_{i+1} \sigma_i \sigma_{i+1} - \left(\frac{t}{1+t^2}\right)^2 \sigma_{i+1}. \tag{13}$$

A further restriction is possible. If we require that both sides of (13) are zero, the Temperley-Lieb algebra [3] is obtained

$$e_i^2 = e_i = e_i^\dagger, \qquad e_i e_j = e_j e_i \quad \text{if} \quad |i-j| \geq 2 \qquad e_i e_{i\pm 1} e_i = \frac{1}{Q} e_i \tag{14}$$

where $Q = (t + t^{-1})^2$. Jones deduced the quantization of $Q \in \{4\cos^2 \pi/m\} \cup [4, \infty)$, $m = 4, 5, \ldots$. This algebra (14) was obtained from the matrices describing the horizontal and vertical interaction of the Q-state Potts model.

6. A Quantum Group Symmetry

A remarkable property of the $RLL = LLR$ relation (7) is that if two L matrices fulfill it and they commute among themselves, their product fulfills the relation too. Therefore the monodromy matrix obeys it:

$$R_{12}(\lambda - \mu) M_N^{(1)}(\lambda) M_N^{(2)}(\mu) = M_N^{(1)}(\mu) M_N^{(2)}(\lambda) R_{12}(\lambda - \mu). \tag{15}$$

Taking the trace leads to (6). But (15) says more. Upon writing $M_N = \begin{pmatrix} A & B \\ C & D \end{pmatrix}$ an algebra among the matrices A_λ, B_λ, C_λ, D_λ results, which is highly nontrivial. For example

$$A_\lambda B_\mu = \frac{a(\mu - \lambda)}{b(\mu - \lambda)} B_\mu A_\lambda - \frac{c(\mu - \lambda)}{b(\mu - \lambda)} B_\lambda A_\mu. \tag{16}$$

If one tries to find a type of highest weight representations of this algebra, a "vacuum" state $|\Omega\rangle$ can be found. The term $B_\lambda A_\mu$ would yield new states applied to $B_{\lambda_1} \dots B_{\lambda_m}|\Omega\rangle$. If we require the vanishing of these terms, Bethe's conditions on the parameters $\lambda_1, \dots, \lambda_m$ result. Filling up holes and particles allows for solutions of the Bethe equations in the limit $N \to \infty$. In this way a partial diagonalization of the transfer matrix can be obtained.

Relation (15) can be generalized. We fix the R-matrix and study matrices $M = \begin{pmatrix} a & b \\ c & d \end{pmatrix}$ solving (15). Here a, b, c, d have to fulfill a special algebra. With R from (12) we obtain the algebra

$$\begin{aligned} ab &= qba, & ad - da &= (q - q^{-1})bc, & bd &= qdb, \\ ac &= qca, & bc &= cb, & cd &= qdc. \end{aligned} \tag{17}$$

Through (17) the q-deformation A_q of the abelian algebra of functions over the group $GL(2, \mathbb{C})$ is defined. A noncommutative coproduct mapping from A_q to $A_q \otimes A_q$ exists

$$\Delta \begin{pmatrix} a & b \\ c & d \end{pmatrix} = \begin{pmatrix} a & b \\ c & d \end{pmatrix} \otimes \begin{pmatrix} a & b \\ c & d \end{pmatrix}. \tag{18}$$

Δ yields an algebra homomorphism. A co–unit $\mathcal{E}$ exists too

$$\mathcal{E} \begin{pmatrix} a & b \\ c & d \end{pmatrix} = \begin{pmatrix} 1 & 0 \\ 0 & 1 \end{pmatrix}. \tag{19}$$

If one factorizes with respect to $\det_q M = ad - qbc = 1$, we obtain the q-deformation of functions over the group $SL(2, \mathbb{C})$. Since a co–inverse exists too, we obtain a Hopf algebra structure. We may go over to a dual Hopf algebra which becomes the deformation of the universal enveloping algebra of $sl(2, \mathbb{C})$. Its generators obey [4]

$$[J_+, J_-] = (2H)_q, \qquad [H, J_\pm] = \pm J_\pm\,, \tag{20}$$

where the q-deformation of x is defined by $(x)_q = (q^x - q^{-x})/(q - q^{-1})$. It is interesting to note that representations of (20) can be obtained directly from the monodromy matrix. L may be transformed to ($xq^{1/2} = e^{-i(\gamma - \lambda)}$, $q = -e^{-i\gamma}$)

$$\tilde{L} = \begin{pmatrix} xq^{s^3} - \frac{1}{x} q^{-s^3} & x(q - q^{-1})s^- \\ \frac{1}{x}(q - q^{-1})s^+ & xq^{-s^3} - \frac{1}{x} q^{s^3} \end{pmatrix} \longrightarrow x \begin{pmatrix} q^{s^3} & (q - q^{-1})s^- \\ 0 & q^{-s^3} \end{pmatrix} \text{ as } x \to \infty \tag{21}$$

and M_N has a triangular form

$$x^{-N} \cdot M_N \longrightarrow \begin{pmatrix} q^H & (q - q^{-1})J_- \\ 0 & q^{-H} \end{pmatrix} \text{ as } x \to \infty\,. \tag{22}$$

The matrices $J_\pm$ and H become explicitly

$$H = \sum_{j=1}^{N} s_j^3, \qquad J_\pm = \sum_{j}^{N} q^{s_3^1 + \ldots + s_3^{j-1}} s_j^\pm q^{-s_3^{j+1} - \ldots - s_3^N} \tag{23}$$

and become generators of the Lie-algebra $sl(2, \mathbb{C})$ for $q \to 1$.

It is remarkable that the Hamiltonian of the XXZ chain (5) commutes even for $\Delta \neq 1$ with the generators H and $J_\pm$

$$[H_{XXZ}, J_\pm] = [H_{XXZ}, H] = 0. \tag{24}$$

A new type of symmetry results. It is not a Lie-algebra which is behind this symmetry but a quantum group. A number of consequences are currently under discussion. At least many relations between integrable models, quantum spin models, algebraic concepts and field theoretic models result [5].

References

1. L. Faddeev and L. Takhtajan, *Hamiltonian Methods in the Theory of Solitons*, Springer, Berlin, 1987.
2. H. Grosse, *Models in Statistical Physics and Quantum Field Theory*, Springer, Berlin, 1988.
3. H. Grosse, *An Introduction to Integrable Models and Conformal Field Theory* (1990), in: Fields and Particles (H. Mitter and W. Schweiger, eds.), Springer, Berlin.
4. V. Pasquier and H. Saleur, Nucl. Phys **B 330** (1990), 523.
5. H. Grosse, *Progress in Completely Integrable Models of Quantum Field Theory*, IAMP Congress, Leipzig, 1991.

INSTITUT FÜR THEORETISCHE PHYSIK, UNIVERSITÄT WIEN, AUSTRIA

RECKONING OF THE BESSELIAN PATH INTEGRAL

Akira Inomata

Abstract. The radial path integral is a Besselian path integral, which is sufficiently ugly even for the harmonic oscillator but has been evaluated by iterating Weber's integral formula. Since there are renewed interests in calculating the radial path integral, a new way of calculating the Besselian path integral for the n-dimensional harmonic oscillator is reported.

Contribution in honor of the 60th birthday of John R. Klauder

1. Introduction

In the very paper [1] that introduced the path integral approach to quantum mechanics, Feynman demonstrated how to carry out path integration for the propagator in cartesian coordinates, and acknowledged the problem associated with path integration by means of non-cartesian coordinates. However, he suggested that the calculation of the short time action should be independent of the choice of coordinate systems. In 1964, Edwards and Gulyaev [2] showed explicitly a way to construct a correct polar coordinate path integral. In fact, as early as 1955, Ozaki [3] had a result similar to that of [2]. In essence, their conclusions were consistent with Feynman's suggestion. They realized that a path integral, if constructed formally with a Lagrangian written in polar variables, does not lead us to a unique propagator. An unambiguous way to reach a correct polar coordinate path integral in flat space is to transform cartesian variables of each short time action into polar variables.

Even if a polar coordinate path integral is correctly constructed, it is not trivial to carry out path integration in polar variables. For a spherically symmetric system, angular path integration is rather easily performed by using the orthogonality properties of spherical harmonics, whereas the radial path integral which depends on the form of the potential is not easy to calculate. The radial path integration is complex enough even for a harmonic oscillator. If we borrow the words from a recent article of Arovas on anyons [4], the path integral after angular integration "still looks sufficiently ugly so as to be rather intimidating." "However," the article continues, "a truly remarkable result due to Peak and Inomata saves the day." The result of the radial path integration for the harmonic oscillator was published in 1969 [5], and the result mentioned by Arovas is this 1969 version. As a matter of fact, there is a simpler and perhaps more "remarkable" version than the one of 1969, which has been found in relation with the harmonic analysis of path integrals but left unpublished. Now that there are renewed interests in evaluating the radial path integral, it seems useful to report the new version of the old subject by taking advantage of the special occasion of celebrating John Klauder's sixtieth birthday.

2. Besselian Path Integrals

Let us start with Feynman's path integral for the propagator in n ($n \in \mathbf{N}$) dimensions:

$$K(\mathbf{r}'',\mathbf{r}';t'',t') = \lim_{N\to\infty} \int \prod_{j=1}^{\infty} K(\mathbf{r}_j,\mathbf{r}_{j-1};t_j - t_{j-1}) \prod_{j=1}^{N-1} d^n\mathbf{r}_j, \tag{1}$$

with the short time propagator,

$$K(\mathbf{r}_j,\mathbf{r}_{j-1};t_j,t_{j-1}) = \left[\frac{M}{2\pi i\hbar\tau_j}\right]^{n/2} \exp\left[\frac{iM}{2i\hbar\tau_j}(\Delta\mathbf{r}_j)^2 - \frac{i}{\hbar}V(r_j)\tau_j\right]. \tag{2}$$

Here $t' = t_0$, $t'' = t_N$, $\tau_j = t_j - t_{j-1}$, $\mathbf{r}_j = \mathbf{r}(t_j)$, and $r_j = |\mathbf{r}_j|$. The time interval $\tau = t_N - t_0$ is divided into N subintervals, and as N tends to infinity, all τ_j are supposed to approach zero. The isometric subdivition $\tau_j = \tau/N$ is the simplest choice, but is not a necessary choice.

The propagator (1) satisfies the composition rule:

$$\int K(\mathbf{r}'',\mathbf{r};t'',t)K(\mathbf{r},\mathbf{r}';t,t')\,d^n\mathbf{r} = K(\mathbf{r}'',\mathbf{r}';t'',t') \tag{3}$$

and the initial condition:

$$\lim_{t''\to t'} K(\mathbf{r}'',\mathbf{r}';t'',t') = \delta^{(n)}(\mathbf{r}'' - \mathbf{r}'). \tag{4}$$

Following the Ozaki-Edwards-Gulyaev procedure, we write the short time space interval as

$$(\Delta\mathbf{r}_j)^2 = r_j^2 + r_{j-1}^2 - 2r_j r_{j-1}\cos\Theta_j, \tag{5}$$

where Θ_j =arccos$(\mathbf{r}_j \cdot \mathbf{r}_{j-1}/r_j r_{j-1})$. Then we use Gegenbauer's expansion formula,

$$e^{z\cos\Theta} = \left(\frac{2}{z}\right)^{\nu} \Gamma(\nu) \sum_{\ell}^{\infty} (\ell+\nu)\, C_\ell^\nu(\cos\Theta) I_{\ell+\nu}(z), \tag{6}$$

to express the short time propagator in polar coordinates:

$$K(\mathbf{r}_j,\mathbf{r}_j;\tau_j) = \sum_{\ell}^{\infty} K_\ell(r_j,r_{j-1};\tau_j)\mathcal{P}_\ell^{(n)}(\cos\Theta) \tag{7}$$

with the radial short time propagator,

$$K_\ell(r_j,r_{j-1};\tau_j) = \frac{M}{i\hbar\tau_j}\left(r_j r_{j-1}\right)^{-(n-2)/2} \times \exp\left[\frac{iM}{2\hbar\tau_j}(r_j^2 + r_{j-1}^2) - \frac{i}{\hbar}\tau_j V_j\right] I_{\ell+\frac{n-2}{2}}\left(\frac{M}{i\hbar\tau_j} r_j r_{j-1}\right), \tag{8}$$

and the normalized spherical function defined by

$$\mathcal{P}_\ell^{(n)}(\cos\Theta) = \pi^{n/2}\frac{2\ell+n-2}{[\Gamma(n/2)]^2}\,\Gamma\left(\frac{n-2}{2}\right)\,C_\ell^{(n-2)/2}(\cos\Theta). \tag{9}$$

In the above, $I_\nu(z)$ is the modified Bessel function, and $C_\ell^\nu(z)$ is the Gegenbauer function. Furthermore, if $n=2$ in (9), we have to use the following properties of the Gegenbauer function,

$$C_0^0(x) = 1 \qquad (\ell = 0),$$

and

$$\lim_{\nu\to 0}\Gamma(\nu)\,C_\ell^\nu(x) = \frac{2}{\ell}\cos(\ell\arccos x)\ (\ell = 1,2,\dots).$$

Substituting the short time propagator (7) into the path integral (1), and noticing that the change of variables transforms the volume elements as

$$d^n\mathbf{r}_j = r_j^{n-1}\,dr_j\,d\Theta_j,$$

we perform path integration over angular variables. With the aid of the integral relation of the normalized spherical functions [6],

$$\int_{S^{n-1}}\mathcal{P}_\ell^{(n)}(\hat{\mathbf{r}}_a\cdot\hat{\mathbf{r}}_b)\,\mathcal{P}_{\ell'}^{(n)}(\hat{\mathbf{r}}_b\cdot\hat{\mathbf{r}}_c)\,d^{n-1}\hat{\mathbf{r}}_b = \delta_{\ell,\ell'}\,\mathcal{P}_\ell^{(n)}(\hat{\mathbf{r}}_a\cdot\hat{\mathbf{r}}_c), \tag{10}$$

where $\hat{\mathbf{r}} = \mathbf{r}/r$, the angular integration can easily be completed. As a consequence, the full propagator (1) is given in the form of a partial wave expansion [5-7]:

$$K(\mathbf{r}'',\mathbf{r}';t'',t') = \sum_{\ell=0}^{\infty}K_\ell(r'',r';t'',t')\mathcal{P}_\ell^{(n)}(\hat{\mathbf{r}}'\cdot\hat{\mathbf{r}}'') \tag{11}$$

with the ℓ-th partial wave propagator given as a radial path integral,

$$K(r'',r';t'',t') = \lim_{N\to\infty}\int\prod_{j=1}^{N}K(r_j,r_{j-1};t_j-t_{j-1})\prod_{j=1}^{N-1}r_j^{n-1}dr_j, \tag{12}$$

which remains to be computed, contingent on specification of the potential.

At this point, let us specify the spherically symmetric potential to be of the form,

$$V(r) = \frac{1}{2}M\omega^2\,r^2 + \frac{b\hbar^2}{2M\,r^2}, \tag{13}$$

where $b > -(n/2-1)^2$. The value of the potential for a short time interval τ_j may be approximated by

$$V_j = \frac{1}{4}M\omega^2(r_j^2 + r_{j-1}^2) + \frac{b\hbar^2}{2Mr_jr_{j-1}}. \tag{14}$$

Note that $V_j = V(r_j), V_j = V(\bar{r}_j), V_j = V(\tilde{r}_j)$, and $V_j = [V(r_j) + V(r_{j-1})]/2$ are all equivalent in path integration. Here, $\bar{r}_j = (r_j + r_{j-1})/2$ and $\tilde{r}_j = \sqrt{r_j r_{j-1}}$.

By the asymptotic recombination trick, the inverse-square potential part may be absorbed by the modified Bessel function. Use of the asymptotic formula of Edwards and Gulyaev [2] for the modified Bessel function,

$$I_\nu(z) \sim \frac{1}{\sqrt{2\pi z}} \exp\left[z - \frac{\nu^2 - \frac{1}{4}}{2z}\right], \tag{15}$$

enables us to combine the inverse-square potential with the Bessel function as

$$\exp\left[-\frac{ib\hbar}{2M\, r_j r_{j-1}}\right] I_{\ell+\frac{n-2}{2}}\left(\frac{M}{2i\hbar\tau_j} r_j r_{j-1}\right) \doteq I_{\lambda(\ell)}\left(\frac{M}{2i\hbar\tau_j} r_j r_{j-1}\right) \tag{16}$$

with

$$\lambda(\ell) = \left[\left(\ell + \frac{n-2}{2}\right)^2 + b\right]^{1/2}. \tag{17}$$

Substitution of (13) and (15) into (8) yields

$$K_\ell(r_j, r_{j-1}; \tau_j) = \frac{M}{i\hbar\tau_j} (r_j r_{j-1})^{-(n-2)/2} \\ \times \exp\left[\frac{iM}{2\hbar\tau_j}\left(1 - \frac{1}{2}\omega^2\tau_j^2\right)(r_j^2 + r_{j-1}^2)\right] I_{\lambda(\ell)}\left(\frac{M}{i\hbar\tau_j} r_j r_{j-1}\right). \tag{18}$$

The radial path integral with this short time propagator, involving the modified Bessel function, is indeed "sufficiently ugly" but has been calculated in a rather laborious manner [5]. Our present goal is to exhibit an improved way of performing the radial path integration with the modified Bessel functions.

To this end, we first introduce a new parameter φ defined by

$$\sin\varphi_j = \omega\tau_j \geq 0 \tag{19}$$

Since

$$\cos\varphi_j = \cos[\arcsin(\omega\tau_j)] = 1 - \frac{1}{2}\omega^2\tau_j^2 + O(\tau_j^4), \tag{20}$$

we can rewrite the short time propagator (17) in the form

$$K_\ell(r_j, r_{j-1}; \tau_j) = \frac{M\omega}{i\hbar} (r_j r_{j-1})^{-(n-2)/2} \csc\varphi_j \\ \times \exp\left[\frac{iM\omega}{2\hbar} \cot\varphi_j\, (r_j^2 + r_{j-1}^2)\right] I_{\lambda(\ell)}\left(\frac{M\omega}{i\hbar} r_j r_{j-1} \csc\varphi_j\right). \tag{21}$$

According to the rules of thumb in path integral calculus, all contributions from the terms of $O(\tau_j^{1+\epsilon})$ $(\epsilon > 0)$ can be ignored in the short time propagator. The above parametrization generates no extra terms other than those of $O(\tau_j^{1+\epsilon})$. Thus, (21) is equivalent to (33) without ambiguity.

Next, for $0 < \eta$, $\eta' < \infty$, $0 < \varphi < \pi$ and Re $\nu > -1$, we define the following function,

$$v_\nu(\eta, \eta'; \varphi) = -i \csc\varphi \, \exp[i(\eta + \eta') \cot\varphi] \, I_\nu\left(-2i\sqrt{\eta\eta'} \csc\varphi\right), \tag{22}$$

which we shall refer to as the v-function. This function as a function of η diverges at $\eta = 0$ unless $\mathrm{Re}\,\nu \geq 0$. However, $\sqrt{\eta\eta'}v_\nu(\eta)$ is regular for the entire range of η, i.e., $0 < \eta < \infty$ insofar as $\mathrm{Re}\,\nu > -1$. It has the following interesting property,

$$\lim_{\varphi\to 0} v_\nu(\eta, \eta'; \varphi) = \frac{1}{2}(\eta\eta')^{-1/4}\delta(\sqrt{\eta} - \sqrt{\eta'}). \tag{23}$$

If we use Weber's formula [8, p.395] or its modified version [5]

$$\int_0^\infty \exp(i\alpha r^2) I_\nu(-iar) I_\nu(-ibr)\, r\, dr = \frac{i}{2\alpha} \exp\left[-\frac{i}{4\alpha}(a^2 + b^2)\right] I_\nu\left(-\frac{ab}{2\alpha}\right), \tag{24}$$

valid for $\mathrm{Re}\,\alpha > 0$ and $\mathrm{Re}\,\nu > -1$, then we can easily verify the following simple composition formula,

$$\int_0^\infty v_\nu(\eta'', \eta; \varphi'') v_\nu(\eta, \eta'; \varphi)\, d\eta = v_\nu(\eta'', \eta'; \varphi'' + \varphi). \tag{25}$$

This composition formula can readily be extended to the multi-integral formula,

$$v_\nu\left(\eta_N, \eta_0; \varphi\right) = \int_0^\infty \prod_{j=1}^{N} v_\nu\left(\eta_j, \eta_{j-1}; \varphi_j\right) \prod_{j=1}^{N-1} d\eta_j \tag{26}$$

where

$$\varphi = \sum_{j=1}^{N} \varphi_j.$$

To carry out the radial path integration, we may recognize an advantage of the use of the v-function defined by (22). In terms of the v-function, the short time propagator (21) becomes as simple as

$$K_\ell(r_j, r_{j-1}; \tau_j) = \frac{M\omega}{\hbar}\left(r_j r_{j-1}\right)^{-(n-2)/2} v_\lambda(\eta_j, \eta_{j-1}; \varphi_j) \tag{27}$$

if the following identifications are made:

$$\eta_j = \frac{M\omega}{2\hbar} r_j^2 \qquad \varphi_j = \arcsin(\omega\tau_j). \tag{28}$$

Applying the composition formula (25), we can easily verify

$$\begin{aligned}\int_0^\infty & K_\ell(r_{j+1}, r_j; \tau_{j+1}) K_\ell(r_j, r_{j-1}; \tau_j) r_j^{n-1}\, dr_j \\ &= \frac{M\omega}{\hbar}\left(r_{j+1} r_{j-1}\right)^{-(n-1)/2} v_{\lambda(\ell)}\left(\eta_{j+1}, \eta_{j-1}; \varphi_{j+1} + \varphi_j\right).\end{aligned} \tag{29}$$

Similarly, use of the multi-integral formula (26) of the v-function immediately leads to

$$K(r_N, r_0; \tau) = \frac{M\omega}{\hbar}(r_0 r_N)^{-(n-2)/2}\, v_\lambda(\eta_N, \eta_0; \varphi) \tag{30}$$

or

$$\begin{aligned} &K_\ell(r_N, r_0; \tau) \\ &= \frac{M\omega}{i\hbar}\,(r_0 r_N)^{-(n-2)/2} \csc\varphi\, \exp[i(\eta_0 + \eta_N)\cot\varphi]\, I_\lambda\,(-2i\sqrt{\eta_0\eta_N}\csc\varphi) \end{aligned} \tag{31}$$

where

$$\tau = \sum_{j=1}^{N} \tau_j\,; \qquad \varphi = \sum_{j=1}^{N} \arcsin(\omega\tau_j).$$

In the limit $N \to \infty$ ($\tau_j \to 0$), we see that $\arcsin(\omega\tau_j) \to \omega\tau_j$, and hence that φ reaches the unique limit,

$$\varphi = \omega\tau. \tag{32}$$

Thus, by applying the same limiting process to (31), we arrive at the radial propagator for the ℓ-th partial wave:

$$\begin{aligned} K_\ell(r'', r'; \tau) = &\frac{M\omega}{i\hbar}(r'r'')^{-(n-2)/2}\csc(\omega\tau) \\ &\times \exp[\frac{iM\omega}{2\hbar}\,(r'^{\,2} + r''^{\,2})\cot(\omega\tau)]\, I_{\lambda(\ell)}\left(\frac{M\omega}{i\hbar}r'r''\,\csc(\omega\tau)\right) \end{aligned} \tag{33}$$

where $\lambda(\ell)$ is given by (16). This coincides if $n = 3$ with the earlier "remarkable" result obtained by direct iteration of Weber's integral formula. The previous way [5] of counting of three parameters accumulated from the repeated use of Weber's integral recurrence formula (24) was considerably laborious and by no means elegant. The present calculation of the radial path integral for the harmonic oscillator in an inverse-square potential is certainly simpler and neater, and perhaps more remarkable than the previous version.

3. Concluding Remarks

We have seen that use of the v-function defined by (21) substantially simplifies the radial path integration for the harmonic oscillator in an inverse-square potential. Insofar as the harmonic plus inverse-square potential is concerned, the radial path integration has become as simple as the angular path integration. The reason why the angular integration is simple is that the system is spherically symmetric. The spherical functions representing the angular part of the propagator are indeed the zonal functions associated with the unitary irreducible representations of the symmetry group $SO(n)$ carried by the system. Unfortunately, there is no geometrical symmetry that can simplify the radial integration. However, although we have introduced the v-function independently of any symmetry consideration, we wish to point out that the function can be identified with a matrix element of the operator which represents a group element belonging to the maximal compact subgroup of the $SU(1,1)$ group. Namely, we have

$$v_\nu(\eta, \eta'; \varphi) = \langle J\,\eta \mid \exp(-2i\varphi J_0) \mid J\,\eta' \rangle$$

where J_0 is the generator of the compact subgroup $U(1)$ of $SU(1,1)$ and $\mid J\,\eta\rangle$ is a continuous basis which diagonalizes the Casimir operator of $SU(1,1)$ and a non-compact operator $K_+ = J_0 + J_1$ simultaneously [9,10]. Here, we merely remark that the dynamical Lie algebra (or the spectrum generating algebra) of the radial harmonic oscillator is $su(1,1)$ and that the v-function can be fully accommodated within the irreducible unitary representations of the Lie algebra $su(1,1)$ [11] or an extended framework [6,12] of the irreducible unitary representations of $SU(1,1)$. Therefore, the present treatment of the radial path integral may also be viewed as an application of harmonic analysis based on the $SU(1,1)$ dynamical group. For more details of the $SU(1,1)$ connection, see [6] and [12]. Recently, Shapiro and Vilenkin [13] attempted conversely to interpret Weber's integral formula as a consequence of the composition rule of the $SU(1,1)$ matrix elements given above. As long as the standard unitary irreducible representations determined by Bargmann [14] are adopted, their interpretation is limited only to the case where $\nu \in \mathbf{N}$.

Finally it should be emphasized that the radial path integral (11) for the harmonic oscillator in the inverse-square potential is no longer a standard Gaussian path integral. In the 1948 paper, Feynman demonstrated how to calculate the path integral by using the recurrence property of the Gaussian integral. If a path integral calculable by the Gaussian recurrence is called a Gaussian path integral, then the radial path integral cannot be a Gaussian path integral. By using the one-term asymptotic formula of Edwards and Gulyaev, it is possible to convert formally the radial path integral into a standard Gaussian integral form. However, the resultant Gaussian path integral along a half line cannot explicitly be calculated. In order to perform the radial integration explicitly, we must use the recurrence property of Weber's integral formula (24) or the composition formula for the v-function (25). In this regard, the radial path integral is a Besselian path integral. The radial motion of a Markovian process has indeed been called by mathematicians a "Bessel motion" or a "Bessel process" [15, p.57].

REFERENCES

1. R.P. Feynman, Rev. Mod. Phys. **20** (1948), 367.
2. S.F. Edwards and Y.V. Gulyaev, Proc. Roy. Soc. (London) **A279** (1964), 229.
3. S. Ozaki, *Lectures at Kyushu University*, (unpublished), 1955.
4. D.P. Arovas, *in: Geometric Phase in Physics*, World Scientific, Singapore, 1989.
5. D. Peak and A. Inomata, J. Math. Phys. **10** (1969), 1422; A. Inomata, Benét Research Laboratory Technical Report, WVT-6718 (1967).
6. A. Inomata, H. Kuratsuji and C.C. Gerry, *Path Integrals and Coherent States of SU(2) and SU(1,1)*, World Scientific, Singapore, 1992.
7. M. Böhm and G. Junker, J. Math. Phys. **28** (1987), 1978.
8. G.N. Watson, *A Treatise on the Theory of Bessel Functions, 2nd ed.*, University Press, Cambridge, 1966.
9. G. Lindblad and B. Nagel, Ann. Inst. Henri Poincaré, A: Phys. théor. **13** (1970), 27.
10. N. Mukunda and B. Radhakrishnan, J. Math. Phys. **14** (1973), 254.
11. A.O. Barut and C. Fronsdal, Proc. Roy. Soc. (London) **A 287** (1965), 532.

12. A. Inomata and G. Junker, *in: Classical and Quantum Systems, Proceedings of the 1991 Wigner Symposium*, World Scientific, Singapore.
13. R.L. Shapiro and N.Ya. Vilenkin, *in: Group Theoretical Methods in Physics, vol. I*, VNU Science Press, Utrecht, 1986.
14. V. Bargmann, Ann. Math. **48** (1947), 568.
15. K. Itô and H.P. McKean, *Diffusion Processes and Their Sample Paths*, Springer-Verlag, Berlin, 1965.

DEPARTMENT OF PHYSICS, STATE UNIVERSITY OF NEW YORK AT ALBANY, ALBANY, NEW YORK 12222, USA

THERMODYNAMICS OF QUARKS AND GLUONS

H. LEUTWYLER

ABSTRACT. Thermodynamics interrelates the discontinuities associated with first order phase transitions. I discuss the relation between the jump in the quark condensate and the latent heat.

Contribution in honor of the 60th birthday of John R. Klauder

While visiting the University of Bern in 1961/62 John Klauder gave us a fabulous lecture course on thermodynamics (in German!). Among other things, he used a bottle of beer to awake understanding for the thermodynamics of fluids and gases. With the following discussion of a closely related problem in particle physics, I would like to give him the satisfaction that, even thirty years later, this bottle is bearing fruit. The problem concerns the transition from the hadronic phase of QCD to the quark-gluon plasma phase. I wish to discuss an analogue of the familiar Clausius-Clapeyron relation and use it to argue that, if the transition liberates any latent heat at all, then the amount of it will at most be small. The argument runs as follows.

For massless quarks, the Hamiltonian of QCD is invariant under chiral transformations. The phenomenology of the strong interactions is consistent with this property provided chiral symmetry is spontaneously broken. Indeed, lattice calculations (for reviews, see [1]) indicate that, at low temperatures, the scalar currents $\bar{q}q = \{\bar{u}u, \bar{d}d, \dots\}$ develop nonzero expectation values $< \bar{q}q >_T$ which are referred to as quark condensates and represent order parameters of the spontaneously broken symmetry. Chiral perturbation theory [2–4] shows that $< \bar{q}q >_T$ melts as the temperature rises. One generally expects that, if the temperature reaches a critical value T_c, chiral symmetry is restored. Lattice results confirm this expectation. Presumably, colour is liberated at the same temperature [1,5]. The transition from the low temperature phase where the symmetry is spontaneously broken to the high temperature phase for which the thermal state is invariant under chiral rotations may be first or second order. Lattice results [6–13] do not yet allow to clearly discriminate between these two possibilities, particularly because it is difficult to deal with light quarks on a lattice. Analytical investigations of the issue also have to be taken with a substantial grain of salt because they invoke simplified models [14–19].

Suppose now that the transition is of first order, i.e. that the equation of state $P = P(T)$ exhibits a discontinuity in the derivative $\partial P/\partial T$. The energy density $\epsilon = T\partial P/\partial T - P$

then also contains a discontinuity, referred to as latent heat,

$$\text{disc}\,\epsilon = T_c\,\text{disc}\,\frac{\partial P}{\partial T}. \tag{1}$$

I wish to draw attention to a relation connecting the discontinuity in the quark condensate with the latent heat[1]. To derive this relation I first note that the operator $\bar{q}q$ occurs in the quark mass term of the Hamiltonian,

$$H = H_0 + \int d^3x \sum_{q=u,d,\ldots} m_q \bar{q}q. \tag{2}$$

The thermal expectation value of $\bar{q}q$ therefore represents the response of the partition function to a change in the quark mass,

$$< \bar{q}q >_T = -\frac{1}{V}\frac{\partial}{\partial m_q}\ln Z \tag{3}$$

where V is the four-dimensional euclidean volume. At sufficiently large volume, the partition function grows exponentially, $\ln Z \to V(P - \epsilon_0)$, where P is the pressure and ϵ_0 is the energy density of the vacuum. In the large volume limit, the relation (3) therefore takes the form

$$< \bar{q}q >_T = \frac{\partial \epsilon_0}{\partial m_q} - \frac{\partial P}{\partial m_q} = < 0 \mid \bar{q}q \mid 0 > - \frac{\partial P}{\partial m_q}. \tag{4}$$

On general grounds, the pressure is a continuous function of T. One might be tempted to infer that the derivative of the pressure with respect to the quark mass is also continuous across the phase transition. This is not the case, however, because the critical temperature changes with the quark mass, $T_c = T_c(m_u, m_d, \ldots)$. In the vicinity of T_c, the pressure is given by

$$P = P_c + (T - T_c)\frac{\partial P}{\partial T} + \ldots \tag{5}$$

where the coefficient of the term linear in $T - T_c$ takes different values above and below the critical point. Evaluating the derivative with respect to the quark mass at $T = T_c$, one obtains

$$\frac{\partial P}{\partial m_q} = \frac{\partial P_c}{\partial m_q} - \frac{\partial T_c}{\partial m_q}\frac{\partial P}{\partial T}. \tag{6}$$

The discontinuity in $\partial P/\partial T$ therefore generates a discontinuity in the quark condensate, given by

$$\text{disc} < \bar{q}q >_T = \frac{1}{T_c}\frac{\partial T_c}{\partial m_q}\text{disc}\,\epsilon. \tag{7}$$

In the following, I first discuss the qualitative content of this relation. Some numerical estimates for the quantities occurring therein are given towards the end of the paper.

[1] An early exploration of this connection can be found in one of the pioneering papers on the deconfinement phase transition [20].

1. The result (7) is analogous to the familiar relation attributed to Clausius and Clapeyron [21] which concerns phase transitions from a fluid to a gas. In that case, the critical temperature depends on the pressure under which the transition takes place, $T_c = T_c(P)$. The relation connects the discontinuity in the volume per molecule, v, to the latent heat, i.e. to the discontinuity in the entropy per molecule, s

$$\frac{\partial T_c}{\partial P} = \frac{v_g - v_f}{s_g - s_f} = \frac{\text{disc } v}{\text{disc } s}. \tag{8}$$

In the relation (7), the role of the thermodynamic variable P is taken over by the quark mass and the corresponding conjugate variable – the volume – is therefore replaced by the quark condensate.

2. The relation (7) may provide a convenient way to measure the latent heat on a lattice: the occurrence of a jump in the quark condensate appears to be more readily identified than a discontinuity in the derivative of the pressure.

3. In the chiral limit, the trace of the energy momentum tensor is given by

$$\Theta^\mu_\mu = -G^2 + < 0 \mid G^2 \mid 0 > \tag{9}$$

where G^2 is proportional to the square of the gluon field strength,

$$G^2 \equiv \frac{(-\beta)}{2g^3} G^a_{\mu\nu} G^{a\mu\nu}. \tag{10}$$

The temperature dependence of the gluon condensate (for an analysis of the gluon condensate on a lattice, see [22]) is therefore given by the thermal expectation value of Θ^μ_μ, i.e. by $\epsilon - 3P$. In particular, the discontinuity in the gluon condensate coincides with the latent heat,

$$\text{disc} < G^2 >_T = -\text{disc } \epsilon. \tag{11}$$

The relation (7) therefore states that, in the chiral limit, the change in the critical temperature generated by a change in the quark mass is given by the ratio of the jump in the quark condensate to the one in the gluon condensate.

4. If the quark masses are turned on, T being kept fixed, pressure and energy density decrease. The sensitivity of the equation of state to the quark masses is qualitatively different in the two phases.

In the high temperature phase, the partition function is invariant under a chiral rotation of the quark mass matrix $\mathcal{M}$. The expansion of Z in powers of $\mathcal{M}$ does therefore not contain a linear term: for $T > T_c$, the breaking of chiral symmetry only shows up at $O(\mathcal{M}^2)$, through a contribution proportional to tr $(\mathcal{M}^+\mathcal{M})$. In particular, increasing the masses m_u, m_d from zero to their physical values at fixed m_s only produces negligibly small effects of order $m_u^2 + m_d^2$.

The low temperature phase is more sensitive to the asymmetries generated by the quark mass matrix, because they are amplified by spontaneous symmetry breaking. For $T < T_c$, the expansion of the partition function in powers of $\mathcal{M}$ does contain a linear term related to the order parameter $< \bar{q}q >_T$. The same term is responsible for the pion mass which affects pressure and energy density significantly, particularly at low T. The same mechanism also amplifies the breaking of SU(3): the excitation energy of the lowest strange energy level, $M_K \simeq 495$ MeV, is considerably larger than $M_\pi \simeq 140$ MeV. Accordingly, the strange quark degrees of freedom remain partially frozen even if the temperature approaches the critical point. The critical temperature must therefore be more sensitive to m_u, m_d than to m_s,

$$\frac{\partial T_c}{\partial m_u} \simeq \frac{\partial T_c}{\partial m_d} > \frac{\partial T_c}{\partial m_s}. \tag{12}$$

The relation (7) then implies that the discontinuity in $< \bar{s}s >_T$ is smaller than the one in $< \bar{u}u >_T$, in qualitative agreement with the estimates for the relative magnitude of these condensates at $T = 0$ [23].

5. Since m_u and m_d are small, the difference between the physical value of T_c and the value for $m_u = m_d = 0$ is approximately given by

$$\Delta T_c = m_u \frac{\partial T_c}{\partial m_u} + m_d \frac{\partial T_c}{\partial m_d}. \tag{13}$$

According to Gell-Mann, Oakes and Renner [24], the vacuum expectation value of the operator $m_u \bar{u}u + m_d \bar{d}d$ is determined by $F_\pi^2 M_\pi^2 \simeq 20$ MeV/fm^3. The relation (7) therefore implies

$$\frac{\Delta T_c}{T_c} = \left| \frac{\text{disc} < \bar{u}u >_T}{< 0 \mid \bar{u}u \mid 0 >} \right| \left| \frac{F_\pi^2 M_\pi^2}{\text{disc}\ \epsilon} \right| \tag{14}$$

6. Finally, I discuss some crude numerical estimates for the various quantities occurring in the above relations, based on the Hagedorn model [25,26]. At low temperatures, a hadronic gas in thermal equilibrium almost exclusively contains pions, more massive particles, such as K, η, ρ, ω, ... being rare on account of the Boltzmann factor. Since chiral symmetry suppresses the interactions of low energy Goldstone bosons both among themselves and with massive hadrons, the gas can approximately be described as a collection of free particles. In this approximation, the abundance of the various particle species is determined by their masses. Evaluating the density with the mass spectrum of the data tables, one finds that at low temperatures the gas is dilute and the internal motion is barely activated: at $T = 120$ MeV e.g., the mean volume per particle is 12 fm^3 and five out of six particles are pions, kaons or etas, i.e. are in the ground state of their internal motion. The density of the gas however rapidly increases with the temperature: particle number and energy per unit volume approximately grow with $n \propto T^{5.5}$ and $\epsilon \propto T^{7.3}$, respectively. At $T = 140$ MeV, the mean volume per particle is 5 fm^3; the (π, K, η)-component accounts for 70 % of all particles, 22 % are excited mesons and 8 % are baryons. Since massive particles carry a substantial amount of rest energy, their share of the energy density is more significant: at $T = 140$ MeV, about half of the energy is stored in the pseudoscalars, one third in excited mesons such as ρ, ω, $K^\star$, ... and one sixth in baryons.

It would be strange if the phase transition took place from a configuration of isolated hadrons in their ground state to a configuration of liberated quarks and gluons. The above numbers indicate a lower limit for the critical temperature,

$$T_c \gtrsim 140 \text{ MeV}. \tag{15}$$

Note also that, below this range, chiral perturbation theory provides a coherent quantitative picture for the properties of the hadronic phase, predicting, in particular, that the quark condensate melts by about 20 % if the temperature reaches 140 MeV [4].

If m_u and m_d are set to zero, the pions loose their mass. The pion density therefore increases, by about 35% if $T = 140$ MeV. Within the dilute gas approximation used above, the change in the abundance of the other particles is also determined by the shift occurring in their mass. The change in M_K and M_η can be estimated with current algebra. In the case of the proton, the mass shift is given by the σ-term, known from πN phenomenology: $\sigma \simeq 45$ MeV [27]. This amounts to $\Delta M_p/M_p \simeq 5\%$, comparable to the change occurring in the pion decay constant, where chiral perturbation theory predicts $\Delta F_\pi/F_\pi \simeq 6\%$ [28,29]. Assuming that the proton value of $\Delta M/M$ is typical for all of the particles except the Goldstone bosons, one finds that, in the limit $m_u, m_d \to 0$, the number density of the gas grows by about 30%. For the volume per particle to remain the same, the temperature must be lowered by $\Delta T/T \simeq 5\%$. [Note that the size of the particles grows if the quarks become lighter – a comparable ratio of diameter to mean distance is therefore reached already at a lower temperature.] Assuming $T_c = 140$ MeV, the relation (14) then implies a remarkably small value for the latent heat,

$$\text{disc}\ \epsilon < 0.4\ \text{GeV/fm}^3. \tag{16}$$

If the critical temperature is higher, the value of $\Delta T_c/T_c$ is smaller, because the pions then represent a less important fraction of the gas – the amplification of m_u, m_d due to spontaneous symmetry breaking diminishes, together with the order parameter. In fact, in the free gas model described above, the temperature dependence of $\Delta T/T$ is approximately the same as for the quark condensate. The estimate (16) is therefore not sensitive to T_c. It is clear, however, that if the gas becomes dense, the model underlying this estimate at best represents a rough guide.

The discontinuity in $< G^2 >_T$ is to be compared with the value of the gluon condensate at $T = 0$. In the notation used above, the estimate given in ref. [30] amounts to

$$< 0 \mid G^2 \mid 0 > \simeq \frac{9}{8} \times 0.012\ \text{GeV}^4 \simeq 1.8\ \text{GeV/fm}^3. \tag{17}$$

The results obtained in the recent literature [31] are even larger. The upper limit (16) therefore indicates that the discontinuity in the gluon condensate is small compared to the value of the condensate at $T = 0$.

With due reservations for the crude nature of these estimates, I conclude that if at all the transition is first order then the latent heat is expected to be small. For a value like disc $\epsilon = 2$ GeV/fm^3, the Clausius-Clapeyron relation (14) would imply $\Delta T_c/T_c < 1\%$, difficult to reconcile with the fact that the properties of the hadronic phase are rather sensitive to M_π.

Acknowledgement

I am very much indebted to Peter Hasenfratz for clarifying discussions and useful remarks.

References

1. For recent reviews, see S. Gottlieb, in: Proc. Lattice '90, Nucl. Phys. B (Proc. Suppl.) **20** (1991), 247; F. Karsch, in: Quark-Gluon-Plasma (R.C. Hwa, ed.), World Scientific, Singapore, 1991; D. Toussaint, Progress in Lattice QCD-91, preprint University of Arizona, AZPH-TH/91-33.
2. P. Binétruy and M.K. Gaillard, Phys. Rev. **D32** (1985), 931.
3. J. Gasser and H. Leutwyler, Phys. Lett. **B184** (1987), 83; **B188** (1987), 477; H. Neuberger, Phys. Rev. Lett. **60** (1988), 889.
4. P. Gerber and H. Leutwyler, Nucl. Phys. **B321** (1989), 387.
5. A. Barducci, R. Casalbuoni, S. De Curtis, R. Gatto and G. Pettini, Phys. Lett. **B244** (1990), 311.
6. C. DeTar and J.B. Kogut, Phys. Rev. Lett. **59** (1987), 399; Phys. Rev. **D36** (1987), 2828.
7. S. Gottlieb, W. Liu, D. Toussaint, R.L. Renken and R.L. Sugar, Phys. Rev. Lett. **59** (1987), 1513, 1881; Phys. Rev. **D41** (1990), 662.
8. A. Gocksch, P. Rossi and U. Heller, Phys. Lett. **B205** (1988), 334.
9. K.D. Born, S. Gupta, A. Irbäck, F. Karsch, E. Laermann, B. Petersson and H. Satz, Phys. Rev. Lett. **67** (1991), 302; R.V. Gavai, S. Gupta, A. Irbäck, F. Karsch, S. Meyer, B. Petersson, H. Satz and H.W. Wyld, Phys. Lett. **B241** (1990), 567.
10. A. Ukawa, Nucl. Phys. B (Proc. Suppl.) **10A** (1989), 66; M. Fukugita, M. Okawa and A. Ukawa, Phys. Rev. Lett. **63** (1989), 1768; Nucl. Phys. **B337** (1990), 181.
11. F.R. Bown, F.P. Butler, H. Chen, N.H. Christ, Z. Dong, W. Schaffer, L.I. Unger and A. Vaccarino, Phys. Rev. Lett. **65** (1990), 2491; **67** (91), 1062.
12. J.B. Kogut, D.K. Sinclair and K.C. Wang, Phys. Lett. **B263** (1991), 101.
13. C. Bernard, M. Ogilvie, T. DeGrand, C. DeTar, S. Gottlieb, A. Krasnitz, R. Sugar and D. Toussaint, preprint (1991).
14. A. Barducci, R. Casalbuoni, S. De Curtis, R. Gatto and G. Pettini, Phys. Lett. **B231** (1989), 463; **B240** (1990), 429; Phys. Rev. **D41** (1990), 610.
15. R.D. Pisarski and F. Wilczek, Phys. Rev. **D29** (1984), 338; S. Midorikawa, H. So and S. Yoshimoto, Z. Phys. **C34** (1987), 307.
16. E.M. Ilgenfritz and E.V. Shuryak, Nucl. Phys. **B319** (1989), 511.
17. T. Hatsuda and T. Kunihiro, Phys. Lett. **B198** (1987), 126; T. Hatsuda, in: Proc. QCD '90, Nucl. Phys. B (Proc. Suppl.) **23** (1991).
18. Z. Frei and A. Patkós, Phys. Lett. **B222** (1989), 469; **B247** (1990), 381.
19. J. Cleymans, A. Kocić and M.D. Scadron, Phys. Rev. **D39** (1989), 323.
20. P. Hasenfratz, F. Karsch and I.O. Stamatescu, Phys. Lett. **B133** (1983), 221.
21. B.P.E. Clapeyron, Journal de l'Ecole Polytechnique (Paris) **14** (1834), 153; R. Clausius, Annalen der Physik **79** (1850), 368, 500.
22. M. Campostrini, G. Curci, A. DiGiacomo and G. Paffuti, Z. Phys. **C32** (1986), 377; M. Campostrini and A. DiGiacomo, Phys. Lett. **B197** (1987), 403; Su. H. Lee, Phys. Rev. **D40** (1989), 2484; R.J. Furnstahl, T. Hatsuda and Su. H. Lee, Phys. Rev. **D42** (1990), 1744; C. Adami, T. Hatsuda and I. Zahed, Phys. Rev. **D43** (1991), 921.
23. V.M. Belyaev and B.L. Ioffe, Sov. Phys. JETP **57** (1983), 716; L. Reinders, H. Rubinstein and S. Yazaki, Phys. Lett. **B120** (1983), 209; L. Reinders and H. Rubinstein, Phys. Lett. **B145** (1984), 108.
24. M. Gell-Mann, R. J. Oakes and B. Renner, Phys. Rev. **175** (1968), 2195.
25. R. Hagedorn, Suppl. Nuovo Cim. **3** (1965), 147; in: Quark Matter '84, Lecture Notes in Physics, **211**, Springer, Berlin, 1985; R. Fiore et al., Nuovo Cim. **88A** (1985), 301.
26. H. Bebie, P. Gerber, J. Goity and H. Leutwyler, preprint University of Bern, BUTP-91/14.
27. J. Gasser, H. Leutwyler and M. E. Sainio, Phys. Lett. **B253** (1991), 252.
28. J. Gasser and H. Leutwyler, Ann. Phys. **158** (1984), 142.
29. J. Donoghue, J. Gasser and H. Leutwyler, Nucl. Phys. **B343** (1990), 341.
30. M.A. Shifman, A. I. Vainshtein and V.I. Zakharov, Nucl. Phys. **B147** (1979), 385; V. Novikov et al., Nucl. Phys. **B191** (1981), 301.
31. G. Launer, S. Narison and R. Tarrach, Z. Phys. **C26** (1984), 433; R.A. Bertlmann, C.A. Dominguez, M. Loewe, M. Perrottet and E. de Rafael, Z. Phys. **C39** (1988), 231; C. A. Dominguez and J. Solà, Z. Phys. **C40** (1988), 63; J. Bordes, V. Giménez and J.A. Peñarrocha, Phys. Lett. **B201** (1988), 365;

B223 (1989), 245; S. Narison, QCD Spectral Sum Rules, Lecture Notes in Physics **26** (1989), World Scientific, Singapore.

INSTITUTE FOR THEORETICAL PHYSICS, UNIVERSITY OF BERN, SIDLERSTRASSE 5, CH-3012 BERN, SWITZERLAND

ON Q-ANALOGS OF COHERENT STATES

M. A. LOHE AND L. C. BIEDENHARN

ABSTRACT. We review the concept of a q-deformation in connection with quantum groups and a q-deformed Heisenberg-Weyl algebra. With these concepts we construct q-analogs of the canonical coherent states and the associated q-harmonic oscillator states. We briefly formulate q-analog vector coherent states and with them construct representations of the quantum group $\mathrm{SU_q(2)}$.

Dedicated to John Klauder at the occasion of his sixtieth birthday

1. Introduction

One of the major interests of John Klauder over the years has been the theory and applications of coherent states, culminating in the book by Klauder and Skagerstam [1] in 1985. We feel it is therefore appropriate in our contribution to the Klauder *Festschrift* that we indicate how the currently interesting concept of "q-deformations" can be used for yet another application of the general concept of coherent states.

To carry out this survey we first discuss q-deformations (and quantum groups) in Section 2 and then in Section 3 introduce q-coherent states (coherent states for the q-deformed Heisenberg-Weyl algebra). In Section 4, we develop another variation on q-coherent states that is important in the representation theory of quantum groups.

2. q-Deformations

Although the concept of a q-analog, for the hypergeometric functions, arose in the mid-nineteenth century (Heine [2]), the significance and generality of this concept has only recently been brought out through the discovery of the so-called "quantum groups". The precise definition of a quantum group is rather forbidding, and, in fact, the quickest way to understand this new structure is by example. Let us consider the quantum group $SU_q(2)$, which is a deformation of the familiar quantal angular momentum group $SU(2)$.

$SU_q(2)$ is generated by the three operators $J_{\pm}^q, J_z^q$ satisfying the commutation relations

$$[J_z^q, J_{\pm}^q] = \pm J_{\pm}^q\,, \tag{2.1}$$

$$[J_+^q, J_-^q] = \frac{q^{J_z^q} - q^{-J_z^q}}{q^{\frac{1}{2}} - q^{-\frac{1}{2}}}\,, \quad q \in I\!R^+. \tag{2.2}$$

These defining relations for $SU_q(2)$ differ from those of ordinary angular momentum in two ways:

(1) The commutator in eq. (2.2) is not $2J_z$ as usual, but an infinite series (for generic q) involving ***all odd powers***: $(J_z^q)^1, (J_z^q)^3, \ldots$. Each such power is a linearly independent operator in the enveloping algebra; accordingly, the Lie algebra of $SU_q(2)$ is *not* of finite dimension.

(2) For $q \to 1$, the right hand side of eq. (2.2) $\to 2J_z$. Thus we recover in the limit the usual Lie algebra of $SU(2)$. The differences noted in (a) and (b) are expressed by saying that the quantum group $SU_q(2)$ is a *deformation* of the enveloping algebra of $SU(2)$.

Let us introduce a notation for the "q-integers" that play a rôle in quantum groups. Define the q-integer $[n]_q$ by:

$$\begin{aligned}[n]_q &\equiv \frac{q^{\frac{n}{2}} - q^{-\frac{n}{2}}}{q^{\frac{1}{2}} - q^{-\frac{1}{2}}} \\ &= \underbrace{q^{(n-1)/2} + q^{(n-3)/2} + \ldots + q^{-(n-1)/2}}_{n \text{ terms}}\,, \quad n \in Z\!\!\!Z \end{aligned} \tag{2.3a}$$

and

$$[n]_q! \equiv [n]_q[n-1]_q \ldots [1]_q\,. \tag{2.3b}$$

These q-integers, $[n]_q$, obey the rule: $[-n]_q = -[n]_q$, with $[0]_q = 0$ and $[1]_q = 1$.

Note that the defining relations, eqs. (2.1) and (2.2) of the quantum group $SU_q(2)$ are invariant to $q \leftrightarrow q^{-1}$. This invariance has been incorporated in the definition (2.3), so that $[n]_q = [n]_{q^{-1}}$. (This—with the use of steps of *unity* for powers of q in eq. (2.3)—accounts for the convention using $q^{\frac{1}{2}}$ in eq. (2.3).) We remark that the q-analogs defined by Heine did not obey the $q \leftrightarrow q^{-1}$ symmetry, and, in fact, the q-factorial in (2.3b) is related to the standard q-analog of Euler's factorial function [3] by:

$$[n]_q! = q^{-n(n-1)/4}\Gamma_q(n+1)\,. \tag{2.4}$$

The concept of a ***q-integer***, mentioned above, is well-defined under the ***addition law***:

$$\begin{aligned}[m]_q \oplus [n]_q &\equiv q^{-\frac{n}{2}}[m]_q + q^{\frac{m}{2}}[n]_q\,. \\ &= [m+n]_q\,, \quad m, n, \in Z\!\!\!Z \end{aligned} \tag{2.5}$$

which establishes an isomorphism: $\{[n]_q\} \leftrightarrow \mathbb{Z}$.

Anticipating that $2J_z^q$ has integer eigenvalues, we can rewrite eq. (2.2) in an equivalent operator form:

$$[J_+^q, J_-^q] \;=\; [2J_z^q]_q\,. \qquad (2.2')$$

In order to understand the meaning of the eqs. (2.1) and (2.2) it is natural to look for representations of the operators $J_\pm^q$, J_z^q as finite-dimensional matrices. For the quantal angular momentum group, there is a standard way to do this: one uses the *Jordan-Schwinger map* [4], which maps the 2×2 matrices $\{J_\pm, J_z\}$ of the fundamental irrep into boson operators.

Is there a q-analog to the Jordan-Schwinger map? There is indeed as is described in detail in [5,6] and developed further in [7,8,9] . The basic idea is to construct q-analogs to the boson operators. Introduce the *q-creation operator* a^q, its Hermitian conjugate the *q-destruction operator* $\overline{a}^q$, and the *q-boson vacuum ket vector* $|0\rangle$ defined by the equation

$$\overline{a}^q|0\rangle \;=\; 0\,. \qquad (2.6)$$

Instead of the Heisenberg relation, postulate the algebraic relation:

$$\overline{a}^q\, a^q - q^{1/2} a^q\, \overline{a}^q \;=\; q^{-N^q/2}\,, \qquad (2.7)$$

where N^q is the Hermitian *number operator* satisfying

$$[N^q, a^q] = a^q\,, \quad [N^q, \overline{a}^q] = -\overline{a}^q\,, \text{ with } N^q|0\rangle \equiv 0. \qquad (2.8a,b,c)$$

This algebra is a deformation of the Heisenberg-Weyl algebra, which is recovered in the limit $q \to 1$. (Note that the q-number operator N^q is now no longer the operator $a\,\overline{a}$ as in the Heisenberg case.)

Orthonormal ket vectors corresponding to states of n q-quanta are given by:

$$|n\rangle_q \;\equiv\; ([n]_q!)^{-\frac{1}{2}}(a^q)^n|0\rangle, \qquad (2.9a)$$

$$\text{with}: \quad N^q|n\rangle_q \;=\; n|n\rangle_q\,. \qquad (2.9b)$$

It is now easy to define a q-analog for the algebra of the generators of the quantum group $SU_q(2)$. In the language of q-boson operators, one defines a pair of *mutually commuting q-bosons* a_i^q and $\overline{a}_i^q$ for $i = 1, 2$. That is, for each i : a_i^q and $\overline{a}_i^q$ obey equations $(2.6 - 2.9)$ and, in addition, the relations:

$$\text{for } i \neq j: \quad [a_i^q, a_j^q] = [\overline{a}_i^q, \overline{a}_j^q] = [a_i^q, \overline{a}_j^q] = 0. \qquad (2.10)$$

The generators $\{J_+^q, J_-^q, J_z^q\}$ of $SU_q(2)$ are then realized by

$$J_+^q = a_1^q\overline{a}_2^q, \quad J_-^q = a_2^q\overline{a}_1^q, \quad J_z^q = \tfrac{1}{2}(N_1 - N_2). \qquad (2.11a,b,c)$$

It can be verified, using (2.6–2.10), that these generators satisfy under commutation the quantum algebra of $SU_q(2)$, eqs. (2.1, 2.2).

We shall not pursue quantum groups further, (see, however, Section 4) since our main interest, coherent states, is more directly connected to the deformed Heisenberg-Weyl algebra.

Remarks. As in the case of the boson operators, one can equivalently regard the operator a^q as effecting multiplication by the complex variable z, but now $\overline{a}^q$ is represented not by differentiation, but by the finite difference operator D_q, defined by

$$D_q f(z) = \frac{f(zq^{\frac{1}{2}}) - f(zq^{-\frac{1}{2}})}{z(q^{\frac{1}{2}} - q^{-\frac{1}{2}})}, \tag{2.12}$$

for suitable functions $f(z)$. Symbolically, we can write $a^q = z$, $\overline{a}^q = z^{-1}[zd/dz]_q$.

Another realization of the q-boson operators can be obtained in which a^q acts as a finite sum operator, performing for $q \to 1$ the operation of integration. Define, for $q < 1$ and f an entire function,

$$(a^q f)(z) = \int f(z)\, d_q z \equiv \sum_{r=0}^{\infty} f\left(zq^{r+\frac{1}{2}}\right)\left(zq^r - zq^{r+1}\right) \tag{2.13a}$$

$$(\overline{a}^q f)(z) = D_q\, z\, D_q\, f(z), \tag{2.13b}$$

then the operators a^q and $\overline{a}^q$ again satisfy (2.7) and (2.8) with $N = zd/dz$, and symbolically we have $a^q = [N]_q^{-1} z$, $\overline{a}^q = z^{-1}[N]_q^2$.

The operator D_q together with q-integration and associated q-extensions to classical functions are not new to quantum groups, having been studied early in this century by Jackson [10]. (In fact the theory of q-functions dates back to Heine [2], as already indicated.) The subject has been developed extensively more recently by Andrews [11], Askey and Wilson [12], Koornwinder [13], Milne [14], and Gustafson [15] particularly the properties of q-analog (basic) hypergeometric functions (see also the monographs by Exton [16], and Gasper and Rahman [17].) Although these developments stand apart from the topic of symmetries and their application to quantum mechanics, there is a link which is provided by the concept of quantum groups. As already indicated, q-boson operators can be used to construct representations of quantum groups, which effectively means that operations of q-differentiation or q-integration are being employed to construct quantum group generators and basis states of representation spaces. Furthermore, properties of q-analog functions, in particular basic hypergeometric functions, appear in the theory of tensor operators for quantum groups, for example, the q-Clebsch-Gordan coefficients for $SU_q(2)$ can be expressed [18,19] in terms of ${}_3\phi_2$ functions and the q-$6j$ and q-Racah coefficients appear as ${}_4\phi_3$ functions [19]. As another example, tensor operators for $SU_q(2)$ in a certain representation can be expressed in terms of ${}_2\phi_1$ functions, and symmetries and algebraic properties of these tensor operators can be related to well-known properties of basic hypergeometric functions [20].

In the constructions to follow, we will find it useful to introduce the q-exponential function defined by: ($\alpha \in \mathbb{C}$)

$$\exp_q(\alpha a^q)|0\rangle \equiv \sum_{n=0}^{\infty} \frac{(\alpha a^q)^n}{[n]!}|0\rangle\,, \tag{2.14}$$

which has the property:

$$\overline{a}^q \exp_q(\alpha a^q)|0\rangle = \alpha \exp_q(\alpha a^q)|0\rangle\,. \tag{2.15}$$

This q-exponential is a q-analog of the classical exponential function, although as such it is not unique; it is, however, invariant under $q \to q^{-1}$.

Exton [16] has defined a family of q-exponential functions (not invariant under $q \to q^{-1}$) by

$$e_q^{\lambda}(z) = \sum_{n=0}^{\infty} \frac{z^n q^{\lambda n(n-1)}}{[n]_q!}\,. \tag{2.16}$$

Properties of these functions are related to those of $\exp_q$, for we may write symbolically,

$$e_q^{\lambda}(z) = \exp_q\left(zq^{2\lambda N}\right) \tag{2.17}$$

where $N = zd/dz$. Another noteworthy property of $\exp_q$ is

$$\exp_q(X)\exp_q(Y) = \exp_q(X+Y)\,, \tag{2.18}$$

where X, Y are "quantum coordinates", satisfying $XY = q^{-1}YX$ (eq. (2.18) follows from an application of the q-binomial theorem.)

A further property of $\exp_q$ which we require is that of *continuity*; this follows from the inequality $[n]_q \geq n$ for all $q > 0$, so that each term in (2.14) is bounded by the corresponding term for the exponential function. In fact, $\exp_q$ is an entire function and so is continuous.

3. Coherent States

Klauder and Skagerstam [1] have identified two key characteristics of coherent states:

(1) *continuity* of the coherent state $|z\rangle$ as a function of z,
(2) *resolution of unity*:

$$\mathbf{1} = \int |z\rangle\,\langle z|\,d\mu(z), \tag{3.1}$$

where the integration takes place with respect to a positive measure $d\mu(z)$.

The best known example of coherent states, which certainly satisfy these two characteristics, are the canonical coherent states generated by creation and annihilation operators a and $\overline{a}$. These canonical coherent states are defined by [1,21]

$$|z\rangle \equiv e^{-|z|^2/2}\, e^{za}\, |0\rangle \tag{3.2a}$$

$$= e^{-|z|^2/2} \sum_{n=0}^{\infty} \frac{z^n}{\sqrt{n!}}\, |n\rangle \tag{3.2b}$$

where $|n\rangle$ denotes the orthonormal vectors generated by the creation operator a.

We can immediately write down q-coherent states $|z\rangle_q$ by replacing the boson operator in (3.2) by its q-boson analog, together with the exponential by the q-exponential function $\exp_q$:

$$|z\rangle_q = \left(\exp_q(|z|^2)\right)^{-\frac{1}{2}} \exp_q(za^q)|0\rangle \tag{3.3a}$$

$$= \left(\exp_q(|z|^2)\right)^{-\frac{1}{2}} \sum_{n=0}^{\infty} \frac{z^n}{\sqrt{[n]_q!}}\, |n\rangle_q. \tag{3.3b}$$

These states satisfy:

$$\overline{a}_q\, |z\rangle_q = z\, |z\rangle_q \tag{3.4}$$

showing that the q-coherent state $|z\rangle_q$ is an eigenstate of the annihilation operator $\overline{a}^q$ with eigenvalue z and, since $z = {}_q\langle z|\overline{a}^q|z\rangle_q$ (assuming the states $|z\rangle_q$ are normalized), the label z is the mean of $\overline{a}^q$ in the state $|z\rangle_q$. The definition (3.3a) is not a unique q-extension of (3.2), for we could have chosen any one of the family e_q^λ of exponential functions given in (2.16); this would introduce explicit q-factors in equations such as (3.4). We outline below how a q-harmonic oscillator model in fact leads naturally to such q-coherent states. The states (3.3a) were considered in [5] and subsequently also in [9,22,23,24]. In fact, as with many q-analogs of classical and quantum concepts, some q-generalizations were obtained before the appearance of quantum groups [25,26].

Let us now consider the two characteristic properties of coherent states, continuity and completeness. The continuity properties of $|z\rangle_q$, as a function of z, follow immediately from continuity of $\exp_q$. The resolution of unity within the Hilbert space, in terms of the states $|z\rangle_q$, has been considered by Gray and Nelson [22] and also Bracken et al. [24]. The q-analog of Euler's formula for $\Gamma(x)$ is required, and is expressed in terms of the q-integration defined in (2.13):

$$\int_0^\zeta \exp_q(-x)x^n d_qx = [n]_q! \tag{3.5}$$

where ζ is the largest zero of $\exp_q(x)$ (note that, unlike e^x, $\exp_q(x)$ alternates in sign as $x \to -\infty$). A natural restriction is $|z|^2 < \zeta$ and then, with the help of (3.5), the resolution of unity can be derived [22],

$$\mathbf{1} = \int |z\rangle_q \, {}_q\langle z| \, d\mu(z) \tag{3.6}$$

where the measure $d\mu(z)$ is given by

$$d\mu(z) = \tfrac{1}{2\pi} \exp_q(|z|^2) \exp_q(-|z|^2) d_q|z|^2 d\theta \,, \tag{3.7}$$

where $\theta = \arg(z)$. It follows from (3.6) that an arbitrary state can be expanded in terms of the states $|z\rangle_q$; in fact, q-coherent states are overcomplete, for an arbitrary q-coherent state is non-orthogonal to $|z\rangle_q$, for any z.

Coherent states arise naturally within the framework of the harmonic oscillator, by defining boson operators from position and momentum operators, Q, P:

$$a = \tfrac{1}{\sqrt{2}}(Q - iP), \qquad \overline{a} = \tfrac{1}{\sqrt{2}}(Q + iP) \,. \tag{3.8}$$

Conversely, we can use these formulas to define momentum and position operators and so, given q-boson operators, these formulas also provide convenient q-analog definitions of q-momentum and position operators [5]. Alternatively, we can define a q-harmonic oscillator by starting with Schrödinger's equation and replacing the derivative by a finite difference operator. Instead of the q-symmetric operator D_q defined in (2.12), we use the following q-derivative,

$$\nabla_q f(x) = \frac{f(xq) - f(x)}{x(q-1)} \,, \tag{3.9}$$

and the q-harmonic oscillator states are determined by the equation

$$\tfrac{1}{2}(-\nabla_q^2 + qx^2)\psi(x) = E\psi(x). \tag{3.10}$$

Effectively, we have chosen q-momentum and q-position operators Q_q, P_q satisfying

$$qQ_qP_q - P_qQ_q = i \tag{3.11}$$

with the realization $Q_q = x, \; P_q = -i\nabla_q$.

Solutions of the finite difference equation (3.10) have been given by several authors [27,28], and involve q-extensions of the Hermite polynomials. The ground state ψ_0 is given by

$$\psi_0(x) = \sum_{n=0}^{\infty} \frac{(-)^n q^{-n^2/2} x^{2n}}{[2n]_q!!} \tag{3.12}$$

where $[2n]_q!! = [2n]_q[2n-2]_q \ldots [2]_q$. Upon using the identity $[2n]_q = [2]_q[n]_{q^2}$ we can identify the function (3.12) as one of the family of q-exponential functions given by (2.16), with $\lambda = -\frac{1}{4}$ and q replaced by q^2. In general the eigenstates ψ_n are labelled by an integer n, and the energy levels are $E_n = \frac{1}{2}[2n+1]_q$. (For comparison, in the model defined in [5], the levels are $\frac{1}{2}([n+1]_q + [n]_q) = \frac{1}{2}[2n+1]_{q^{1/2}}$). The eigenstates ψ_n take the form

$$\psi_n(x) = H_n^q(x)\psi_0(xq^{-n/2}) \tag{3.13}$$

where ψ_0 is given by (3.12) and $H_n^q(x)$ denotes a q-extension of the classical Hermite polynomial, with the explicit formula:

$$H_n^q(x) = \sum_{r=0}^{n} \frac{C_r x^r q^{-(2n+1)r/4}}{[r]_q!} \tag{3.14}$$

where the coefficients C_r are given (for even or odd r) by

$$\begin{aligned} C_{2m} &= (-)^m q^{(2n+1)m/2}[2n]_q[2n-4]_q \ldots [2n-4m+4]_q \\ C_{2m+1} &= (-)^m q^{(2n+1)m/2}[2n-2]_q[2n-6]_q \ldots [2n-4m+2]_q \,. \end{aligned} \tag{3.15}$$

From the explicit eigenstates one can identify q-boson operators which step between the eigenstates $\psi_n(x)$, from which one can form the q-coherent states of this model of the q-harmonic oscillator [28].

4. Vector Coherent States

Coherent states can be generalized in many ways, but one such generalization, to vector coherent states, is particularly useful for the direct and explicit construction of representations of symmetry algebras. We outline this construction for $SU(2)$ and its q-extension to $SU_q(2)$, in a way which can be generalized to all the quantum unitary groups. VCS (vector coherent) states are described in detail in the monograph by Hecht [29] (in which further references to the large body of previous work may be found). The q-extension to $U_q(n)$ appears in [30].

Let us consider first $SU(2)$; the normalized VCS state $|jm\rangle_{\text{VCS}}$ is defined by

$$|jm\rangle_{\text{VCS}} = \langle jj|e^{aJ_+}|jm\rangle|0\rangle \otimes |jj\rangle \tag{4.1}$$

where, in standard notation, $|jm\rangle$ is an $SU(2)$ basis vector having angular momentum j and z-projection m. In this expression we have formed a matrix element by using angular momentum coherent states, which themselves are formed from the exponential of the boson creation operator a acting on the vacuum $|0\rangle$, whereas the raising generator J_+ acts on the basis vector $|jm\rangle$. Only one term in the exponential series actually contributes to this matrix element, which therefore reduces to a monomial in a acting on $|0\rangle$ (see (4.3) below). The action of the generators on these VCS states is defined by

$$\Gamma(J_m)|jm\rangle_{\rm VCS} = \langle jj|e^{zJ_+}J_m|jm\rangle|0\rangle \otimes |jj\rangle \tag{4.2}$$

where $\Gamma(J_m)$ denotes the representative of the $SU(2)$ generator J_m, acting in the space of VCS states. The operators $\{\Gamma(J_m)\}$ form a representation of $SU(2)$, i.e. Γ is a homomorphism from the $SU(2)$ generators $\{J_m\}$ to a set of operators formed from a and $\overline{a}$ which satisfy the commutation relations of $SU(2)$. The VCS states in fact carry a representation of the group, which of course extends to the algebra. The actual calculation of $\Gamma(J_m)$ is performed by commuting J through the exponential, so that it acts on the vector $\langle jj|$, using the commutation relations of $SU(2)$. The matrix element in (4.1) is readily evaluated, giving

$$|jm\rangle_{\rm VCS} = \left[\frac{(2j)!}{(j+m)!}\right]^{\frac{1}{2}} \frac{a^{j-m}}{\sqrt{(j-m)!}}|0\rangle \otimes |jj\rangle \tag{4.3}$$

and the VCS generators are found to be:

$$\begin{aligned} \Gamma(J_+) &= \overline{a}, \quad \Gamma(J_z) = \mathcal{J}_z - N\,, \\ \Gamma(J_-) &= a(2\mathcal{J}_z - N) \end{aligned} \tag{4.4}$$

where $N = a\overline{a}$, and $\mathcal{J}_z$ is the $SU(2)$ generator acting on the (fixed) one dimensional vector $|jj\rangle$ and so can be replaced by its eigenvalue j.

The q-generalization of this construction is immediate and can be used to determine details of the quantum group representations. The q-VCS states are, by analogy with (4.1),

$$|jm\rangle_{q-\rm VCS} = \langle jj|\exp_q(a^q J_+^q)|jm\rangle|0\rangle \otimes |jj\rangle \tag{4.5}$$

which can be regarded as involving q-analogs of angular momentum coherent states. The action of the algebra is also analogous to that defined in (4.2) for $SU(2)$. By explicit evaluation of the matrix elements in (4.5) one finds:

$$|jm\rangle_{q-\rm VCS} = \left[\frac{[2j]_q!}{[j+m]_q!}\right]^{\frac{1}{2}} \frac{(a^q)^{j-m}}{\sqrt{[j-m]_q!}}\,|0\rangle \otimes |jj\rangle\,, \tag{4.6}$$

and the q-VCS generators are realized by

$$\begin{aligned} \Gamma(J_+^q) &= \overline{a}^q, \quad \Gamma(J_z^q) = \mathcal{J}_z^q - N \\ \Gamma(J_-^q) &= a^q[2\mathcal{J}_z^q - N]_q\,. \end{aligned} \tag{4.7}$$

It can be verified directly that the commutation relations (2.1, 2.2) are satisfied (to do this one requires $a^q\bar{a}^q \to [N]_q$ and the identity $[n+1]_q[n']_q - [n]_q[n'+1]_q = [n'-n]_q$). From (4.6, 4.7) we may determine the explicit matrix elements of the quantum group generators. This VCS construction realizes all unitary irreps of the quantum group $SU_q(2)$. Remarkably, (for q generic real) all such irreps are finite dimensional (dimension $= (2j+1)$, $j = 0, \frac{1}{2}, \ldots$) and in one-to-one correspondence with the usual $SU(2)$ irreps.

The q-VCS construction generalizes to the quantum group $U_q(n)$, by employing Gel'fand-Weyl basis states in place of the $SU_q(2)$ states $|jm\rangle$ and by replacing the argument $a^q J_+^q$ in (4.5) by a sum over a subalgebra of raising generators and q-boson operators. Details are provided in [30], where the explicit realization of the quantum group generators corresponding to (4.7) are given.

REFERENCES

1. J. R. Klauder and B. S. Skagerstam, *Coherent States. Applications in Physics and Mathematical Physics*, World Scientific, Singapore, 1985.
2. E. Heine, J. reine und angew. Math. **34** (1847), 285.
3. L. J. Slater, *Generalized Hypergeometric Functions*, Cambridge University Press, Cambridge, 1966.
4. L. C. Biedenharn and J. D. Louck, *Angular Momentum in Quantum Physics, Encyclopedia of Mathematics and Its Applications* **8** , Addison-Wesley, Reading, MA, 1981.
5. L. C. Biedenharn, J. Phys. A. Math. Gen. **22** (1989), 873.
6. A. J. Macfarlane, J. Phys. A. Math. Gen. **22** (1989), 4581.
7. L. C. Biedenharn and M. A. Lohe, *Some Remarks on Quantum Groups— A New Variation on the Theme of Symmetry*, From Symmetries to Strings: Forty Years of Rochester Conferences, World Scientific, 1990, pp. 189-206.
8. C.-P. Sun and H.-C. Fu, J. Phys. A. Math. Gen. **22** (1989), 983.
9. P. Kulish and E. Damashinsky, J. Phys. A. Math. Gen. **23** (1990), 415.
10. F. H. Jackson, Messenger Math. **38** (1909), 57-61; 62-64; Quart. J. Pure Appl. Math. **41** (1910), 193-203.
11. G. E. Andrews, *q-Series: Their Development and Application in Analysis, Number Theory, Combinatorics, Physics, and Computer Algebra (Conference Board of the Mathematical Sciences* **66** *)*, American Mathematical Society, Providence, RI, 1986.
12. R. Askey and J. A. Wilson, Memoirs Amer. Math. Soc. **319** (1985).
13. T. H. Koornwinder, Nederl. Acad. Wetensch. Proc. Ser. **A92** (1989), 97.
14. S. C. Milne, Advances in Math. **72** (1988), 59-131.
15. R. A. Gustafson, SIAM J. Math. Anal. **18** (1987), 294.
16. H. Exton, *q-Hypergeometric Functions and Applications, Ellis Horwood Series*, Wiley, New York, 1983.
17. G. Gasper and M. Rahman, *Basic Hypergeometric Series, Encyclopedia of Mathematics and its Applications* **35** , Cambridge University Press, Cambridge, 1990.
18. L. Vaksman, Sov. Math. Dokl. **39** (1989), 467.

19. L. C. Biedenharn and M. A. Lohe, *Symmetries of Quantum Group Coupling Coefficients,*, Differential Geometry, Group Representations, and Quantization, Lecture Notes in Physics **379** , pp. 193.

20. L. C. Biedenharn and M. A. Lohe, *Quantum Groups and Basic Hypergeometric Functions*, Quantum Groups, Proceedings of Argonne Workshop, World Scientific, Singapore, 1991, pp. 123.

21. A. Perelomov, *Generalized Coherent States and Their Applications*, Springer, Berlin, 1986.

22. R. W. Gray and C. A. Nelson, J. Phys. A. Math. Gen. **23** (1990), 945.

23. C. Quesne, Phys. Lett. **153A** (1991), 303.

24. A. J. Bracken, D. S. McAnally, R. B. Zhang and M. D. Gould, J. Phys. A. Math. Gen. **24** (1991), 1379.

25. M. Arik and D. D. Coon, J. Math. Phys. **17** (1976), 524.

26. A. Jannussis, G. Brodimas, D. Sourlas, A. Streclas, P. Siafaricas, L. Papaloucas and N. Tsangas, Hadronic J. **5** (1982), 1923.

27. J. A. Minahan, *The q-Schrodinger Equation*, University of Florida preprint, August 1990.

28. N. M. Atakishiev and S. K. Suslov, Theor. and Math. Phys. **8** (1991)), 1055; Teor. i Mat. Fiz. **87** (1991), 154.

29. K. T. Hecht, *The Vector Coherent State Method and Its Application to Problems of Higher Symmetries*, Lecture Notes in Physics **290**, 1987.

30. L. C. Biedenharn and M. A. Lohe, *An Extension of the Borel-Weil Construction to the Quantum Group $U_q(n)$*, Comm. Math. Phys. (to appear).

DEPARTMENT OF PHYSICS, DUKE UNIVERSITY, DURHAM, NORTH CAROLINA, U.S.A.
PERMANENT ADDRESS FOR M.A. LOHE: NORTHERN TERRITORY UNIVERSITY, P.O. BOX 40146, CASUARINA NT 0811, AUSTRALIA

DYNAMICAL ENTROPY OF QUANTUM SYSTEMS AND THEIR ABELIAN COUNTERPART

HEIDE NARNHOFER AND W. THIRRING

ABSTRACT. We calculate the dynamical entropy of quasifree automorphisms by showing that their abelian counterpart already gives an optimal abelian model.

Birthday Present for John R. Klauder

1. Introduction

The dynamical entropy is an important tool in classical ergodic theory [1,2]. Its generalization to nuclear C^*-algebras [3,4,5] is more complicated, its definition depending on the notion of an abelian model for an algebra and a state over it. In this paper we want to make this aspect more intuitive by showing that in the cases of interest the ergodic information is carried by a maximal abelian subalgebra corresponding to a classical version of the system [6]. In this case the quantum dynamical entropy is simply its classical version for this abelian skeleton of the system.

In Sect. 2 we review the salient points of the quantum dynamical entropy; we illustrate in Sect. 3 the general theory by quasifree automorphisms of the CAR algebra. The classical version of the system of infinitely many fermions is studied in Sect. 4. Its classical phase space is identified with the power set of Z^{2d} and endowed with a Borel structure. Although algebraically isomorphic to a maximal abelian subalgebra of the quantum system it shows the same entropic features.

2. Entropies on Noncommutative Algebras

If we make a finite subset I of N a probability space by giving each point $j \in I$ a measure μ_j, $\sum_{j\in I} \mu_j = 1$, then everybody agrees that the entropy of this measure is

$$S_\mu = -\sum_{j\in I} \mu_j \ln \mu_j \, . \tag{2.1}$$

For the quantum mechanical generalization it is expedient to phrase this in an algebraic language by considering a finite abelian algebra $\mathcal{M}$ with minimal orthogonal projections P_j, $j \in I$, and a state over $\mathcal{M}$ with $\omega(P_j) = \mu_j$. Then we say that the entropy of $\mathcal{M}$ in the state ω is $S_\omega(\mathcal{M}) = S_\mu$. The function $\eta : [0,1] \to \mathcal{R}^+$, $x \to \eta(x) = -x \ln x$ is strictly concave and subadditive. This makes $S_\omega(\mathcal{M})$ monotonic in the sense that it decreases by going to subalgebras $\mathcal{A} \subset \mathcal{M}$. A general element of $\mathcal{M}$ can be written $m = \sum_{j\in I} c_j P_j$, $c_j \in C$ and if for a partition $I = \cup_\alpha I_\alpha$ we have $c_j = c_\alpha \forall j \in I_\alpha$, then these m's form a subalgebra $\mathcal{A}$. Thus the elements of $\mathcal{A}$ can be written as $a = \sum_\alpha c_\alpha P_\alpha$, $P_a = \sum_{j\in I_\alpha} P_j$ and $\omega_{|\mathcal{A}}$, the restriction of ω to $\mathcal{A}$ is

$$\omega_{|\mathcal{A}}(P_\alpha) = \sum_{j\in I_\alpha} \mu_j =: \mu_\alpha .$$

Thus it is natural to call

$$S_{\omega|\mathcal{A}}(\mathcal{A}) = \sum_\alpha \eta(\mu_\alpha) \tag{2.2}$$

the entropy of the subalgebra. The subadditivity of η implies the monotonicity

$$S_{\omega|\mathcal{A}}(\mathcal{A}) \leq S_\omega(\mathcal{M}) . \tag{2.3}$$

In spite of that, $\omega_{|\mathcal{A}}$ together with a particular map $E : \mathcal{M} \to \mathcal{A}$ carry the full information of ω in the sense that

$$\omega = \omega_{|\mathcal{A}} \circ E . \tag{2.4}$$

This map E is a positive linear map defined by

$$E(P_j) = P_\alpha \frac{\omega(P_j)}{\omega(P_\alpha)} \quad \text{if } j \in I_\alpha . \tag{2.5}$$

In addition to (2.4) it satisfies

$$E(a_1 m a_2) = a_1 E(m) a_2 \quad \forall\, a_i \in \mathcal{A},\, m \in \mathcal{M} \tag{2.6}$$

and is called the canonical conditional expectation. From (2.6) follows that E is a left inverse of the canonical embedding $i : \mathcal{A} \to \mathcal{M} : P_\alpha \to \sum_{j\in I_\alpha} P_j$,

$$E \circ i = 1_{|\mathcal{A}} . \tag{2.7}$$

These nice properties get lost when we turn to the nonabelian case. Let $\mathcal{M}$ be a finite matrix algebra M^n $= n \times n$ matrices. Any state ω of $\mathcal{M}$ is characterized by a density matrix $\rho, 0 \leq \rho \leq 1$, tr $\rho = 1$ and von Neumann proposed his famous formula

$$S_\omega(\mathcal{M}) = Tr\, \eta(\rho) . \tag{2.8}$$

To motivate (2.8) one might reduce the nonabelian case to the previous one by considering maximal abelian subalgebras $\mathcal{A}$ of $\mathcal{M}$. Since they are much smaller than $\mathcal{M}$, the linear dimensions being n^2 and n, one may think that they contain less information and that because of the monotonicity (2.3) their entropy is much less. Actually, exactly the opposite happens and we have (for the proof see [7,8])

Theorem (2.9)

$$S_\omega(\mathcal{M}) = \inf_{\mathcal{A}} S_{\omega|\mathcal{A}}(\mathcal{A})$$

where the inf is taken over all maximal abelian subalgebras $\mathcal{A} \subset \mathcal{M}$. For instance, it is a well–known fact that the pure state "spin up" for the algebra $\mathcal{M} = \{1, \overrightarrow{\sigma}\}$ gives a state $(1/2, 1/2)$ for the abelian subalgebra $\mathcal{A} : \{1, \sigma_x\}$. Thus $S_{\omega|\mathcal{A}}(\mathcal{A}) = \ln 2 > S_\omega(\mathcal{M}) = 0$.

Since ρ is hermitian it can be written $\rho = \sum_j \rho_j P_j$ where the P_j are one–dimensional projections orthogonal to each other. Thus we have

$$S_\omega(\mathcal{M}) = -\sum_j p_j \ln p_j$$

and we see that the inf in (2.9) is reached by the algebra $\mathcal{A}$ generated by the P_j. In case of degeneracy this $\mathcal{A}$ may not be abelian and we are led to the general

Definition (2.10)

The centralizer $\mathcal{M}_\omega \subset \mathcal{M}$ is defined by

$$\mathcal{M}_\omega = \{m \in \mathcal{M} : [m, \rho] = 0\}.$$

Remark (2.11)

If ρ is faithful, it has the canonical form $\rho = e^{-H}$ and $\mathcal{M}_\omega$ is the subalgebra of elements invariant under the "modular automorphism" $\sigma_\omega : m \to e^{iHt} m\, e^{-iHt}$ of $\mathcal{M}$.

We then have

Theorem (2.12)

Assume $\mathcal{M}$ is finite dimensional. Then the inf in (2.9) is reached iff $\mathcal{A}$ is a maximal abelian subalgebra of $\mathcal{M}_\omega$ (which at the same time is also maximal abelian in $\mathcal{M}$).

The next difference to the abelian case is the absence of a conditional expectation $E : M \to \mathcal{A}$ for most subalgebras. We have

Theorem (2.13)

Given ω we have a conditional expectation $E_\omega : \mathcal{M} \to \mathcal{A}$, $\omega = \omega_{|\mathcal{A}} \circ E_\omega iff \mathcal{A} \subset \mathcal{M}_\omega$ [9].

Remark (2.14)

The tracial state $\tau(m) = \frac{1}{n}$ Tr m has trivial modular automorphism and thus $\mathcal{M}_\tau = \mathcal{M}$. Therefore for any subalgebra $\mathcal{A} \subset \mathcal{M}$ there is a positive linear map $E_\tau : \mathcal{M} \to \mathcal{A}$ with $E_\tau(a_1 m a_2) = a_1 E_\tau(m) a_2$. But this map will not carry all the states, in general, $\omega \neq \omega_{|\mathcal{A}} \circ E$.

According to these considerations it seems advisible in quantum ergodic theory to map the noncommutative system with E onto a maximal abelian subalgebra of $\mathcal{M}_\omega$ and there to continue with classical ergodic theory. Since most of the features used so far also generalize to infinite systems Emch [10] and others [11] proposed this approach. Unfortunately it does not fit at all in our general program where we are interested in equilibrium (= KMS) states and σ_ω is the time evolution of $\mathcal{M}$. Now it is a minimal requirement for chaotic systems that they are ergodic in the sense that there are no nontrivial constants of motion, that is $\mathcal{M}_\omega = 1 \cdot C$ and E_ω maps $\mathcal{M}$ onto a one–dimensional algebra. To get more flexibility we introduced (with A. Connes [5]) the following notion:

Definition (2.15)

An abelian model for $(\mathcal{M},\omega)$ is a positive map P into an abelian algebra $\mathcal{B}$ with state μ such that $\omega = \mu \circ P$.

Remarks (2.16)

1. (2.15) can also be phrased by saying that ω is the image of μ under the adjoint map P^* of the duals: $\mathcal{B}^* \to \mathcal{M}^*$ as $(P^*\mu)(m) = \mu(P(m)) = \mu \circ P(m)$.
2. If the P_j are the minimal projections of $\mathcal{B}$ and we identify them with the pure states $\overline{P}_j$ over $\mathcal{B} : \overline{P}_j(P_k) = \delta_{ik}$, then $\mu = \sum_j \overline{P}_j \mu_j$ and $\omega = \sum_j \mu_j P^*(\overline{P}_j) =: \sum_j \mu_j \omega_j$. Thus an abelian model corresponds to a decomposition of ω into the states ω_j and to any decompositon we can associate an abelian model.

Since the notion (2.15) is so general one cannot expect a relation between $S_\omega(\mathcal{M})$ and $S_\mu(\mathcal{B})$. In fact $\mathcal{B} = 1 \cdot C$, $\mu = 1$, $P = \omega$ would qualify and make $S_\mu(\mathcal{B}) = 0$. But if $\mathcal{B}$ is maximal abelian $\subset \mathcal{M}_\omega$, $P = E_\omega$ satisfies (2.5) and in finite dimensions gives $S_\omega(\mathcal{M}) = S_\mu(\mathcal{B})$. Thus one might conjecture that $S_\omega(\mathcal{M})$ is the sup of $S_\mu(\mathcal{B})$ over all abelian models but, because of trivial reasons, this requires a further correction. Even if ω is pure such that $S_\omega(\mathcal{M}) = 0$ one could take $\mathcal{B} = C \times C$, $\mu = (\mu_1, 1-\mu_1)$, $P(m) = \omega(m)P_1 + \omega(m)P_2$ such that $S_\mu(\mathcal{B})$ can be anything between 0 and ln 2. What goes wrong in this model is that the corresponding decomposition of ω is $\omega = \mu_1\omega + (1-\mu_1)\omega$ and does not carry any information about the decomposability of ω. Furthermore, though the $\overline{P}_j$ are pure, the $P^*(\overline{P}_j) = \omega_j$ need not be satisfied, and one has to correct for the lack of complete information in the $P^*(\overline{P}_j)$. Therefore one will associate with an abelian model an information gain

$$S_{P^*\mu}(\mathcal{M}) - \sum_j \mu_j S_{P^*\overline{P}_j}(M) = \sum_j \mu_j S(P^*\mu | P^*\overline{P}_j)$$

where $S(\omega|\nu)$ is the relative entropy. Thus one gets as first generalization of (2.8):

Definition (2.17)

$$S_\omega(\mathcal{M}) = \sup \sum_j S(P^*\mu | P^*\overline{P}_j)\mu_j$$

where the sup is taken over all abelian models $\mathcal{M} \xrightarrow{P} \mathcal{B}$ with $\omega = P^*\mu$ and $\mathcal{B}$ spanned by the minimal projection P_j , $\mu_j = \mu(P_j)$.

Remarks (2.18)

1. (2.17) is a generalization of (2.8) since if states are described by density matrices ρ and σ, then $S(\rho,\sigma) = Tr\ \rho(\ln \rho - \ln \sigma)$ leads back to (2.8) by taking for the $P^*(\overline{P}_j)$ one–dimensional projections of ρ. But the relative entropy is still defined for infinite systems where states cannot be described by density matrices.
2. The information gain of an abelian model may be thought of in the following way. The P_j are certain intervals for the values of the observables $m \in \mathcal{M}$ and the μ_j are the probabilities that they are assumed in the state $\omega = \sum_j \mu_j \omega_j$. By the knowledge in which component we are, we have an information gain $S_\omega(\mathcal{M}) - S_{\omega_j}(\mathcal{M})$.

$$\sum_j \mu_j S(P^*\mu | P^*\overline{P}_j) = \sum_j \mu_j (S_\omega(\mathcal{M}) - S_{\omega_j}(\mathcal{M}))$$

is the mean of these quantities weighted by the probabilities with which they occur in ω.

The fact that (2.17) is expressed as a sup rather than an inf as in (2.9) has been used by us [12] to give a definition of the entropy of a subalgebra $\mathcal{A}$ which satisfies the monotonicity (2.3) also in the nonabelian case and thus differs from $S_{\omega|\mathcal{A}}(\mathcal{A})$.

Definition (2.19)

$$H_\omega(\mathcal{A}) = \sup \sum_j \mu_j S(P^*(\mu) \circ i | P^*(\overline{P}_j) \circ i)$$

where $i : \mathcal{A} \to \mathcal{M}$ is the canonical embedding and the sup is taken over all abelian models $M \xrightarrow{P} \mathcal{B} = \{P_j\}$ with $\omega = \mu \circ P$. Further on, we shall also employ the notation $H_\omega(i)$.

Remarks (2.20)

1. $H_\omega(\mathcal{A})$ does not only depend on $\omega_{|\mathcal{A}}$ but represents the maximal information gain about $\mathcal{A}$ by a model $\mathcal{M} \xrightarrow{P} \mathcal{B}$.
2. The monotonicity $H_\omega(\overline{\mathcal{A}}) \geq H_\omega(\mathcal{A})$ for $\overline{\mathcal{A}} \supset \mathcal{A}$ follows from $S(\mu|\nu) \geq S(\mu \circ j | \nu \circ j)$ for any completely positive map and in particular for the embedding $\mathcal{A} \xrightarrow{j} \overline{\mathcal{A}}$.
3. $H_\omega(\mathcal{A}) \leq S_{\omega|\mathcal{A}}(\mathcal{A})$ since $S_{\omega|\mathcal{A}}(\mathcal{A})$ is the same expression (2.19) with the sup over abelian models $\mathcal{A} \xrightarrow{P} \mathcal{B}_0$. Since any model $\mathcal{M} \xrightarrow{P} \mathcal{B}$ gives a model $\mathcal{A} \xrightarrow{P_{|\mathcal{A}}} \mathcal{B}_0$ the sup is taken over a larger set for $S_{\omega_{|\mathcal{A}}}(\mathcal{A})$.

With the tools developed so far we can solve the problem of generalizing the dynamical entropy for noncommutative $\mathcal{M}$. There we need the union of two partitions which algebraically means two sets of projections P_j, Q_k with $1 = \sum_j P_j = \sum_k Q_k$. In the abelian case it is given by the projections $\{P_j Q_k\}$; but if they do not commute this product is not even hermitian. A way out of this dilemma was given by Connes and Størmer [3] who suggested to consider an information content $H(\mathcal{A}_1, \mathcal{A}_2, \ldots, \mathcal{A}_n)$ of several finite dimensional subalgebras $\mathcal{A}_i$. Their prescription was taylormade for the case where $\mathcal{M}$ has a tracial state and thus there are always conditional expectations $\mathcal{M} \xrightarrow{E_i} \mathcal{A}_i$ with $\omega = \omega_{|\mathcal{A}_i} \circ E_i$. But even without this assumption we can consider an abelian model which incorporates all $\mathcal{A}_i$ according to the scheme:

$$\begin{array}{ccccccc} \mathcal{A}_1 & & & & & & \mathcal{B}_1 \\ & i_1 & & & & E_1 & \\ & \searrow & & & & \nearrow & \\ \vdots & \to & & \mathcal{M} \xrightarrow{P} \mathcal{B} & & \to & \vdots \\ & i_n & & & & E_n & \\ & \nearrow & & & & \searrow & \\ \mathcal{A}_n & & & & & & \mathcal{B}_n \end{array} \tag{2.21}$$

where the i_k are the embeddings, E_k conditional expectations $\mathcal{B} \xrightarrow{E_j} \mathcal{B}_k \subset \mathcal{B}$. All these maps are supposed to carry the states according to

$$\omega_{|\mathcal{A}_k} = \omega \circ i_k = \mu \circ P \circ i_k = \mu_{|\mathcal{B}_k} \circ E_k \circ P \circ i_k .$$

To associate to this model an information gain about the algebras $\mathcal{A}_i$ we introduce the notion of the entropy defect of a model. It measures how much the abelian entropy $S_\mu(\mathcal{B})$ differs from the information gain determined previously.

Definition (2.22)

The entropy defect of $\mathcal{A} \xrightarrow{P} \mathcal{B} = \{P_k\}$ is

$$s_\mu(P) = S_\mu(\mathcal{B}) - \sum_j \mu_j \, S(P^*(\mu)|P^*(\overline{P}_j)).$$

The model (2.21) provides abelian models $A_k \overset{E_k \circ P \circ i_k}{\rightarrow} \mathcal{B}_k$ for the subalgebras $\mathcal{A}_k$ and it turns out that a quantity with the properties we have been looking for is obtained by correcting $S_\mu(\mathcal{B})$ with the entropy defects of the subalgebras and taking the sup over all models (2.21)[5].

Definition (2.23)

$$H_\omega(i_1, i_2, \ldots, i_n) = \sup_{\mu \circ P = \omega} [S_\mu(\mathcal{B}) - \sum_k s_{\mu|\mathcal{B}_k}(E_k \circ P \circ i_k)].$$

Indeed using the various properties of the relative entropy one can show [5]

Proposition (2.24)

a) If $\Theta_j : \mathcal{A}'_j \to \mathcal{A}_j \supset \mathcal{A}'_j$ are embeddings, then

$$H_\omega(i_1 \circ \Theta_1, \ldots, i_n \circ \Theta_n) \le H_\omega(i_1, \ldots, i_n).$$

b) If $\Theta : \mathcal{A} \to \mathcal{A}$ is a completely positive map, then

$$H_\omega(\Theta \circ i_1, \ldots, \Theta \circ i_n) \le H_\omega(i_1, \ldots, i_n).$$

c) $H_\omega(i_1, i_2, \ldots, i_n)$ depends only on the set $\{i_1, i_2, \ldots, i_n\} = X$. This means it is a symmetric function, $H(i,i) = H(i)$, etc.

d) With the notation of c) we have

$$\max\{H_\omega(X), H_\omega(Y)\} \le H_\omega(X \cup Y) \le H_\omega(X) + H_\omega(Y).$$

If ω is a tracial state (2.23) agrees with entropy defined by Connes and Størmer and if the $\mathcal{A}$'s are abelian and generate the algebra $\bigvee_k \mathcal{A}_k$, then

$$H_\omega(i_1, \ldots, i_n) = S_\omega(\bigvee_k \mathcal{A}_k).$$

This can be seen by using the $\mathcal{A}_k$ as their own models, i.e. $\omega = \mu, E_k \circ P \circ i_k = id_{|\mathcal{A}_k}, \mathcal{B} = \bigvee_k \mathcal{A}_k$. In this case $(E_k \circ P \circ i_k)^*$ is the identity map and $\sum_j \mu_j S(\mu|\overline{P}_j) = S_\mu(\mathcal{A})$, so the entropy defects vanish.

The properties (2.24) guarantee the existence of the following limit:

Definition (2.25)

Let $\sigma : \mathcal{M} \to \mathcal{M}$ be an automorphism of an AF C^*−algebra with $\omega \circ \sigma = \omega$ and $\mathcal{A} \subset \mathcal{M}$ finite dimensional, $\mathcal{A} \xrightarrow{i} \mathcal{M}$ the embedding. Then

$$h_\omega(\mathcal{A}, \sigma) = \lim_{k \to \infty} \frac{1}{k} H_\omega(i, \sigma \circ i, \sigma^2 \circ i, \ldots, \sigma^{k-1} \circ i)$$

is the dynamical entropy of the subalgebra $\mathcal{A}$ and

$$h_\omega(\sigma) = \sup_{\mathcal{A}=\text{fin.dim.}} h_\omega(\mathcal{A}, \sigma)$$

the dynamical entropy of $(\mathcal{M}, \omega, \sigma)$.

Proposition (2.26)

The invariance and scaling properties shown in the classical theory

(i) $h_{\omega \circ \Theta}(\Theta^{-1} \sigma \Theta) = h_\omega(\sigma) \, \forall \, \Theta \in Aut\mathcal{M}$

(ii) $h_\omega(\sigma^n) = |n| h_\omega(\sigma) \, \forall \, n \in \mathbf{Z}$

(iii) $h_{\omega \otimes \nu}(\sigma \otimes \tau) \geq h_\omega(\sigma) + h_\nu(\tau)$

hold also for the generalization (2.25).

Since the entropy functions H are defined as sup's over abelian models the question arises how to find good abelian models for finite dimensional subalgebra $\mathcal{A}_j$. The first thought are maximal abelian subalgebras $\hat{\mathcal{A}}_j \subset \mathcal{A}_j$ and in certain situations one can prove that they actually model optimally.

Proposition (2.27)

If the $\mathcal{A}_j$, $j = 1, \ldots, n$ are mutually commuting finite subalgebras of $\mathcal{M}$ containing maximal abelian subalgebras $\tilde{\mathcal{A}}_j \subset \mathcal{M}_\omega$, $\hat{i}_j : \hat{\mathcal{A}}_j \to \mathcal{A}_j$ the embedding, then

$$H_\omega(i_1 \circ \hat{i}_1, i_2 \circ \hat{i}_2, \ldots, i_n \circ \hat{i}_n) = H_\omega(i_1, i_2, \ldots, i_n) = S(\omega_{|\bigvee_{j=1}^n \hat{\mathcal{A}}_j}).$$

Proof: Use the $\hat{\mathcal{A}}_j$ as abelian models according to the scheme

$$\hat{\mathcal{A}}_j \xrightarrow{\hat{i}_j} \mathcal{A}_j \xrightarrow{i_j} \mathcal{M} \xrightarrow{E} \bigvee_{j=1}^{n} \hat{\mathcal{A}}_j \xrightarrow{E_j} \hat{\mathcal{A}}_j .$$

The conditional expectations E, E_j carry the state ω since the $\hat{\mathcal{A}}_j$ are contained in the centralizer $\mathcal{M}_\omega$. Thus we have the abelian situation where the $\hat{\mathcal{A}}_j$ are their own models which means that the entropy defects $s_{\omega_{|\mathcal{A}_j}}(E_j \circ E \circ i_j \circ \hat{i}_j)$ vanish. Furthermore, because of (2.24) we have

$$H_\omega(\hat{\mathcal{A}}_1, \hat{\mathcal{A}}_2, \ldots, \hat{\mathcal{A}}_n) \leq H_\omega(\bigvee_j \hat{\mathcal{A}}_j, \bigvee_j \hat{\mathcal{A}}_j, \ldots, \bigvee_j \hat{\mathcal{A}}_j) = H_\omega(\bigvee_j \hat{\mathcal{A}}_j) \leq S(\omega_{|\bigvee_j \hat{\mathcal{A}}_j})$$

and we see that his model gives already the maximal value: Since it has vanishing entropy defects we can continue on the left with

$$S(\omega_{|\bigvee_j \hat{\mathcal{A}}_j}) \leq H_\omega(\hat{\mathcal{A}}_1, \hat{\mathcal{A}}_2, \ldots, \hat{\mathcal{A}}_n)$$

and thus we have equality throughout. Since the $\bigvee_j \hat{\mathcal{A}}_j$ are maximal abelian in $\bigvee_j \mathcal{A}_j$ we can continue with (2.9) to conclude

$$S(\omega_{|\bigvee_j \hat{\mathcal{A}}_j}) \geq S(\omega_{|\bigvee_j \mathcal{A}_j}) \geq H_\omega(\mathcal{A}_1, \mathcal{A}_2, \ldots, \mathcal{A}_n) \geq H_\omega(\hat{\mathcal{A}}_1, \hat{\mathcal{A}}_2, \ldots, \hat{\mathcal{A}}_n) = S(\omega_{|\bigvee_j \hat{\mathcal{A}}_j})$$

and so all these quantities coincide.

If $\mathcal{M}_\omega$ is trivial there does not exist a subalgebra $\mathcal{A}(\neq z \cdot 1)$ with a conditional expectation $E : \mathcal{M} \to \mathcal{A}$ and $\omega = \omega_{|\mathcal{A}} \circ E$. However, if ω is faithful, one can always construct a completely positive map which carries the state and whose difference from a conditional expectation can be estimated [5]. Then the continuity properties of the H–functions insure that one can get a pretty good abelian model with them.

Definition (2.28)

Let $\mathcal{M}_1$ and $\mathcal{M}_2$ be two von Neumann algebras with faithful normal states ϕ_1 and ϕ_2. Given a completely positive map $\gamma : \mathcal{M}_1 \to \mathcal{M}_2$ with $\phi_1 = \phi_2 \circ \gamma$. Its adjoint $\gamma^\dagger : \mathcal{M}_2 \to \mathcal{M}_1$ is defined with the modular automorphisms $\sigma_t^{\phi_k} \in Aut\, \mathcal{M}_k$ of ϕ_k by

$$\phi_1(\sigma^{\phi_1}_{i/2}(x_1)\gamma^\dagger(x_2)) = \phi_2(\gamma(x_1)\sigma^{\phi_2}_{-i/2}(x_2)) \quad , \quad x_k \in \mathcal{M}_k.$$

Proposition (2.29)

(i) $\gamma^\dagger$ is completely positive with $\phi_2 = \phi_1 \circ \gamma^\dagger$.

(ii) $(\gamma \circ \gamma')^\dagger = \gamma'^\dagger \circ \gamma^\dagger$, $\gamma^{\dagger\dagger} = \gamma$.

(iii) If γ is the inclusion for $\mathcal{M}_1 \subset \mathcal{M}_2$ and $\mathcal{M}_1 \subset \mathcal{M}_{2_{\phi_2}}$, then $\gamma^\dagger$ is the conditional expectation.

(iv) If γ is an automorphism $\gamma^\dagger = \gamma^{-1}$.

Remark (2.30)

When we use von Neumann algebras we think $\Pi_\omega(\mathcal{M})''.\sigma \in textAut\, \mathcal{M}$ with $\omega \circ \sigma = \omega$ can be extended to $\Pi_\omega(\mathcal{M})''$; and $(\mathcal{M}, \omega, \sigma)$ and $(\Pi_\omega(\mathcal{M})'', \overline{\omega}, \overline{\sigma})$, with $\overline{\omega}, \overline{\sigma}$ the extensions, give the same dynamical entropy [5].

Since $\gamma^\dagger$ carries the state we can construct abelian models for abelian commuting subalgebras $\mathcal{A}_j$ with the adjoints $i_j^\dagger$ of the inclusions. If the elements of $\mathcal{A}_j$ are nearly invariant under σ_t^ω one can show that the $i_j^\dagger$ are not too different from canonical expectations and we will see in the next section that they have small entropy defects. In any case we can use them as abelian models and get (with (2.9))

Proposition (2.31)

Let $\mathcal{A}_j$, $j = 1, \ldots, n$ be pairwise commuting subalgebras of $\mathcal{M}$ and $\mathcal{A}$ the algebra they generate, further $\mathcal{A}_j \xrightarrow{i_j} \mathcal{M}$ the inclusions. Then, even if the maximal abelian subalgebras $\hat{\mathcal{A}}_j \subset \mathcal{A}_j$, are not in $\mathcal{M}_\omega$ we have

$$H_\omega(i_1, \ldots, i_n) \geq S(\omega_{|\mathcal{A}}) - \sum_{j=1}^n s_{\omega_{|\mathcal{A}_j}}(\hat{i}_j^\dagger \circ i_j^\dagger \circ i_j \circ \hat{i}_j)$$

where $\hat{i}_j : \hat{\mathcal{A}}_j \to \mathcal{A}_j$ are the inclusions.

Since $S(\omega_{|\mathcal{A}})$ does not depend on the choice of the $\hat{\mathcal{A}}_j$ we have to seek them so as to minimize the entropy defects.

3. Dynamical Entropy for Quasifree Automorphism

Already in [3] the shift on a lattice system was considered as example to calculate its dynamical entropy, here restricted to the tracial state. In [5] we generalized the considerations to arbitrary states [IX.1] but with a condition on the state that is not so easy to check: We repeat the statement.

Theorem (3.1)
Assume $\mathcal{M} = \otimes_{k=-\infty}^{+\infty} M_k^l$ where M^l is an $l \times l$ matrix. $\sigma M_k^l = M_{k+1}^l$. Define $\mathcal{M}_I = \otimes_{k\in I} M_k^l \subset \mathcal{M}$. Take $I = [-s, n+s]$ and consider the state ϕ on $\mathcal{M}$ with $\phi \circ \sigma = \phi$. Let $\psi = \phi_{|\mathcal{M}_I}$ and assume that the modular automorphisms are such that $\forall \varepsilon > 0 \, \exists \, s, n_0$ with

$$\|\sigma_{i/2}^{\phi} \circ \sigma_{-i/2}^{\psi} Q - Q\| \leq \varepsilon \|Q\| \quad \forall Q \in \bigotimes_{k=0}^{n} M_k^l \,, \forall n > n_0 \,.$$

Then

$$H_\phi(\sigma) = s(\phi) = \lim_{n\to\infty} \frac{S(\phi_{|\mathcal{M}_I})}{|I|} \,.$$

Remark (3.2)
$h_\phi(\sigma) \leq s(\phi)$ follows by the subadditivity

$$\begin{aligned} h_\phi(\sigma, \mathcal{M}_{(0,n)}) &= \lim_{k\to\infty} \frac{1}{k} H(\mathcal{M}(0,n), \ldots, \mathcal{M}_{(k,k+n)}) \\ &\leq \lim_{k\to\infty} \frac{1}{k} H(\mathcal{M}_{(0,n+k)}) \\ &\leq \lim_{k\to\infty} \frac{1}{k} S(\phi_{|\mathcal{M}_{(0,n+k)}}) \end{aligned}$$

and

$$h_\phi(\sigma) = \lim_{n\to\infty} h_\phi(\sigma, \mathcal{M}_{(0,n)}) \,.$$

The condition on the modular automorphism is needed to control the entropy defect and thus the validity of the abelian model. Though it is for sure not an optimal condition it was good enough for quasifree automorphisms provided the state was quasifree and its momentum distribution twice differentiable on the torus [5,A.3]. The estimates are rather involved though they show that Theorem 3.1 is at least not empty. We will give an easier proof in the following section, also generalizing the result to a larger class of states and a larger class of quasifree automorphisms.

Theorem (3.3)
Consider the CAR algebra over the Hilbert space $\mathcal{H} = L^2((-\infty, \infty), dp)$. Consider the quasifree automorphism $\sigma^\alpha \, a(f) = a(U^\alpha f)$. Assume that $U^\alpha = e^{i\alpha\omega(p)}$ in momentum representation of f and $\omega(p)$ is apart from a discrete number of steps a C^1–function.
Consider the quasifree state defined by

$$\phi(a^\dagger(f)\, a(g)) = \int_{-\infty}^{+\infty} dp\, f(p) \overline{g}(p)\, \rho(p)$$

where ρ satisfies the same assumptions as ω. Then

$$h_\phi(\sigma) = -\int_{-\infty}^{+\infty} dp\, |\omega'(p)|\, [\rho(p) \ln \rho(p) + (1-\rho(p)) \ln(1-\rho(p))] \,.$$

Remark (3.4)

The same statement can be found in [13]. There, not only the superadditivity (2.26(iii)), but even additivity of h for tensor products was assumed. Whether it holds or not is still an open problem. We will circumvent the use of this property in the following proof.

Nondegeneracy of the spectrum of the one–particle Hamiltonian was assumed to simplify the notation of the result. How degeneracy can be dealt with can be found in [14], here restricted to discrete spectrum of $\rho(p)$. How to combine the two methods is fairly evident. In addition [14] deal with U that have singular continuous spectrum and they show that such σ always give zero dynamical entropy. Therefore the problem of how to calculate the dynamical entropy of quasifree automorphisms in quasifree states can be considered as settled.

We turn to the proof of (3.3) which we do in various steps.

Lemma (3.5)

Let $\mathcal{A} \subset \mathcal{M}$ be abelian with minimal projections $P_I, \sum_I P_I = 1, d = \text{card}\{I\} < \infty$. If ϕ is faithful, σ_t^ϕ its modular automorphism, and

$$\varepsilon := \max_I \left(1 - \frac{\phi(P_I \sigma^\phi_{i/2} P_I)}{\phi(P_I)}\right)$$

then

$$H_\phi(\mathcal{A}) \geq S(\phi_{|\mathcal{A}}) - \varepsilon \ln 2d/\varepsilon.$$

Proof: In (2.28) take $M_1 = \mathcal{A}$, $M_2 = M$, γ =inclusion, $\phi_1 = \phi_{|\mathcal{A}}$, $\phi_2 = \phi$; $\sigma^{\phi_1} = 1$ since $\mathcal{A}$ is abelian. Then

$$\gamma^\dagger \circ \gamma(P_I) = \gamma^\dagger(P_I) = \sum_{I'} P_{I'} \phi(P_I \sigma^\phi_{i/2} P_{I'})/\phi(P_{I'}) \equiv \sum_{I'} \mu_{II'} P_{I'}/\mu_{I'} \,.$$

The abelian model $\mathcal{A} \xrightarrow{\gamma} M \xrightarrow{\gamma^\dagger} \mathcal{A}$ gives

$$H_\phi(\mathcal{A}) \geq S(\phi\circ\gamma) - \sum_I \mu_I S((\gamma^\dagger \circ \gamma)^* \overline{P}_I) = -\sum_I \mu_I \ln \mu_I - \sum_{I,I'} \mu_{II'} \ln \mu_{II'}/\mu_{I'} \,.$$

Now $\sum_I \mu_{II'}/\mu_{I'} = 1$ and $\varepsilon = \max_I \|(\gamma^\dagger \circ \gamma)^* \overline{P}_I - \overline{P}\|$ where $\overline{P}_I(P_{I'}) = \delta_{II'}$ is a pure state. As $S(\overline{P}_I) = 0$ we get from [15,§2]

$$\sum_I \frac{\mu_{II'}}{\mu_{I'}} \ln \frac{\mu_{II'}}{\mu_{I'}} < \varepsilon \ln 2d/\varepsilon \; \forall I'.$$

Since $\sum_{I'} \mu_{I'} = 1$ this completes the proof of the lemma.

We shall first concentrate on the space translation on the lattice algebra $\mathcal{M}$ generated by a_x with the CAR relations $[a_x, a^*_{x'}]_+ = \delta_{x,x'}$, $x, x' \in \mathbf{Z}$. We shall map it to the CAR algebra over $L^2((0,1), dp)$ by

$$a(f) = \sum_{x \in \mathbf{Z}} \int_0^1 dp\, e^{ipx} f(p)\, a_x \,.$$

The quasifree state ϕ shall be written in the form of a Fermi distribution

$$\phi(a^*(f)\, a(g)) = \int_0^1 dp \frac{\overline{f}(p)\, g(p)}{1 + e^{\varepsilon(p)}} \,.$$

$\mathcal{M}$ is isomorphic to the subalgebra $\mathcal{M}_1$ generated by $a(f)$, $f \in L^2([0, 1/n], dp)$, $n \in N$, the restriction of ϕ to $\mathcal{M}_1$ being

$$\phi(a^*(f)\, a(g)) = \int_0^{1/n} dp \frac{f(p)\, g^*(p)}{1 + e^{\varepsilon(p)}} \,.$$

Therefore $\sigma^{\phi}_{i/2}(a(f)) = a(g)$, $g(p) = e^{\varepsilon(p)/2} f(p)$ and $\sigma^{\phi}_{i/2}(a^*(f)) = a^*(\hat{g})$, $\hat{g}(p) = e^{-\varepsilon(p)/2} f(p)$. We shall now calculate $H_\phi(\mathcal{A}_1, \sigma^n \mathcal{A}_1, \dots, \sigma^{kn} \mathcal{A}_1)$ where $\mathcal{A}_1$ is generated by $P_1 = a^*(\sqrt{n})\; a(\sqrt{n})$ and $1 - P_1$. The translates

$$\sigma^{kn} P_1 = a^*(e^{2\pi inkp} \sqrt{n})\, a(e^{2\pi inkp} \sqrt{n}) \equiv P_{1,k} \quad , \quad k \in Z$$

commute with each other since the translated wave functions are orthogonal to each other. As the $\mathcal{A}_{1,k} = \sigma^{nk} \mathcal{A}_1$ are commuting abelian algebras we may use

$$\mathcal{A}_{1,j} \xrightarrow{i_j} \mathcal{B}_1 = \bigvee_{i=1}^{k} \mathcal{A}_{1,i} \xrightarrow{\gamma} M \xrightarrow{\gamma^\dagger} \mathcal{B}_1 \xrightarrow{E_i} \mathcal{A}_i$$

as abelian model. Here the i_j, γ are inclusions and the $E_j = i_j^\dagger$ conditional expectations. This model gives

$$H_\phi(\mathcal{A}_{1,o}, \mathcal{A}_{1,1}, \dots, \mathcal{A}_{1,k}) \geq S(\phi_{|\mathcal{B}_1}) - \sum_{j=1}^{k} s_{\phi_{|\mathcal{A}_{1,j}}}(\gamma_j^\dagger \circ \gamma_j) \quad , \quad \gamma_j = \gamma \circ i_j \,.$$

To calculate the entropy defects we have to evaluate as in the lemma ($f(p) = e^{2\pi injp} \sqrt{n}$, g and $\hat{g}$ as defined above)

$$\begin{aligned}
\phi(P_{1,j}\, \sigma^{\phi}_{i/2}\, P_{1,j}) &= \phi(a^*(f)\, a(f)\, a^*(\hat{g})\, a(g)) = \\
&= \phi(a^*(f)\, a(f))\, \phi(a^*(\hat{g})\, a(g)) + \phi(a^*(f)\, a(g))\, \phi(a(f)\, a^*(\hat{g})) = \\
&= n^2 \int_0^{1/n} dp\, dp' \left[\frac{1}{1 + e^{\varepsilon(p)}} \, \frac{1}{1 + e^{\varepsilon(p')}} + \frac{e^{\varepsilon(p)/2}}{1 + e^{\varepsilon(p)}} \, \frac{e^{-\varepsilon(p')/2}}{1 + e^{\varepsilon(p')}} \right] = \\
&= n \int_0^{1/n} \frac{dp}{1 + e^{\varepsilon(p)}} - \frac{n^2}{2} \int_0^{1/n} dp\, dp' \frac{(e^{\varepsilon(p)/2} - e^{\varepsilon(p')/2})^2}{(1 + e^{\varepsilon(p)})(1 + e^{\varepsilon(p')})} \,.
\end{aligned}$$

If $\varepsilon(p) \in C^1([0,1/n])$ then

$$(e^{\varepsilon(p)/2} - e^{-\varepsilon(p')/2})^2 \le \frac{1}{8n^2} \sup_p |\varepsilon'(p) e^{\varepsilon(p)/2}|^2 =: \frac{c}{8n^2}$$

and

$$\phi(a^*(f)\, a(f)\, a^*(\hat{g})\, a(g)) \ge \phi(a^*(f)\, a(f)) - \frac{c}{16n^2}\,.$$

Thus $\varepsilon \le c/16n^2$ and lemma (3.5) tells us (with $d=2$)

$$s_{\phi|\mathcal{A}_{1,j}}(\gamma_j^\dagger \circ \gamma_j) \le \frac{c}{8n^2} \ln 8n/\sqrt{c} \quad \forall \;\; j.$$

Thus

$$h_{\phi,\sigma^n}(\mathcal{A}_1) \ge s_1(\phi) - \frac{c}{8n^2} \ln 8n/\sqrt{c}$$

where $s_1(\phi)$ is the entropy density of $\mathcal{M}_1$. Here we used that $\bigvee_j \sigma^{jn}\mathcal{A}_1$ is maximal abelian in $\mathcal{M}_1$ and thus by virtue of (2.9)

$$\lim_{k\to\infty} \frac{1}{k} S(\phi_{|\bigvee_{j=1}^k \sigma^{jn}\mathcal{A}_1}) \ge s_1(\phi)\,.$$

It is a standard formula that the entropy density of a free Fermi gas is

$$s_1(\phi) = n \int_0^{1/n} dp\, [\eta(\rho(p)) + \eta(1-\rho(p))]$$

with $\rho(p) = (1+e^{\varepsilon(p)})^{-1}$. Because of the strong subadditivity of the entropy $\lim k \to \infty$ is actually $\inf k$. The same procedure works for the CAR algebras $\mathcal{M}_j$ where $f \in L^2([\frac{j-1}{n}, \frac{j}{n}], dp)$, $1 \le j \le n$. If we denote the corresponding abelian algebras by $A_{j,k}$ we note that they commute for all j and k, and that ϕ and σ factor in the j. Thus, from the facts that h is superadditive (2.26(iii)) and the entropy density is subadditive, we conclude

$$h_{\phi,\sigma^n}(\mathcal{A}_1 \otimes \mathcal{A}_2 \otimes \cdots \otimes \mathcal{A}_n) \ge \sum_{j=1}^n h_{\phi,\sigma^n}(\mathcal{A}_j) \ge \sum_{j=1}^n s_j(\phi) - \frac{c}{8n} \ln 8n/\sqrt{c}$$

and with $\mathcal{A}^{(n)} = \otimes_{j=1}^n \mathcal{A}_j$

$$h_\phi(\sigma) = \frac{1}{n} h_\phi(\sigma^n) \ge \frac{1}{n} h_{\phi,\sigma^n}(\mathcal{A}^{(n)}) \ge \frac{1}{n}\sum_{j=1}^n s_1(\phi) - \frac{c}{8n^2} \ln 8n/\sqrt{c} \quad \forall \;\; n\,.$$

But for $n \to \infty$, $\frac{1}{n}\sum_{j=1}^n s_j(\phi)$ is the entropy density

$$-\int_0^1 dp\, [\rho(p) \ln \rho(p) + (1-\rho(p)) \ln(1-\rho(p))]\,.$$

Since it cannot be exceeded by $h_\phi(\sigma)$ we have proved the

Lemma (3.6)

Let $\sigma : a(f) \to a(e^{ip} f)$ be the shift of CAR algebra $\mathcal{M}$ over $L^2([0,1], dp)$. Then, for all $\rho(p) \in C^1$ we have

$$h_\phi(\sigma) = -\int_0^1 dp\,[\rho(p)\ln\rho(p) + (1-\rho(p))\ln(1-\rho(p))].$$

Corollary 1: Let $[0,1] = \bigcup_{\alpha=1}^{k} I_\alpha$, where I_α are disjoint invervals, so that $L^2([0,1], dp) = \oplus_\alpha L^2(I_\alpha)$, and thus in a natural way $\mathcal{M}^e \supset \otimes_\alpha \mathcal{M}_\alpha^e$, $\phi_{|\mathcal{M}^e} = \otimes_\alpha \phi_\alpha$, $\sigma = \otimes_\alpha \sigma_\alpha$. Let $\rho(p)$ be C^1 in all I_α. Then Lemma (3.6) still holds.

Proof: First note that $\sigma^n : f(p) \to e^{2\pi i n p} f(p)$ is unitary in $L^2([0,1], dp)$ not only for $n \in Z$ but can be extended to $n \in \mathbf{R}$. The relation $h_\phi(\sigma^n) = |n|\, h_\phi(\sigma)$ can be extended first to n rational and then by continuity to $n \in \mathbf{R}$ [13]. To demonstrate this claim we first note that the map $R \to \mathcal{B}(\mathcal{H}) : x \to e^{2\pi i x p}$ is strongly continuous and thus $R \to \mathcal{M} : x \to \sigma^x(a)$ is norm continuous. If $\mathcal{A} \subset \mathcal{M}$ is finite-dimennsional then for any $\delta > 0\ \exists\, n_0 \in N$ such that $\forall x \in [0,1]$, $n > n_0\ \exists\, l \in \{0,1,\ldots,n\}$ with $\sigma^x\mathcal{A} \overset{\delta}{\subset} \sigma^{l/n}\mathcal{A}$. ($\mathcal{B} \overset{\delta}{\subset} C$ means that $\forall\, b \in \mathcal{B}^1\ \exists\, c \in C^1$, $\|c-b\| < \delta$.) Since $\mathcal{M}$ is hyperfinite and $\mathcal{A}$ finite–dimensional there exists, for all $n \in N$, a finite-dimensional $\mathcal{B} \subset \mathcal{M}$ with $\sigma^{l/n}\mathcal{A} \overset{\delta}{\subset} \mathcal{B}$ for $l \in \{0,1,\ldots,n\}$. Together we have $\sigma^{r+x}\mathcal{A} \overset{2\delta}{\subset} \sigma^r\mathcal{B}\,\forall\, r \in Z$, $x \in [0,1)$.
This implies by the standard properties of the $H-$functionals

$$h_\phi(\sigma^{1-\varepsilon}) = n\, h_\phi(\sigma^{(1-\varepsilon)/n}) = \lim_{r\to\infty}\frac{1}{r}H_\phi(\mathcal{A}, \sigma^{(1-\varepsilon)/n}\mathcal{A}, \ldots, \sigma^{nr(1-\varepsilon)/2}\mathcal{A}) \le$$

$$\le \lim_{r\to\infty}\frac{1}{r}H(\mathcal{B}, \sigma\mathcal{B}, \ldots, \sigma^{\overline{r}}\mathcal{B}) + \eta(\delta)$$

where $\overline{r}$ is the next integer after $r(1-\varepsilon)$ and $\lim_{\delta\to 0}\eta(\delta) = 0$. Thus $h_\phi(\sigma^{1-\varepsilon}) \le (1-\varepsilon)h_\phi(\sigma)$ for small real ε and $h_\phi(\sigma^x) = |x| H_\phi(\sigma)$ extends from the rationals to the reals. Now $L^2([0,1])$ can be unitarily tranformed into $L^2([0,\delta])$, $[0,\delta]$ some I_α, with $Ue^{2\pi i p}U^{-1} = e^{2\pi i p/\delta}$. U induces an isomorphism between the CAR algebras which transforms σ_α into $\hat{\sigma}_\alpha = \sigma_\alpha^{1/\delta}$. Upon rescaling the variables, we see

$$h_{\phi_\alpha}(\hat{\sigma}_\alpha) = \frac{1}{\delta}h_{\phi_\alpha}(\sigma_\alpha) = -\int_0^1 dp\,[\eta(\rho(p\delta)) + \eta(1-\rho(p\delta))]$$

or

$$h_{\phi_\alpha}(\sigma_\alpha) = -\int_{I_\alpha} dp\,[\rho(p)\ln\rho(p) + (1-\rho(p))\ln(1-\rho(p))].$$

Using the superadditivity (2.26(iii)) of h for tensor products and $\mathcal{M}^e \supset \otimes_\alpha \mathcal{M}_\alpha^e$ we have

$$h_\phi(\sigma) = h_{\phi_{|\mathcal{M}^e}} \ge h_{\otimes_\alpha\phi_\alpha}(\otimes_\alpha\sigma_\alpha) \ge \sum_\alpha h_{\phi_\alpha}(\sigma_\alpha) =$$

$$= -\int_0^1 dp\,[\rho(p))\ln\rho(p) + (1-\rho(p))\ln(1-\rho(p))].$$

Again, since $h_\phi(\sigma)$ cannot exceed the entropy density, we have equality.

Corollary 2: Let $\omega \to p(\omega)$ be a strictly increasing C^1 map $[0,1] \to [p_0, p_1]$ with $\partial p/\partial\omega \in L^\infty([0,1])$, $\omega(p)$ the inverse map. Then the dynamical entropy of the quasifree automorphism τ corresponding to $f(p) \to e^{i\omega(p)} f(p)$ is

$$-\int dp \, \frac{\partial \omega(p)}{\partial p} \left[\rho(p) \ln \rho(p) + (1-\rho(p)) \ln(1-\rho(p))\right].$$

Proof:

$$\hat{f}(\omega) \to \sqrt{\frac{\partial p}{\partial \omega}} \hat{f}(\omega(p)) = f(p)$$

is a unitary map $L^2([0,1], d\omega) \to L^2([p_0, p_1], dp)$ which gives an isomorphism i of the CAR algebras. The states are transformed as

$$\hat{\phi}(a^*(\hat{f})\, a(\hat{g})) = \int d\omega \, \frac{\hat{g}^*(\omega)\hat{f}(\omega)}{1+e^{\hat{\varepsilon}(\omega)}} = \phi(i(a^*(\hat{f})\, a(\hat{g}))) = \phi(a^*(f)\, a(g)) \int dp \frac{g^*(p)f(p)}{1+e^{\varepsilon(p)}}$$

with

$$\varepsilon(p) = \hat{\varepsilon}(\omega(p)).$$

Since the dynamical entropy is an invariant

$$\begin{aligned} h_{\hat{\phi}}(\hat{\tau}) &= -\int d\omega \left[\hat{\rho}(\omega) \ln \hat{\rho}(\omega) + (1-\hat{\rho}(\omega)) \ln(1-\hat{\rho}(\omega))\right] = \\ &= -\int dp \frac{\partial \omega(p)}{\partial p} \left[\rho(p) \ln \rho(p) + (1-\rho(p)) \ln(1-\rho(p))\right] = h_\phi(\tau) \end{aligned}$$

where

$$\rho(p) = \hat{\rho}(\omega(p)).$$

Corollary 3: Let $\mathcal{M}_R$ be the CAR algebra over R, $L^2((-\infty,\infty), dp)$, ϕ a quasifree state corresponding to $\rho(p)$ and σ^x space translations $\sigma_x a(f) = a(e^{iPx} f)$. Then

$$h_\phi(\sigma^x, \mathcal{M}_R) = -|x| \int_{-\infty}^{+\infty} dp \left[\rho(p) \ln \rho(p) + (1-\rho(p)) \ln(1-\rho(p))\right] \equiv |x| \int_{-\infty}^{+\infty} dp \, s(p).$$

Proof: Assume that the integral exists. Then

$$\int_{-\infty}^{+\infty} dp \, s(p) = \int_{-N\pi}^{N\pi} s(p)\, dp - \varepsilon$$

and (3.2) tells us

$$h_\phi(\sigma^x, \mathcal{M}_R) \le |x| \int_{-\infty}^{+\infty} dp \, s(p).$$

On the other hand.

$$
\begin{aligned}
&h_\phi(\sigma^x,\mathcal{M}_R)\geq \\
&\geq h_{\phi_{[-\infty,-N\pi]}}(\sigma^x,\mathcal{M}_{(-\infty,-N\pi)})+h_{\phi_{[-N\pi,N\pi]}}(\sigma^x,\mathcal{M}_{(-N\pi,-N\pi)})+h_{\phi_{[N\pi,\infty]}}(\sigma^x,\mathcal{M}_{[N\pi,\infty]}) \\
&\geq |x|\int_{-N\pi}^{N\pi} s(p)\,dp
\end{aligned}
$$

where in the last step we used again the scaling automorphism

$$\gamma a(fp)) = \frac{1}{\sqrt{N}}a(f(\frac{p}{N})) \qquad \text{and} \qquad \gamma^{-1}\sigma^x\gamma = \sigma^{xN}\,.$$

Corollary 4: Let $\mathcal{M}$ be the CAR algebra over some interval (R or finite). Let σ be an automorphism $\sigma a(f) = a(Uf)$ with $U = e^{i\omega(p)}$ and $\omega(p)$ a piecewise C^1 function. Let ϕ be a quasifree state corresponding to $\rho(p), \rho(p)$ piecewise C^1. Then

$$h_\phi(\sigma^x) = |x|\int_{-\infty}^{+\infty} dp\left|\frac{d\omega(p)}{dp}\right| s(p)\,.$$

Proof: If $\omega(p)$ is monotonic and C^1, we have already given the proof. Assume now that $\omega(p)$ is C^1 and monotonically increasing for $p < \alpha$ and monotonically decreasing for $p > \alpha$. Generalization to other $\omega(p)$ that are piecewise monotonic is evident. We introduce the automorphism γ:

$$\gamma a(f) = a(\Theta(p-\alpha)f(p)) + a^\dagger(\Theta(\alpha-p)\overline{f}(p))\,.$$

This automorphism makes σ conjugate to

$$\gamma\sigma\gamma a(f) = a(e^{i\omega(p)}\Theta(p-\alpha)f(p)) + a(e^{-i\omega(p)}\Theta(\alpha-p)f(p))\,.$$

Now $\omega(p)$ is monotonically increasing apart from a step at $p = \alpha$. If $\omega(p) = k\pi, k$ integer, this step disappears in the unitary operator and we can replace $\omega(p)$ by a monotonically increasing continuous $\overline{\omega}(p)$ that implements the same automorphism. If $(k-1)\pi < \omega(p) < k\pi$ we could rescale it and use the linearity of the dynamical entropy. But this can fail if we have to use different scalings for different steps. There is another alternative. We imbed $\mathcal{M}$ in $\overline{\mathcal{M}}$ by

$$g(p) = \begin{cases} f(p) & p < \alpha \\ 0 & \alpha < p < \alpha + 1 \\ f(p-1) & p > \alpha + 1 \end{cases}$$

On this $\overline{\mathcal{M}}$ we define the automorphism

$$\overline{\sigma}a(g) = a(\overline{U}g)$$

$$(\overline{U}g)(p) = \begin{cases} e^{i\omega(p)}g(p) & p < \alpha \\ e^{i\overline{\omega}(p)}g(p) & \alpha < p < \alpha + 1 \\ e^{-i\omega(p-1)+i2k\pi}g(p) & p > \alpha + 1 \end{cases}$$

with $\overline{\omega}(p) \in C^1$, $\overline{\omega}(\alpha) = \omega(\alpha)$ and $\overline{\omega}(\alpha+1) = -\omega(\alpha) + 2\pi k$.

We extend ϕ on $\overline{\mathcal{M}}$ by putting with some $\varepsilon \in (0,1)$

$$\overline{\rho}(p) = \begin{cases} \rho(p) & p < \alpha \\ \varepsilon & \alpha < p < \alpha + 1 \\ \rho(p-1) & p > \alpha + 1 \end{cases}$$

Then

$$\begin{aligned} h_\phi(\sigma^x, \mathcal{M}) \le h_\phi(\overline{\sigma}, \overline{\mathcal{M}}) &= \\ = |x| \int dp\, \overline{s}(p) &= \\ = |x| \int dp\, s(p) &- \varepsilon \ln \varepsilon - (1-\varepsilon)\ln(1-\varepsilon). \end{aligned}$$

On the other hand,

$$h_\phi(\sigma^x, \mathcal{M}) \ge h_{\phi(-\infty,\alpha)}(\sigma^x) + h_{\phi(a,\infty)}(\sigma^x) = |x| \int dp\, s(p).$$

Since ε is arbitrary we have equality. In the same way discontinuities in ω can be handled. This proves Corollary 4. Theorem (3.3) is the summary of the Corollaries 1–4.

4. The Classical Fermi Gas

In one dimension the one–particle phase space discretized with $h = 1$ is $\mathbf{Z}^2 = \{(x,v)\}$. The phase space Ω of the infinite "classical" Fermi gas is the power set of $\mathbf{Z}^2$ (compare [6]). Its element are the subsets $K \subset \mathbf{Z}^2$ and the points of K are the coordinates of the particles. There is not more than one particle per cell so that Fermi statistics is incorporated. Its mathematical expression is $\#\{K \cap O\} \le \#O \ \forall\, O \subset \mathbf{Z}^2$. Next, we endow Ω with a Borel structure by specifiying the algebra $\mathcal{M}$ of measurable functions. $\mathcal{M}$ is generated by the characteristic functions of cylindrical sets

$$P_{x,v}(K) = \begin{matrix} 1 & \text{if } (x,v) \in K \\ 0 & \text{if } (x,v) \notin K\,. \end{matrix} \tag{4.1}$$

The $P_{x,v}$ are minimal projections in $\mathcal{M}$. The translation $T(x,v) = (x+1, v)$ of the one–particle phase space induces an automorphism σ in $\mathcal{M}$ in a natural way

$$\sigma(P_{x,v}) = P_{x+1,v}\,.$$

Next, we need states ω on $\mathcal{M}$ which are invariant under T, and the ones without spatial correlations and a velocity distribution $0 \le f \le 1$ are of the form

$$\omega(P_{(x,v)}) = f(v)\,,$$

$$\omega(P_{(x,v)}P_{(x',v')}) = \omega(P_{(x,v)})\,\omega(P_{(x',v')}) \quad \text{unless} \quad x = x'\,,\ v = v'.$$

To construct a partition of Ω we need orthogonal projections. With the notation $n_{x,v} \in \{0,1\}$,

$$P_{n_{x,v}} = \begin{matrix} P_{x,v} & \text{if } n_{x,v} = 1 \\ 1 - P_{x,v} & \text{if } n_{x,v} = 0 \end{matrix} \tag{4.2}$$

and

$$\mathcal{P}_{L,V}(n) = \prod_{x=-L}^{L} \prod_{v=-V}^{V} P_{n_{x,v}} \tag{4.3}$$

where n is the index set $\{n_{x,v} : |x| \leq L, |v| \leq V\}, \#n = 2^{(2L+1)(2V+1)}$, we get

$$\mathcal{P}_{L,V}(n)\mathcal{P}_{L,V}(n') = \delta_{n,n'}\mathcal{P}_{L,V}(n),$$
$$\sum_n \mathcal{P}_{L,V}(n) = 1.$$

Thus the $P_{L,V}$ give a partition of Ω and its entropy in the state ω is

$$\begin{aligned} S_\omega(\{\mathcal{P}_{L,V}\}) &= -\sum_n \omega(\mathcal{P}_{L,V}(n)) \ln \omega(\mathcal{P}_{L,V}(n)) \\ &= -\sum_n \prod_{x=-L}^{L} \prod_{v=-V}^{V} \omega(P_{n_{x,v}}) \sum_{x'=-L}^{L} \sum_{v'=-V}^{V} \ln \omega(P_{n_{x',v'}}) \\ &= -\sum_{x=-L}^{L} \sum_{v=-V}^{V} [f(v) \ln f(v) + (1-f(v)) \ln(1-f(v))]. \end{aligned} \tag{4.4}$$

From this expression one can easily calculate the dynamical entropy of the shift σ. Since for $L \to \infty$, $V \to \infty$ the operators in $P_{L,V}$ generate all of $\mathcal{M}$ we may employ the Kolmogorov–Sinai theorem, but first we need a slight extension of (4.4). If O is a finite subset of $\mathbf{Z}^2$ then (4.4) obviously generalizes to

$$S_\omega(\mathcal{P}_O) = -\sum_{(x,v)\in O} (f(v) \ln f(v) + (1-f(v)) \ln(1-f(v))). \tag{4.5}$$

Therefore, taking for instance $O = \{(x,v) : |x| \leq L \text{ and } |v| < V\}$,

$$\begin{aligned} h_\omega(\sigma) &= \lim{}_{O\uparrow \mathbf{Z}^2} \lim{}_{n\to\infty}[S_\omega(\{\mathcal{P}_{\bigvee_{i=o}^{n} T^i O}\})] - S_\omega(\{\mathcal{P}_{\bigvee_{i=1}^{n} T^i O}\})] \\ &= -\sum_{v=-\infty}^{+\infty} [f(v) \ln f(v) + (1-f(v)) \ln(1-f(v))]. \end{aligned} \tag{4.6}$$

If we consider a time evolution where the particles move independently from one another, $\sigma(P_{x,v}) = P_{x+1,v}$ is replaced by $\tau(P_{x,v}) = P_{x+\delta(v),v}$, $\delta(v) \in \mathbf{Z}$ being an effective velocity. Again we have to calculate

$$\begin{aligned} h_\omega(\tau) &= \lim{}_{O\uparrow \mathbf{Z}^2} \lim{}_{n\to\infty}[S_\omega(\{\cup_{i=0}^{n} \tau^i \mathcal{P}_O\}) - S_\omega(\{\cup_{i=1}^{n} \tau^i \mathcal{P}_O\})] \\ &- \sum_{v=-\infty}^{+\infty} |\delta(v)|[f(v) \ln f(v) + (1-f(v)) \ln(1-f(v))] \end{aligned} \tag{4.7}$$

where the limit is already obtained for any of the set

$$O = \{(x,v) | c(v) \leq x \leq |\delta(v)| - 1 + c(v),\ c(v) \in \mathbf{Z}\}.$$

5. The Classical System as an Abelian Skeleton of the Quantum System

We can imbed the classical phase space of Sect. 4 into the CAR algebra in the following way, essentially repeating the construction in Sect. 3:

We denote by $f_{x,v}$ the function $\sqrt{n}e^{2\pi ipnx}\Theta(\frac{v+1}{n}-p)\Theta(p-\frac{v}{n})$, $\{x,v\}\in\mathbf{Z}^2$. These functions are orthogonal and form a basis of $L^2(R,dp)$. Since they are strictly localized in momentum space they decay in x space only like $1/|x|$. In spite of this fuzzyness they provide a strict isomorphism to the classical system of Sect. 4 which was strictly local in x and $p\equiv v$ (within an uncertainty of h). We identify $a^\dagger(f_{x,v})\,a(f_{x,v})$ with $P_{x,v}$. The automorphism $\sigma^y a(f)=a(e^{2\pi ipy}f)$ gives $\sigma^{yn}\,a^\dagger(f_{x,v})\,a(f_{x,v}) = a^\dagger(f_{x+y,v})\,a(f_{x+y,v})$. The operators $a^\dagger(f_{x,v})a(f_{x,v})$ and $a(f_{x,v})a^\dagger(f_{x,v})$, $(x,v)\in\mathbf{Z}^2$, form a maximal abelian subalgebra of the CAR algebra. This maximal abelian subalgebra serves as its own abelian model, and this abelian model is already optimal provided the maximal abelian algebra belongs to the centralizer of the modular automorphism (the centralizer is also invariant as a set under σ). This follows from the general relation

$$H_{\omega,\mathcal{M}}(\mathcal{A})=H_{\omega,\gamma\mathcal{M}}(\mathcal{A})$$

if γ is a conditional expectation $\mathcal{M}\xrightarrow{\gamma}\mathcal{M}_\omega$ with $\omega\circ\gamma=\gamma$. This situation arises if $\rho(p)$ is constant for $\frac{v}{n}<p<\frac{v+1}{n}$. We notice that under these circumstances the dynamical entropy of [5] coincides with the one propsed by Emch [10].

If $\rho(p)$ becomes a continuously varying function ($\rho'\neq 0$), the modular automorphism becomes asymptotically abelian and the centralizer becomes trivial. Because of this triviality the definition of [10] breaks down. But [5] still allows to use for abelian models maximal abelian subalgebras on which the appropriately scaled shift acts as an automorphism. The modular automorphism acts on this algebra as an isomorphism whose effect decreases with increasing n so that the model improves with increasing n.

Finally, we can also give an interpretation to $\tau a(f) = a(e^{i\varepsilon(p)}f)$ and its dynamical entropy. Let us first assume that it is possible to divide the p-space into intervals such that there is an increasing sequence of $n\in\mathbf{Z}$ and a function $k:\mathbf{Z}/n\to\mathbf{Z}^+$ with

$$\varepsilon\left(\frac{v+1}{n}\right)-\varepsilon\left(\frac{v}{n}\right)=\frac{k(v)}{t}\quad\text{for some } t\in\mathbf{R}.$$

Then the

$$f_{l,v}(p)=\exp\frac{2\pi i\varepsilon(p)lt}{k(v)}\sqrt{\frac{|\varepsilon'(p)|}{k(v)}}\Theta(\frac{v+1}{n}-p)\Theta(p-\frac{v}{n})$$

form a basis of $L^2(R,dp)$. The projection $a^\dagger(f_{l,v})a(f_{l,v})$ and $a(f_{l,v})a^\dagger(f_{l,v})$ with $0\le l<k(v)$ together with their translates τ^{ks}, $s\in\mathbf{Z}$, give a maximal abelian subalgebra that in the limit $n\to\infty$ becomes an optimally abelian model where τ^t acts as a shift with step size depending on v.

If $\varepsilon(p)$ does not satisfy the above condition, we can approximate it by an $\bar{\varepsilon}_n(p)$ with the desired property such that $\|\tau(a(f))-\bar{\tau}_n(a(f))\|\to 0$ for $n\to\infty$. Again $(\overline{\mathcal{A}}_n)$ is a maximal abelian subalgebra that serves as an abelian model together with $\bar{\tau}_n$.

In any case, we notice that though we are dealing with a noncommutative situation, the essential effect of mixing and memory loss stems from an abelian subalgebra. This abelian subalgebra is only under special circumstances invariant under the automorphism. Otherwise we can find an approximative automorphism on this abelian subalgebra such that the abelian model we use in Sect. 4 carries itself a K−dynamics. This seems to be the case for a large class of systems (systems that are K−entropic and also K-algebraic dynamical systems). Whether it is a more general property that entropic K−systems demand abelian K−subsystems or approximately abelian K−subsystems (remember they demand asymptotic abelianness [16]), or whether our imagination is able to give good models only when it is allowed to work on the familiar ground of abelian algebras remains an open problem.

References

1. Kolmogorov A. N., Dokl. Akad. Nauk **119** (1958), 861.
2. Sinai Ya, Dokl. Akad. Nauk **124** (1959), 768.
3. Connes A., Størmer E, Acta Math. **134** (1975), 289.
4. Connes A., C. R. Acad. Sci. Paris **t 301 I** (1985), 1.
5. Connes A., Narnhofer H., Thirring W., Commun. Math. Phys. **112** (1987), 691.
6. Gaspard P, Hudetz T, in preparation.
7. Wehrl A, Rev. Mod. Phys. **50** (1978), 221.
8. Thirring W, *A course in Mathematical Physics IV*, Springer, New York (1983).
9. Takesaki M, *Theory of Operator Algebras I*, Springer, New York (1979).
10. Emch G. G., Commun. Math. Phys. **49** (1976), 191.
11. Kümmerer B., Schröder W, Commun. Math. Phys **90** (1983), 251.
12. Narnhofer H., Thirring W., Fizika **17** (1985), 257.
13. Narnhofer H., Thirring W., Lett. Math. Phys. **14** (1987), 89.
14. Størmer E., Voiculescu D., Commun. Math. Phys. **133** (1990), 521.
15. Narnhofer H., Thirring W., Lett. Math. Phys. **15** (1988), 261.
16. Benatti F., Narnhofer H., Commun. Math. Phys. **136** (1991), 231.

Institut für Theoretische Physik, Universität Wien, AUSTRIA

COHERENT STATES AND SQUEEZED STATES, SUPERCOHERENT STATES AND SUPERSQUEEZED STATES

Michael Martin Nieto

Abstract. This article reports on a program to obtain and understand coherent states for general systems. Most recently this has included supersymmetric systems. A byproduct of this work has been studies of squeezed and supersqueezed states. To obtain a physical understanding of these systems has always been a primary goal. In particular, in the work on supersymmetry an attempt to understand the role of Grassmann numbers in quantum mechanics has been initiated.

Contribution in honor of the 60th birthday of John R. Klauder

1. Introduction

In 1926, Schrödinger published a paper [1] which described what we today call the "coherent states." This paper was separate from his fundamental series on "Quantization as an Eigenvalue Problem." Schrödinger sent a copy of this paper to Lorentz in response to Lorentz's objection to using wave packets to represent particles (since the packets must spread out with time) [2].

From what we would call a knowledge of the generating function for the Hermite polynomials, Schrödinger was able to show that in a harmonic oscillator potential, a general Gaussian wave with the width of the ground state could have arbitrary energy and momentum, follow the classical motion of a classical particle in the potential, and not change its shape with time. This insight eventually became formulated in one of the standard ways to define the coherent states (of the harmonic oscillator), as minimum uncertainty coherent states (MUCS).

In the early 1960s, stimulated by the classic work of Klauder [3,5], Sudarshan [4,5], and Glauber [6], there was a renewed interest in these states in the context of quantum optics. By using the boson operator techniques of these authors, the coherent states can be defined as displacement operator coherent states (DOCS) or as annihilation operator coherent states (AOCS). The displacement operator method uses symmetry, or group theory, techniques.

For the harmonic oscillator all three methods are equivalent, but for other systems they are not, in general. After reviewing these methods for the harmonic oscillator, I will discuss extensions of them to general potentials (MUCS) and arbitary symmetries

(DOCS). In these contexts, the place of the "squeezed states" of the harmonic oscillator then follows.

Although my own work for bosonic systems has strongly emphasized the physically intuitive minimum-uncertainty approach, when I and my colleagues came to consider supercoherent states, we reached the conclusion that the best approach there was to use the more abstract DOCS method. This method provided an explicitly defined mathematical approach which we hoped would yield physical insight into what such a bosonic-fermionic system means, something that, *a priori*, was opaque.

In the final sections I will review this work for coherent states, announce some new results for squeezed states, and discuss some physical conclusions which can be inferred from this work. This includes an indication of what the role of Grassmann numbers in quantum mechanics might be.

2. Coherent states

The harmonic oscillator Hamiltonian,

$$H = \frac{1}{2m}p^2 + \frac{1}{2}m\omega^2 x^2, \tag{1}$$

is quadratic in the operators, x and p, which classically vary as $\sin(\omega t)$ and $\cos(\omega t)$. The commutation relation of the associated quantum operators ($\hbar = 1$)

$$[x, p] = i, \tag{2}$$

defines an uncertainty relation

$$(\Delta x)^2(\Delta p)^2 \geq 1/4. \tag{3}$$

> *I: (MUCS). The minimum uncertainty coherent states for the harmonic oscillator potential can be defined as those states which minimize the uncertainty relation (3), subject to the added constraint that the ground state is a member of the set.*

Those states which minimize the uncertainty relation (3) are

$$\psi(x) = [2\pi\sigma^2]^{-1/4} \exp\left[-\left(\frac{x - x_0}{2\sigma}\right)^2 + ip_0 x\right], \tag{4}$$

$$\sigma = s\sigma_0 = s/(2m\omega)^{1/2}. \tag{5}$$

When $s = 1$, these Gaussians have the width of the ground state, so they are the coherent states. The states are labeled by two parameters, $x_0 = \langle x\rangle$ and $p_0 = \langle p\rangle$.

Now consider the displacement operator approach, which was first championed by Klauder [7]. Consider the oscillator algebra defined by a, a^+, a^+a, and I. The displacement operator is the unitary exponentiation of the elements of the factor algebra, spanned by a and a^+:

$$D(\alpha) = \exp[\alpha a^+ - \alpha^* a] = \exp\left[-\frac{1}{2}|\alpha|^2\right] \exp[\alpha a^+] \exp[-\alpha^* a], \tag{6}$$

where the last equality comes from using a Baker-Campbell-Hausdorff relation.

II: (DOCS). The displacement operator coherent states are obtained by applying the displacement operator $D(\alpha)$ on an extremal state, i.e., the ground state.

Specifically, this yields

$$D(\alpha)|0\rangle = \exp[\alpha a^+ - \alpha^* a]|0\rangle = \exp\left[-\frac{1}{2}|\alpha|^2\right] \sum_n \frac{\alpha^n}{\sqrt{n!}}|n\rangle \equiv |\alpha\rangle, \tag{7}$$

where $|n\rangle$ are the number states. With the identifications

$$Re(\alpha) = [m\omega/2]^{1/2} x_0\,, \qquad Im(\alpha) = p_0/[2m\omega]^{1/2}, \tag{8}$$

these are the same as the MUCS.

The third definition I mention here is

III: (AOCS). The annihilation operator coherent states are the eigenstates of the destruction operator:

$$a|\alpha\rangle = \alpha|\alpha\rangle. \tag{9}$$

(For the harmonic oscillator, III can be shown to follow from II.)

3. Coherent states for general potentials and symmetries

At the end of his 1926 paper [1], Schrödinger predicted that similar (coherent) states could be constructed for the hydrogen atom, but that it would be difficult. (He never returned to the problem.) Some 50 years later, Pete Carruthers asked me if I thought it could be done for arbitrary potentials. In striving to answer Pete's question Mike Simmons, Vincent Gutschick, and myself developed our generalization of the minimum-uncertainty method.

Consider an arbitrary classical Hamiltonian and find those classical variables, $X_c(x_c, p_c)$ and $P_c(x_c, p_c)$, that vary as $\sin(\omega t)$ and $\cos(\omega t)$. Our first *ansatz* was that the classical Hamiltonian would be quadratic in these variables. (For the many systems we studied, it was.) Now turn these classical variables into quantum operators, X and P, by turning x_c and p_c into the quantum operators x and p and then appropriately symmetrizing them in the functionals X and P. These operators define a commutation relation and hence an uncertainty relation

$$[X, P] = iG\,, \qquad (\Delta X)^2(\Delta P)^2 \geq \frac{1}{4}\langle G\rangle^2. \tag{10}$$

The state which minimizes the uncertainty relation (10) is given by the solution to the eigenvalue equation

$$\left(X + \frac{i\langle G\rangle}{2(\Delta P)^2}P\right)\psi_{mus} = \left(\langle X\rangle + \frac{i\langle G\rangle}{2(\Delta P)^2}\langle P\rangle\right)\psi_{mus}. \tag{11}$$

Note that of the four parameters $\langle X\rangle, \langle P\rangle, \langle P^2\rangle$, and $\langle G\rangle$, only three are independent because they satisfy the equality in Eq. (10).

Our second *ansatz* is our definition for

> *I: General MUCS for Arbitrary Potentials. The coherent states are the states ψ_{mus}, subject to the restriction that the ground state solution of the Schrödinger equation be a member of the set. (It always turned out to work.) This means that the three independent parameters are now reduced to two.*

As indicated, this definition worked for every solvable potential we tried. (In WKB approximation it holds in general.) The results are described in a series of papers [8], and a summary article is reprinted in the book by Klauder and Skagerstam [9]. We even produced a 16 mm color-sound film on the time evolution of these coherent states. It was later made into a video [10]. (Unfortunately, I have not found it on the shelves at any Blockbuster Video Store.) As hoped for, these states can be shown to follow the classical motion. They disperse with time, as they have to, since the eigenenergies are in general not commensurate. The variation of decoherence time from system to system can also be understood. These states maintain their coherence as well as or better than those from other methods. In the end this is not too surprising since they were physically designed to do so. It is this physically intuitive basis for these states which is one of their advantages.

Numerous times John Klauder expressed an interest in this program, and his "Doubting Thomas" penetrating questions led to specific results on at least two occasions: i) the numerical comparison of the time evolution of our states with continuous representation states, and ii) an explicit demonstration of the resolution of the identity [11].

Continuing, the generalization of the DOCS method to other symmetries is clear. Its application to arbitrary Lie groups has been discussed by many people [7,12].

> *II: General DOCS for Arbitrary (Lie) Symmetries. The general DOCS are those states obtained by applying the displacement operator, which is the unitary exponentiation of the elements of the factor algebra, on the extremal state. This is the extremal weight vector for noncompact groups. (I'd call it the ground state but John would wince.)*

One important advantage of this method is that it presents a well-defined mathematical procedure to obtain the states.

4. Squeezed states

The squeezed states of the harmonic oscillator are very easy to obtain from the MUCS point of view [13]. Look back to Eqs. (4-5). Simply let $s \neq 1$, and you have the "squeezed states." That is, they are minimum uncertainty Gaussians whose widths are not necessarily that of the ground state. This is a continuous three-parameter set of states.

Similarly, the generalization of squeezed states for arbitrary potentials works the same. These states are one step back from the general MUCS. They are the three parameter set of states defined in Eq. (11), ψ_{mus}.

The displacement operator squeezed states for the harmonic oscillator are more complicated. Consider the "unitary squeeze operator"

$$S(z) = \exp[\frac{1}{2}za^+a^+ - \frac{1}{2}z^*aa] \tag{12}$$

$$= \exp[\frac{1}{2}e^{i\phi}(\tanh r)a^+a^+]\left(\frac{1}{\cosh r}\right)^{(\frac{1}{2}+a^+a)} \exp[-\frac{1}{2}e^{-i\phi}(\tanh r)aa], \tag{13}$$

where Eq. (13) is obtained from a BCH relation [14]. The squeezed states equivalent to the ψ of Eqs. (4-5) are obtained by operating on the ground state by

$$T(\alpha, z)|0\rangle = D(\alpha)S(z)|0\rangle \equiv |\alpha, z\rangle, \tag{14}$$

$$z \equiv re^{i\phi}, \qquad r = \ln s. \tag{15}$$

[ϕ is a phase which defines the starting time, $t_0 = (\phi/2\omega)$, and s is the wave-function squeeze of Eq. (5).] But here one sees a difference. $S(z)$ by itself can be considered to be the displacement operator for the group SU(1,1) defined by

$$K_+ = \frac{1}{2}a^+a^+, \qquad K_- = \frac{1}{2}aa, \qquad K_0 = \frac{1}{2}(a^+a + \frac{1}{2}), \tag{16}$$

so that the $S(z)|0\rangle$ by themselves are the "SU(1,1) coherent states."

It is not clear how one generalizes DOCS squeezed states to other systems, e.g., to the $\cosh^{-2}$ potential. (Obviously, squeezed states represent a more complicated symmetry than coherent states.) But the MUCS generalization is clear.

Bob Fisher, Vern Sandberg, and myself looked at the possiblity of naively generalizing the harmonic oscillator squeezed states to higher order in a and a^+ [15]. It turned out this could not be done. (Once again John's hand crept in, helping us along the way.) This background [13-15] was later an aid to Alan Kostelecký, Rod Truax, and myself, when we came to the problem of "supersqueezed states."

5. Supercoherent states

The introduction of graded Lie algebras was an important milestone in the study of combined internal and space-time symmetries. This led to the development of supersymmetric theories which predict the existence of boson and fermion partner states: e.g., for the photon there is a partner photino, etc.

Up to now none of the supersymmetric partners have been found, so that supersymmetry, if extant, is broken. All evidence for supersymmetry has been found in the low energy regime: e.g., in nuclei [16] in atomic systems [17] and in WKB calculations [18]. But these results, although exciting and indicative, do not unequivocally prove the need for a fundamental supersymmetry. Rather, they are tantalizing hints. The proof would be in the discovery of supersymmetric particles, perhaps at the SSC.

Our own interest in physical manifestations of supersymmetry in atomic systems [17], combined with our interest in coherent states, led Kostelecký, Truax, and myself to consider how one should define "supercoherent states." We came to the conclusion that we should not seek to generalize the minimum-uncertainty method, because just how that should be applied to the fermion sector was unclear.

In the MUCS program, physical intuition had led to the mathematics. But here we felt we should let the mathematics lead us to the physical intuition. That is to say, by using the displacement operator method, but with supergroups, we had a method that was well-defined and which we could use because we had already developed the theory and application of super-BCH relations [19]. This gave us a tool that other workers in the field did not have [20].

Joined by Beata Fatyga, Alan's graduate student, we derived our general supercoherent states [21]. (As before, John Klauder could be found giving advice.) The definition is the same as the DOCS method for general Lie groups given above, only one uses super-groups and their associated (anti)commutation relations. We presented three examples, with physical models for each: i) The super oscillator algebra defines the supersymmetric harmonic oscillator. ii) A supersymmetric quantum-mechanical system was given which has a Heisenberg-Weyl albegra plus another bosonic degree of freedom. This represents an electron in a constant magnetic field. iii) An OSP(1/2) supersymmetry represents the electron-monopole system.

Of the three examples, the first is the simplest and I use it to demonstrate the technique. The super oscillator algebra is defined by

$$[a, a^+] = I, \qquad \{b, b^+\} = I. \tag{17}$$

Using super-BCH relations, one obtains that the super displacement operator is

$$\mathbf{D}(g) = \exp[Aa^+ - \overline{A}a + \theta b^+ + \overline{\theta} b] \tag{18}$$

$$= \left(\exp[-\frac{1}{2}|A|^2]\exp[Aa^+]\exp[-\overline{A}a]\right)\left(\exp[-\frac{1}{2}\overline{\theta}\theta]\exp[\theta b^+]\exp[\overline{\theta}b]\right) \tag{19}$$

$$\equiv D_B(A)\, D_F(\theta). \tag{20}$$

The B and F subscripts denote the fact that our supersymmetric displacement operator can be written as a product of "boson" and "fermion" (more properly, even and odd) displacement operators.

θ and $\overline{\theta}$ are odd Grassmann numbers. They are nilpotent (they only contain a "soul"). They satisfy anticommutation relations among themselves and with the fermion operators b and b^+. A and $\overline{A}$ are complex, even, Grassmann numbers. Because a^+ and a are pure even elements, we associate the "body" of A and $\overline{A}$ with the α and α^* of the ordinary coherent states. The "soul" part of A must be even in the Grassmann numbers. For example, A could be of the form

$$A = \alpha + \tau\overline{\theta}\theta, \tag{21}$$

although technically A is not restricted to the two odd basis elements we have.

Explicit calculation yields

$$\mathbf{D}(g)|0,0\rangle = [1 - (1/2)\overline{\theta}\theta]|A,0\rangle + \theta|A,1\rangle \equiv |Z\rangle. \tag{22}$$

The two labels of $|0,0\rangle$in Eq. (22) represent the even (bosonic) and odd (fermionic) sectors. The bosonic sector has the form of an ordinary coherent state $|A\rangle$ and the fermionic sector has zero or one fermions. (I refer the reader to Ref. [21] for further details.) If the bosonic displacement is turned off, then the fermionic displacement produces states with zero and one fermion, but no bosons. (In Sec. 7, I will return to an idea on how to physically interpret these states.)

6. Supersqueezed states

From the previous sections, it is clear that the supersymmetric generalization of the SU(1,1) squeeze operator of Eqs. (12-13) is what is desired. The group involved is the supergroup OSP(2/2). In addition to the SU(1,1) elements of Eq. (16), it has five more:

$$\begin{aligned} M_0 &= \frac{1}{2}(b^+b - \frac{1}{2}), \\ Q_1 &= \frac{1}{2}a^+b^+, \qquad Q_2 = \frac{1}{2}ab, \qquad Q_3 = \frac{1}{2}a^+b, \qquad Q_4 = \frac{1}{2}ab^+. \end{aligned} \tag{23}$$

The commutation relations follow, and so the supersqueeze operator can in principle be written as ($\hat{g}$ is the factor algebra)

$$\mathbf{S}(g) = \exp\left[\sum_{i=1}^{6} \alpha_i \hat{g}_i\right] \tag{24}$$

$$= \prod_{i=1}^{8} \exp[\beta_i g_i]. \tag{25}$$

The above can be solved by using super-BCH relations. They yield twenty coupled differential equations. (The even operators have separate equations for order zero, two and four Grassmann numbers, and the odd operators have separate equations for orders one and three.)

We have just finished solving these equations and are in the process of performing some final calculations [22]. For now it is useful to note that the squeeze operator can again be separated into a product of bosonic and fermionic pieces:

$$\mathbf{S}(g) = S_B(g)\, S_F(g). \tag{26}$$

The fermionic squeeze operating by itself on $|0,0\rangle$ produces the states $|0,0\rangle, |1,1\rangle$, and $|2,0\rangle$.

Therefore, one finally obtains that the supersqueezed states are, in general, of the form

$$\begin{aligned} \mathbf{T}(g)|0,0\rangle &= \mathbf{DS}|0,0\rangle \\ &= D_B(A)D_F(\theta)S_B(g)S_F(g)|0,0\rangle \\ &= [D_B(A)S_B(g)][D_F(\theta)S_F(g)]|0,0\rangle \\ &\equiv \mathbf{T}_B(g)\mathbf{T}_F(g)|0,0\rangle. \end{aligned} \tag{27}$$

The general operator produces states with arbitrary numbers of bosons and zero or one fermion.

7. Discussion

If supersymmetric partners were actually found, then fundamentally one would need to give a physical interpretation to Grassmann numbers. This situation would be similar to the problem of the physical interpretation of imaginary numbers when quantum mechanics was discovered. When Schrödinger described his "coherent states" [1], he thought that the physics was contained in the real part of his wave solutions. The realization that the complex phase had physical information came later.

Recently I addressed this problem [23], although I was by no means the first to do so (see, e.g., the appendix in [24]). Looking at the supercoherent states of Eq. (22), I suggested that the odd part describes the existence of a coherent, massless fermion; i.e., a "photino" with energy E, it being coherent with the various n-photon states in the bosonic sector.

As to the Grassmann numbers, I made the suggestion that the fermion sector "phase" relative to the boson sector is defined by the c in $\theta \equiv c\zeta$, ζ being a Grassmann basis vector and labeling the fermionic part of the state. The probability of finding a supercoherent state that has a bosonic sector coherent with one photino is c^*c, with the $\overline{\zeta}\zeta$ in $\overline{\theta}\theta$ labeling the probability as being for a fermion. Thus, the probability of finding a bosonic sector without a coherent photino is $(1 - c^*c)$, from $(1 - \overline{\theta}\theta)$. Note that one must have $|c| \leq 1$. Then the probabilities for the coherent state having or not having a fermion are both ≤ 1.

The restriction on the value of c is analogous to physical restrictions placed on ordinary quantum mechanics. For example, one demands that all state probabilities $P_n = |a_n|^2 \leq 1$ and one disallows unnormalizable solutions of the Schrödinger equation. (See Ref. [23] for more details.) Even so, I must emphasize that at this point this restriction on c is based on physical intuition rather than on mathematical rigor.

Now that we have our supersqueezed states [22], I hope to pursue this line of thought, taking into account the insight available from studying the Grassmann structure of the supersqueezed states. It appears that a similar restriction can be made for the quantities analogous to c in the supersqueezed states. Just as as was the case for Ref. [23], I expect useful comments to come from John Klauder.

Acknowledgements

Of course, in a review of a program such as this, I must once again acknowledge my colleagues and friends with whom I have enjoyed working over the years.

But most all, this is John Klauder's celebration. As the reader will have noted, the influence of John Klauder can be seen throughout this work. It is an honor, but most of all it is a great personal pleasure, to congratulate John on the occasion of his 60th birthday. I eagerly await our next discussion, perhaps on Grassmann numbers and supersqueezed states, when I can once again hear John say: "But Michael, ..."

References

1. E. Schrödinger, Naturwiss. **14** (1926), 664.
2. K. Przibram (ed.), Letters on Wave Mechanics (Letter # 19), Philosophical Library, NY, 1967.
3. J. R. Klauder, J. Math. Phys. **4** (1963), 1055; 1058; ibid. **5** (1964), 177.
4. E. C. G. Sudarshan, Phys. Rev. Lett. **10** (1963), 277.

5. J. R. Klauder and E. C. G. Sudarshan, *Fundamentals of Quantum Optics*, W. A. Benjamin, New York, 1968.
6. R. J. Glauber, Phys. Rev. Lett. **10** (1963), 84; Phys. Rev. **130** (1963), 2529; ibid. **131** (1963), 2766.
7. J. R. Klauder, J. Math. Phys. **4** (1963), 1058.
8. M. M. Nieto and L. M. Simmons Jr., Phys. Rev. D **20** (1979), 1321, the first of a series of six papers, ending with; M. M. Nieto, L. M. Simmons, Jr. and V. P. Gutschick, Phys. Rev. D **23** (1981), 927.
9. M. M. Nieto, in: Coherent States, Applications in Physics and Mathematical Physics (J. R. Klauder and B.-S. Skagerstam, eds.), World Scientific, Singapore, 1985, pp. 429 and ff.
10. V. P. Gutschick, M. M. Nieto and F. Baker, *Time-Evolution of Coherent States for General Potentials*, 16 mm, color, sound movie, available from Cinesound Co. 915 N. Highland Ave, Hollywood, CA 90038, (Also available in VHS.).
11. V. P. Gutschick and M. M. Nieto, Phys. Rev. D **22** (1980), 403; M. M. Nieto, ibid. **30** (1984), 770.
12. A. O. Barut and L. Girardello, Commun. Math. Phys. **21** (1971), 41; A. M. Perelomov, Commun. Math. Phys. **26** (1972), 222; J. R. Gilmore, *Lie Groups, Lie Algebras, and Some of Their Applications*, John Wiley, New York, 1974.
13. M. M. Nieto, in: Frontiers of Nonequilbrium Statistical Physics (G. T. Moore and M. O. Scully, eds.), Plenum, New York, 1986, pp. 287 and ff.
14. D. R. Truax, Phys. Rev. D **31** (1985), 1988.
15. R. A. Fisher, M. M. Nieto and V. D. Sandberg, Phys. Rev. D **29** (1984), 1107.
16. F. Iachello, Phys. Rev. Lett. **44** (1981), 772; Nucl. Phys. **A370** (1981), 284.
17. V. A. Kostelecký and M. M. Nieto, Phys. Rev. Lett. **53** (1984), 2285; V. A. Kostelecký, M. M. Nieto and D. R. Truax, Phys. Rev. A **38** (1988), 4413, (and references therein).
18. A. Comptet, A. D. Bandrauk and D. K. Campbell, Phys. Lett. B **150** (1985), 159.
19. D. R. Truax, V. A. Kostelecký, and M. M. Nieto, J. Math. Phys. **27** (1986), 354; V. A. Kostelecký, M. M. Nieto and D. R. Truax, J. Math. Phys. **27** (1986), 1419.
20. C. Aragone and F. Zypman, J. Phys. A **19** (1986), 2267; A. B. Balatekin, H. A. Schmitt and B. R. Barrett, J. Math. Phys. **29** (1988), 1634; M. Chaichian, D. Ellinas and P. Prešnajder, CERN preprint CERN-TH-5925/90; H. A. Schmitt and A. Mufti, Phys. Lett. A **148** (1990), 139.
21. B. W. Fatyga, V. A. Kostelecký, M. M. Nieto and D. R. Truax, Phys, Rev. D **43** (1991), 1403.
22. V. A. Kostelecký, M. M. Nieto and D. R. Truax, (in preparation).
23. M. M. Nieto, in: Santa Fe Workshop: Foundations of Quantum Mechanics (T. D. Black, M. M. Nieto, M. O. Scully, R. M. Sinclair and H. S. Pilloff, eds.), World Scientific, Singapore, 1992, pp. 95 and ff.
24. See, e.g. the appendix in: R. Delbourgo and J. A. Henderson, J. Phys. A. **20** (1987), 529.

THEORETICAL DIVISION, LOS ALAMOS NATIONAL LABORATORY, UNIVERSITY OF CALIFORNIA, LOS ALAMOS, NEW MEXICO 87545, U.S.A., EMAIL ADDRESS: MMN@PION.LANL.GOV

MASS AND MIXING ANGLE PATTERNS IN THE STANDARD MODEL AND ITS MINIMAL SUPERSYMMETRIC EXTENSION

PIERRE RAMOND

ABSTRACT. Using renormalization group techniques, we examine several interesting relations among masses and mixing angles of quarks and leptons in the Standard Model of Elementary Particle Interactions as a function of scale. We extend the analysis to the minimal Supersymmetric Extension to determine its effect on these mass relations. For a heavy top quark, and minimal supersymmetry, most of these relations can be made to agree at one unification scale.

Dedicated to John Klauder at the occasion of his sixtieth birthday

Introduction

Most of the parameters in the Standard Model of the Strong, Weak, and Electromagnetic Interactions are to be found in the Yukawa sector of the theory where they serve to parametrize quark and lepton masses, as well as the interfamily mixings of the quarks, and CP violation. Historically, of these thirteen parameters, only one was ever predicted [1], the charmed quark mass, but only after an inspired guess on the value of a strong (*i.e.* presently uncalculable) matrix element.

Theoretical guesses on the nature of physics beyond the Standard Model have been formulated, using as inspiration the idea of Grand Unification [2] which emerged from the observed pattern of the quantum numbers of the elementary particles. When applied in conjunction with the renormalization group [3], this idea has proven extremely fruitful. Recent work indicates that the experimental values of the gauge couplings are such that all three couplings evolve to the same value [4] at shorter distances when supersymmetry is included at SSC scales. Without supersymmetry, the gauge couplings meet two at a time, forming a small GUT triangle in the plot of their evolution as a function of scale.

This encouraging situation, hinting at a Super Grand Unified Theory, should be matched by concomitant simplicity in the other parameters of the theory. To that purpose we present a comparative analysis of possible relations among Yukawa couplings at shorter distances both in the Standard Model itself and in its minimal supersymmetric extension.

In the context of the $SU(5)$ Grand Unified Theory [5], several mass relations were proposed based on simple assumptions on the possible Higgs structure. The first of these (assuming only a **5** Higgs representation) leads to the equality between the τ-lepton and b-quark Yukawas or masses at the GUT scale:

$$m_b = m_\tau \ . \qquad \text{RELATION I}$$

This relation, when folded into the running of the masses with distance is not inconsistent with experiment, due to the fact that QCD provides, through the anomalous dimension of the quark mass, the required factor to bring it in rough agreement at experimental scales [6]. This relation, if applied to the lighter two families, is off by a factor of ten. A new scheme was proposed [7] with a slightly more complicated Higgs structure (using a **45** representation in conjunction with the **5**). It replaces the above with the more complicated relations

$$m_d = 3m_e \ , \qquad \text{RELATION IIa}$$
$$3m_s = m_\mu \ . \qquad \text{RELATION IIb}$$

These are typical of the SU(5) types of model in which the charge $-1/3$ quarks and the charged leptons Yukawa couplings appear with the same quantum numbers.

The situation concerning the mixing angles is equally intriguing. It was noticed long ago that there existed a near numerical equality between the square of the tangent of the Cabibbo angle and the ratio of the down to the strange quark masses (determined from current algebra). This Oakes relation reads

$$\tan\theta_c \approx \sqrt{\frac{m_d}{m_s}} \ . \qquad \text{RELATION III}$$

It has provided the central inspiration in the search for Yukawa matrices. It was found [8] that very general classes of matrices with judiciously chosen textures (*i.e.* zeroes in the right places) reproduced this relation, at least approximately.

In a model based on $SO(10)$ [9] it was found [10] that these three different relations could all be obtained at the same time, with the required texture being enforced naturally by discrete symmetries at the GUT scale. As a consequence of the model, the mixing of the third family with the two lighter ones was dictated exclusively through the Yukawa matrices of the charge 2/3 quarks. Accordingly an Oakes-like relation for the mixing of the second and third families ensued [10]

$$V_{cb} = \sqrt{\frac{m_c}{m_t}} \ ; \qquad \text{RELATION IV}$$

it relates the "23" matrix element of the CKM matrix to the ratio of the charm quark mass to the top quark mass. This relation, if true, presents us with the very exciting possibility of predicting the mass of the top quark. It is known that the top quark mass is somewhere in between 100 and 200 GeV, the lower limit being set by direct experimental searches, the upper by the radiative effect of the top quark mass on the ratio of neutral to charged current processes.

These four relations can all be satisfied if one takes the Yukawa Mixing matrices to be of the form [10], shown here in a specific gauge

$$\mathbf{Y}_u = \begin{pmatrix} 0 & A & 0 \\ A & 0 & B \\ 0 & B & C \end{pmatrix} ,$$

$$\mathbf{Y}_d = \begin{pmatrix} 0 & R & 0 \\ R & S & 0 \\ 0 & 0 & T \end{pmatrix} \quad , \quad \mathbf{Y}_e = \begin{pmatrix} 0 & R & 0 \\ R & -3S & 0 \\ 0 & 0 & T \end{pmatrix} .$$

This form has been recently rediscovered by several groups [11,12], and some of our analysis overlaps with their work. Although derived with specific and sometimes complicated Higgs structures in mind, these relations may well prove sturdier than the theories which generated them. In the following, we first examine these relations in the context of the Standard Model at varying distances all the way down to Planck length. We then extend the analysis to the minimal supersymmetric extension of the Standard Model, and compare the effect of this extension on their relative validity. A more thorough treatment is in preparation [13].

Running the Standard Model to Planck

In this section we make use of the numerical techniques and routines developed in a previous work [14]. We first use experiment to fix the parameters of the Standard Model at lower energies. We then use these values as initial conditions in the renormalization group running to lower length scales, using the $\overline{MS}$ scheme. Our incomplete knowledge of the Standard Model parameters forces us to repeat the analysis for a range of allowed values of the top quark and Higgs masses. In these runs, we take $g_3(M_Z) = 1.191$ and the physical bottom quark mass $M_b = 4.89$ GeV.

Let us summarize the salient features of the renormalization group running in the Standard Model. At the one loop level, the gauge couplings are unaffected by the other couplings in the theory. On the other hand, the Yukawa couplings are affected at one loop by both the gauge and Yukawa coupling constants. Since the top Yukawa coupling is at least as big as the gauge couplings at low energy, it means that the Yukawa runnings are sensitive to mostly the top Yukawa and the QCD gauge couplings. Thus we can expect that the mass and mixing relations we have just described to be sensitive to the value of the top quark mass. On the other hand, the Higgs quartic self–coupling enters the running of the other couplings only at the two loop level, and its effect on the quark and lepton parameters is expected to be small. However its own running is very sensitive to the top quark mass; it can become negative as easily as it can blow up. The former leads to vacuum instability, the latter, called the triviality bound leads to strong self–interaction of the Higgs. The following graphs summarize these bounds for representative values of the top quark mass. It is amusing to note that it is for comparable value of the top and Higgs masses that these bounds are least effective, but it is important to emphasize that a high value of the top with a relatively low value of the Higgs necessarily indicates the presence of new physics within reach of the SSC.

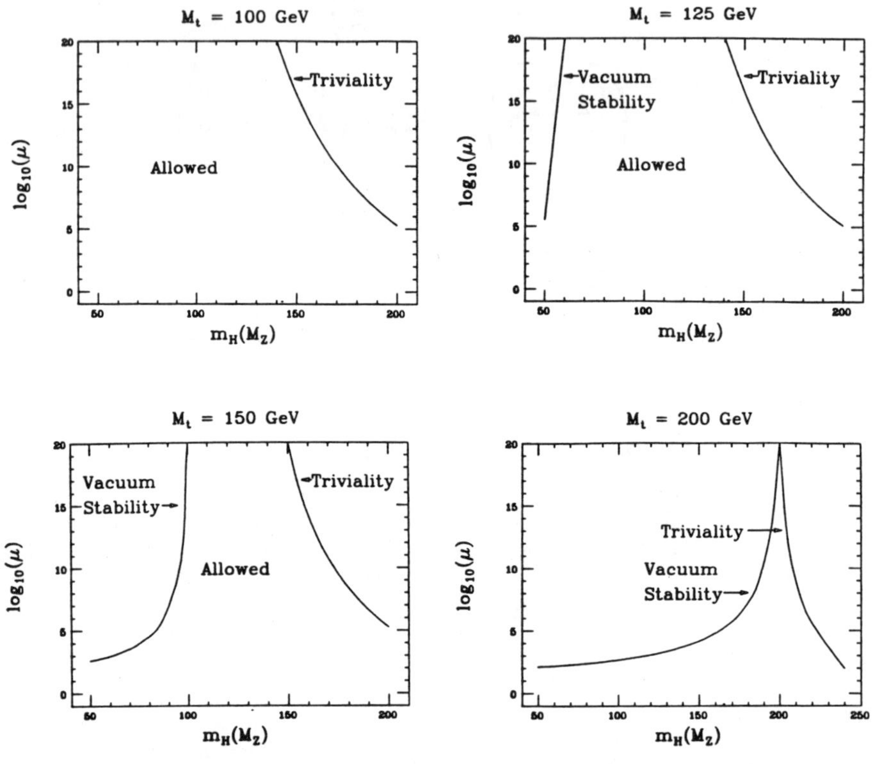

Scales of Expected New Physics in the Standard Model

RELATION (I)

This relation is the most natural one in the $SU(5)$ theory, and it could be expected to be valid at scales where the Standard Model gauge couplings are the closest to one another. We examine its validity for different physical values of the top and Higgs masses in the Standard Model. The results are summarized in the following tables:

Standard Model: $m_t = 100$ GeV; $m_H = 100$ GeV						
Scale (GeV)	6×10^7	10^{11}	10^{13}	10^{15}	10^{17}	10^{19}
$\frac{m_b}{m_\tau}$	1.001	.858	.798	.752	.716	.688

Standard Model: $m_t = 150$ GeV; $m_H = 150$ GeV						
Scale (GeV)	1.5×10^7	10^{11}	10^{13}	10^{15}	10^{17}	10^{19}
$\frac{m_b}{m_\tau}$	1.003	.818	.758	.711	.675	.651

Standard Model: $m_t = 190$ GeV; $m_H = 180$ GeV						
Scale (GeV)	4×10^6	10^{11}	10^{13}	10^{15}	10^{17}	10^{19}
$\frac{m_b}{m_\tau}$	1.001	.763	.699	.649	.611	.581

Their noteworthy feature is that this simplest of the SU(5) relation tends to be valid at an energy scale many orders of magnitude removed from that at which the gauge couplings tend to converge, which is around 10^{15} GeV.

RELATION (II)

We now turn to the more complicated relation between the masses from the two lighter families. Again we run the predictions for different values of the top and Higgs masses. The results are summarized in the following tables

Standard Model: $m_t = 100$ GeV; $m_H = 100$ GeV				
Scale (GeV)	10^{13}	10^{15}	10^{17}	10^{19}
$\frac{m_d}{3m_e}$	1.445	1.376	1.30	1.25
$\frac{3m_s}{m_\mu}$	1.30	1.22	1.17	1.12

Standard Model: $m_t = 150$ GeV; $m_H = 150$ GeV				
Scale (GeV)	10^{13}	10^{15}	10^{17}	10^{19}
$\frac{m_d}{3m_e}$	1.43	1.35	1.29	1.25
$\frac{3m_s}{m_\mu}$	1.28	1.21	1.16	1.12

Standard Model: $m_t = 190$ GeV; $m_H = 180$ GeV				
Scale (GeV)	10^{13}	10^{15}	10^{17}	10^{19}
$\frac{m_d}{3m_e}$	1.42	1.34	1.28	1.23
$\frac{3m_s}{m_\mu}$	1.28	1.21	1.15	1.11

We see that these relations are never satisfied at any scale in the Standard Model, and they are also quite insensitive to the value of the top. The formula for the second family provides the better agreement, but it never gets below 11%. From these tables, one can easily read off the ratio of the determinants of the charge $-1/3$ to charge -1 mass matrices. We note that for the lowest top mass, these two determinants are equal at 10^{18} GeV, while for the highest they converge at 10^{15} GeV.

RELATION (III)

We find the Oakes relation to be quite independent of scale. The reason is that the Cabibbo angle does not run, and the ratio of light quarks is essentially unaffected by QCD, since both are far away from the Pendleton-Ross infrared fixed point. Further one finds that their numerical values are pretty much independent of the value of the top quark mass and of the Higgs mass. However the agreement is spectacular, hovering around the 4% level. We only present a representative example for one case:

Standard Model: $m_t = 100$ GeV; $m_H = 100$ GeV					
Scale (GeV)	10^7	10^{13}	10^{15}	10^{17}	10^{19}
$\tan\theta_C\sqrt{\frac{m_s}{m_d}}$	1.038	1.038	1.038	1.038	1.038

RELATION (IV)

Since this relation involves the top quark mass directly, it could be used to fix its value. On the other side of the equation, the experimental value of the "23" element of the CKM matrix, V_{cb} is known to within $\sim$ 10% only [15]

$$V_{cb} = 0.043 \pm 0.006 \ .$$

In the following we use the central value. It is interesting to see under what conditions this relation can be made to hold [16], in particular if it is satisfied for a top quark anywhere in its allowed range between 100 and 200 GeV. In this case, because of the Pendleton-Ross fixed point the ratio of the two quark masses runs appreciably in the infrared region. For a top quark in its lower range, 100 – 150 GeV, this relation fails over all scales, so we start with a 190 GeV top in the following tables. The value of V_{cb} at all scales is obtained by running the angles; the particular value used for the CP–violating phase produces no appreciable difference in our results.

Standard Model: $m_t = 190$ GeV; $m_H = 180$ GeV					
Scale (GeV)	10^{11}	10^{13}	10^{15}	10^{17}	10^{19}
$\sqrt{\frac{m_c}{m_t}}$	.0565	.0559	.0555	.0552	.0548
V_{cb}	.0502	.0513	.0519	.0529	.0534

Standard Model: $m_t = 200$ GeV; $m_H = 195$ GeV					
Scale (GeV)	10^{13}	10^{14}	10^{15}	10^{17}	10^{19}
$\sqrt{\frac{m_c}{m_t}}$	.0536	.0533	.0530	.0525	.0519
V_{cb}	.0527	.0534	.0539	.055	.0561

We see that it is only for a very heavy top quark that this relation can be fulfilled. Of course things get better if we use the largest experimentally allowed value of V_{cb}. In particular, for a yet heavier top quark mass, the region of agreement spans more scales.

Putting all these results together, it is hard to arrive at a unified picture in the context of the Standard Model. The scale at which one relation tends to be satisfied does not coincide with that at which the other is valid. Still, the disagreement is never too large, which makes us hope that small course corrections in the running of the parameters may make most if not all of these relations hold simultaneously at a unified or similar scales. It is remarkable that for a top quark at the upper reaches of its allowed range, the long life of the bottom quark lends plausibility to the $SO(10)$–inspired relation.

Running the Supersymmetric Standard Model to Planck

As is well known, the Standard Model shows no apparent inconsistencies until perhaps the Planck scale, where quantum gravity enters the picture. Thus the nature of the physics that is to be encountered in between our scale and the Planck scale is a matter of theoretical taste. At one extreme, the value of the gauge couplings may be interpreted to infer new phenomena every two orders of magnitude. At the other, there is the possible desert suggested by GUTs; however, the absence of new phenomena over many orders of magnitude cannot be understood (perturbatively) unless one generalizes the Standard Model to be Supersymmetric at an experimentally accessible scale. This particular scenario is bolstered by the fact that with such "low energy" supersymmetry, the three gauge couplings of the Standard Model meet at a scale of $\approx 10^{16}$ GeV at the perturbative value of 1/26 [4]. In the following we restrict ourselves to this particular scheme in investigating the fate of these four relations among masses and mixing angles.

The collapse of the "GUT Triangle" in the supersymmetric extension fixes two scales, one is that at which the gauge couplings unify, the other denotes the threshold of supersymmetry. Minimal supersymmetry [17] implies two Higgs doublets, and eliminates the feisty quartic self–coupling of the Standard Model. Accordingly, even in the limit where only one Higgs is light, there appears an extra parameter, the ratio of the vacuum values of these two doublets, parametrized by an angle β

$$\tan\beta = \frac{v_u}{v_d} \ .$$

In a previous publication [18], it was shown that with supersymmetry it becomes possible to assume that relation (I) is valid at gauge unification. This fixes the angle β in terms of the top quark mass and the mass of the lightest neutral Higgs. These results are displayed for two scales of supersymmetry. For $M_{SUSY} = 1$ TeV, the gauge unification occurs between a low of 6.92×10^{15} GeV and a high of 1.26×10^{16} GeV, corresponding to $g_3 = 1.171$ and $g_3 = 1.197$, respectively. The error bars in the strong coupling allow for a SUSY scale as high as 10 TeV, with unification at 6.46×10^{15} GeV. The values in the table are obtained for the lower g_3.

Super-Standard Model: $M_{SUSY} = 1$ TeV		
$g_3 = 1.171$; $M_{GUT} = 6.92 \times 10^{15}$ GeV		
β (degrees)	M_H(GeV)	M_t(GeV)
50	57	154
60	81	171
70	102.5	183
80	116.5	190.5
85	120	192

Super-Standard Model: $M_{SUSY} = 1$ TeV		
$g_3 = 1.197$; $M_{GUT} = 1.26 \times 10^{16}$ GeV		
β (degrees)	M_H(GeV)	M_t(GeV)
50	60	159
60	84	176
70	105	188
80	119	196
85	123	198

Super-Standard Model: $M_{SUSY} = 10$ TeV		
β (degrees)	M_H (GeV)	M_t(GeV)
50	86	163
60	112.5	180
70	134	191.5
80	148	198
85	151	199.6

We still have one unknown degree of freedom, the angle β. The strategy of this paper is to fix its value by demanding optimum agreement on the remaining relations (II-IV). The results are again given in terms of tables for the central value of $V_{cb} = .044$. We treat three cases, the first where unification takes place at its lowest value. The second table uses the same value of the SUSY scale but the highest unification scale. Finally the third table uses the maximum SUSY value of 10 TeV, consistent with the error on g_3.

Super-Standard Model: $M_{SUSY} = 1$ TeV					
$g_3 = 1.171$; $M_{GUT} = 6.92 \times 10^{15}$ GeV					
β (degrees)	50	60	70	80	85
$\frac{m_d}{3m_e}$	1.543	1.539	1.534	1.53	1.53
$\frac{3m_s}{m_\mu}$	1.38	1.381	1.379	1.378	1.379
$\sqrt{\frac{m_c}{m_t}}$	.0518	.0485	.0463	.0451	.0450
V_{cb}	.0371	.0362	.0369	.0370	.0370

Super-Standard Model: $M_{SUSY} = 1$ TeV					
$g_3 = 1.197$; $M_{GUT} = 1.26 \times 10^{16}$ GeV					
β (degrees)	50	60	70	80	85
$\frac{m_d}{3m_e}$	1.38	1.38	1.38	1.38	1.37
$\frac{3m_s}{m_\mu}$	1.24	1.24	1.23	1.23	1.23
$\sqrt{\frac{m_c}{m_t}}$	.0457	.0458	.0409	.0398	.0397
V_{cb}	.0351	.0351	.0351	.0350	.0350

Super-Standard Model $M_{SUSY} = 10$ TeV					
$g_3 = 1.171$; $M_{GUT} = 6.46 \times 10^{15}$ GeV					
β (degrees)	50	60	70	80	85
$\frac{m_d}{3m_e}$	1.54	1.54	1.54	1.55	1.55
$\frac{3m_s}{m_\mu}$	1.39	1.39	1.38	1.38	1.38
$\sqrt{\frac{m_c}{m_t}}$	.0489	.0456	.0435	.0424	.0422
V_{cb}	.0373	.0372	.0371	.0372	.0372

It is clear from the above that the mass relation between the two lightest families are consistently off over all scales. The same is not true for the mixing angle relation. Using the central value for V_{cb}, and supersymmetry, there is no agreement at the unification scale, although it gets close especially in the second case. Thus it lends credence to the fact that with a higher value of the mixing angle, one could satisfy that relation. We have made several runs with a higher value $V_{cb} = .050$. For instance, we find for the higher value of the strong coupling, $g_3 = 1.197$, and $\beta = 85$ degrees, the following graph

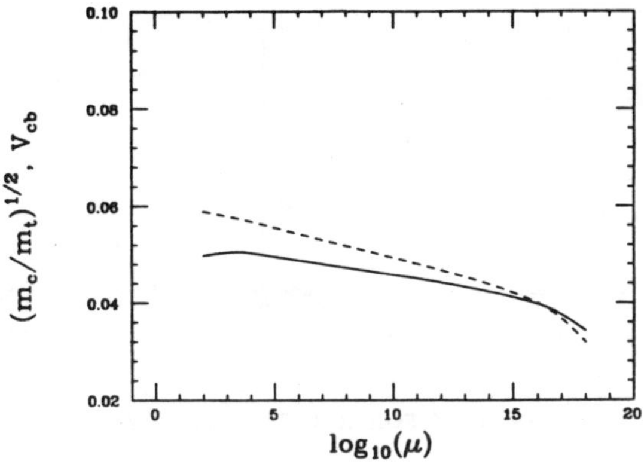

EVOLUTION OF V_{cb} (SOLID LINE) AND $\sqrt{m_c/m_t}$ (DASHED LINE) WITH $V_{cb} = .050$

If we consider the lower value of the strong coupling constant, the two curves meet closer to the Planck scale. Also, by raising the value of V_{cb} we can change the crossing scale. We just show this particular curve as an illustrative example.

From the theoretical point of view, we do not know exactly where the scale at which the $SO(10)$-inspired relation is valid. It could be much higher than M_{GUT}, the scale of unification of the Standard Model's couplings. Accordingly, we now plot V_{cb} as a function of β, assuming that relation (IV) is valid at M_{GUT}, $10M_{GUT}$, and $100M_{GUT}$, and the higher value of g_3.

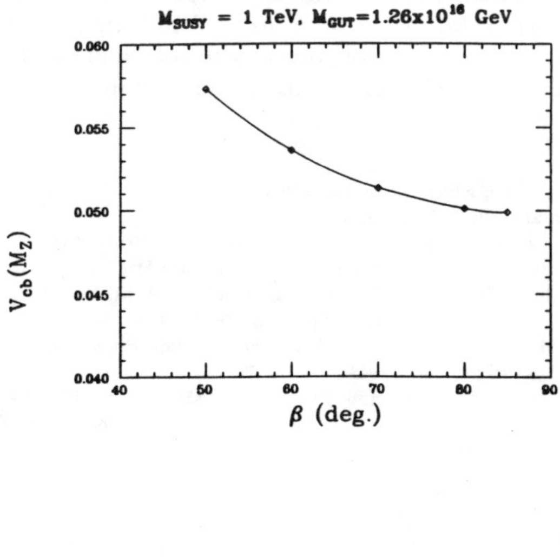

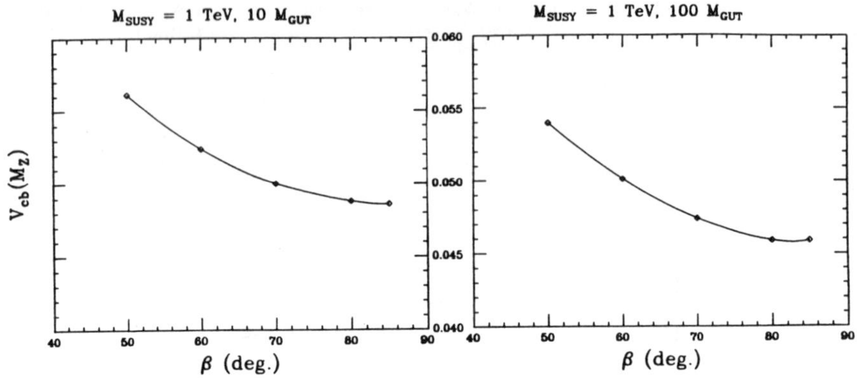

V_{cb} AS A FUNCTION OF β FOR RELATION (IV) VALID AT DIFFERENT SCALES

We conclude that it is not impossible to achieve agreement for three out of the four relations. But, for this to be true, several things must occur: one V_{cb} must be larger than its presently measured value; second the top quark mass must be around 190 GeV (if it is a bit lighter, then agreement dictates that the mixing angle should be larger still); third the Higgs mass should hover around 120 GeV. These conclusions are qualitatively correct if one demands maximum agreement. A similar analysis which recently appeared in the literature has reached similar conclusions [12]. However, it is difficult to arrive at a definite number without an exhaustive analysis of the parameter space. We leave this to a future publication [13].

Acknowledgements

All of this work uses numerical routines developed in collaboration with H. Arason, D. Castaño, B. Keszthelyi, S. Mikaelian, E. Piard and B. Wright. I also wish to thank Professor G.G. Emch for giving me the opportunity to contribute to John's birthday volume. Special thanks are due to D. Castaño for last minute computer heroics.

References

1. M.K. Gaillard and B. Lee, Phys. Rev. **D10** (1974), 897.
2. J.C. Pati and A. Salam, Phys. Rev. **D10** (1974), 275.
3. H. Georgi, H. Quinn and S. Weinberg, Phys. Rev. Lett. **33** (1974), 451.
4. U. Amaldi, W. de Boer and H. Fürstenau, Phys. Lett. **260B** (1991), 447; J. Ellis, S. Kelley and D. Nanopoulos, Phys. Lett. **B260** (1991), 131; P. Langacker and M. Luo, Phys. Rev. **D44** (1991), 817.
5. H. Georgi and S. L. Glashow, Phys. Rev. Lett. **32** (1974), 438.
6. A. Buras, J. Ellis, M. K. Gaillard, and D. Nanopoulos, Nucl. Phys. **B135** (1978), 66.
7. H. Georgi, C. Jarlskog, Phys. Lett. **89B** (1979), 297.
8. H. Fritzsch, Phys. Lett. **B70** (1977), 436; Phys. Lett. **73B** (1978), 317; Phys. Lett. **166B** (1986), 423.
9. H. Fritzsch and P. Minkowski, Ann. Phys. **93** (1975), 193; H. Georgi, in: Particles and Fields (C.E. Carlson, ed.), AIP, New York, 1974.

10. J. Harvey, P. Ramond and D. Reiss, Phys. Lett. **92B** (1980), 309; Nucl. Phys. **B199** (1982), 223.
11. G. Lazarides and Q. Shafi, Nucl. Phys. **B350** (1991), 179.
12. S. Dimopoulos, L. Hall and S. Raby, OSU Preprints DOE-ER-01545-566, 567 (1991).
13. H. Arason, D. Castaño, B. Keszthelyi, S. Mikaelian, E.Piard, P. Ramond and B. Wright, *in preparation.*
14. H. Arason et al., Florida Preprint, UFIFT-HEP-91-33 (1991).
15. Atwood, *private communication.*
16. E. Freire, G. Lazarides and Q. Shafi, Mod. Phys. Lett. **29A** (1990), 2453.
17. For a review, see H.P. Nilles, Phys. Rep. **110** (1984), 1; H. Haber and G. Kane, Phys. Rep. **117** (1985), 75.
18. H. Arason et al, Phys. Rev. Lett. **67** (1991), 2933.

INSTITUTE FOR FUNDAMENTAL THEORY, DEPARTMENT OF PHYSICS, UNIVERSITY OF FLORIDA, GAINESVILLE, FL 32611 USA

HAMILTONIANS WITH CONTACT INTERACTION

HELMUT REEH

ABSTRACT. Hamiltonians with perturbations of vanishing support are discussed. It is shown that there are generalizations to Friedman's theorem. Applications to physics, however, may be limited.

Dedicated to John Klauder at the occasion of his sixtieth birthday

1. Introduction.

Hamilton operators with contact interactions, i.e. with δ-function like potentials occur in models of solid state physics, of nuclear physics and of atomic physics, e.g. the Darwin term of the non relativistic approximation to the Dirac equation ($\pi\ Z\ e^2\hbar^2/_{2m^2c^2} \cdot \delta^3(\vec{x})$, repulsive) and the hyperfine interaction term ($\frac{2}{3}\ \vec{\mu} \cdot \vec{\mathcal{M}}\ 4\ \pi\ \delta^3(\vec{x})$ for s-states; interaction between magnetic moments of electron and nucleus). Another, two dimensional, application is a treatment of the Aharonov-Bohm effect where a flux thread is enclosed in a cylindrical repulsive tube of very small diameter.

Mathematically, Hamiltonians with interaction located in a point were studied in 1961 by Berezin and Faddeev [1] as self-adjoint extensions of $H_0 = -\Delta$, the Laplacean on $\mathbb{R}^3$ restricted at first to functions vanishing around the origin.

In 1972 Friedman [2] studied $H = H_0 + V_n(x,t)$ as self-adjoint operators on $\mathcal{L}^2(\mathbb{R}^d)$ with a bounded self-adjoint multiplication operator $V_n(x,t)$ (the bound may depend on n), $x \in \mathbb{R}^d, t \in I \subset \mathbb{R}^1$ the time. Under certain assumptions concerning the possible time dependence of V_n he showed in case $V_n \geq 0$ as operators that the unitary time evolution $U(t)$ belonging to H converges strongly to the free time evolution e^{itH_0} for $n \to \infty$ if $d \geq 2$ and if the support of $V_n(x,t)$ in $\mathbb{R}^d$ shrinks to 0 as $n \to \infty$. With other words: There are no repulsive δ-function like potentials for the 2 and 3 dimensional case (It is easy to see that for dimension $d = 1$ both attractive and repulsive δ-function perturbations of H_0 may be defined, cf. [3] p. 168. It is also known that for $d \geq 4$ the Laplacean on the punctured space is already essentially selfadjoint, cf. [3] p. 161).

The strategy of Berezin and Faddeev was generalized and extended by Albeverio et al. [4] to perturbations of exactly solvable H (like e.g. the hydrogen Hamiltonian). Recently, Warzog [5] tried to extend it to perturbations of $H_0 = -\Delta + V(x)$ for a certain class of potentials $V(x)$. Also here it turned out that only attractive contact terms are possible.

An elegant proof of Friedman's theorem by path integral methods was given by Simon [6].

In a paper by Gosdzinsky and Tarrach [7] contact perturbations of the Laplacean were studied and relations to notions of quantum field theory were pointed out. These authors also showed that Friedman's result is not optimal: For suitably normalized sequences $V_n(x) \geq 0$ with shrinking supports, the critical point where $H_0 + \lambda V_n$ converges to H_0 is not $\lambda = 0$ but some negative λ_0 (I will return to this point). Explicitly solvable models with contact interaction are also studied by Perez and Coutinho [8] and by Exner [9].

In the following I should like to show by an elementary heuristic argument that a generalization of Friedman's theorem holds. For convenience I discuss only contact perturbations of $H = -\Delta + V(|\vec{x}|)$ for the 3 dimensional case with spherical symmetry.

2. A heuristic consideration.

Consider a particle in a central symmetric potential $V(\vec{x})$, $\vec{x} \in \mathbb{R}^3$. The radial partial wave Hamiltonians $h_\ell = -\frac{d^2}{dr^2} + \ell(\ell+1)\frac{1}{r^2} + V(r), \ell = 0, 1, 2, \ldots$ on smooth functions on the positive real line $\mathbb{R}^+$ with supports not containing the point $r = 0$, for suitable V are essentially self-adjoint if $\ell \geq 1$, whereas for $\ell = 0$ there are different self-adjoint extensions corresponding to different boundary conditions at $r = 0$ for the functions $u(r)$ in the domain of the extension. It is well known from elementary quantum mechanics, that the usual physical boundary condition is $u(0) = 0$ whereas other boundary conditions correspond to perturbations of H by a perturbation concentrated at the origin. These may be studied by the following limiting procedure.

Consider the radial Schrödinger equation for s-waves

$$-u''(r) + V_R^0(r) \cdot u(r) + V_R(r) \cdot u(r) = \lambda\, u(r), \; r \in \mathbb{R}^+ \tag{1}$$

with

$$V_R(r) = \begin{cases} V(r) & \text{for } r > R \\ 0 & \text{for } r \leq R \end{cases}$$

$$V(z) = \frac{\eta}{z^2} - \frac{q}{z} + v(z)$$

with smooth $v(z)$ regular analytic around $z = 0$, $-\frac{1}{4} < \eta < 2$, and

$$V_R^0(r) = a \cdot \frac{1}{R^\alpha}\, \chi\left(\frac{r}{R}\right), \quad \chi(s) = \begin{cases} 1 & s \leq 1 \\ 0 & s > 1 \end{cases}$$

(a proper δ-sequence for $\alpha = 3$ and $R \downarrow 0$). At $r = R$ $u(r)$ and $u'(r)$ are continuous.

It is convenient to distinguish 3 cases: (a) $\eta = q = 0$, (b) $\eta = 0$, $q \neq 0$, (c) $\eta \neq 0$, $-\frac{1}{4} < \eta < 2$.

It is well known that in case (a)

$$-u'' + V(r)\, u = \lambda u \tag{2}$$

has a unique solution for given $u(r_0),\ u'(r_0),\ r_0 \geq 0$.

In case (b) this is only true for $r_0 > 0$, and around $r = 0$ one has the solution

$$\begin{aligned} w(r) &= A \cdot w_1(r) \ + \ B \cdot w_2(r) \\ w_1(r) &= r - \frac{q}{2} r^2 \ + \ \frac{\frac{q^2}{2} + \alpha_0 - \lambda}{6} r^3 + \cdots \\ w_2(r) &= w_1(r)(\log r - 1) \ + \ \sum_{\nu=0}^{\infty} h_\nu \, r^\nu, \\ & \qquad h_0 \ = \ -\frac{1}{q}\,,\ h_1 \ = \ 0\,,\ h_2 \ = \ \frac{3}{4}\,q \ - \ \frac{\alpha_0 - \lambda}{2q}\,,\cdots \end{aligned}$$

$\left(\text{with } v(r) = \sum_{\nu=0}^{\infty} \alpha_\nu r^\nu \text{ around } r = 0\right)$ or

$$\begin{aligned} w(r) &= -B \cdot \frac{1}{q} \ + \ B\ r(\log r - 1) \ + \ Ar + \cdots \\ w'(r) &= B \log r \ + \ A - Bqr \log r - (Aq + \frac{Bq}{2} - 2Bh_2)r + \cdots \end{aligned}$$

such that w is specified by A, B.

In case (c) one has instead

$$w_{1,2}(r) \ = \ r^{\gamma_{1,2}} \sum_{\nu=0}^{\infty} \beta_\nu^{(1,2)}\, r^\nu,\ \beta_0^{(1,2)} = 1,\ \gamma_{1,2} \ = \ \frac{1}{2} \ \pm \ \sqrt{\frac{1}{4} + \eta}\,.$$

The different selfadjoint extensions of H are fixed by the boundary condition of the admitted functions at the origin: in case (a) by the relation between $u'(0)$ and $u(0)$, in the other cases by A and B.

Consider first for the case (a) the selfadjoint operators

$$h_R \ = \ -\frac{d^2}{dr^2} \ + \ V_R^0(r) \ + \ V_R(r)$$

with the physical boundary condition $u(0) \ = \ 0$ and study the limit $R \downarrow 0$. For $0 \leq r \leq R$ the solutions of (1) are

$$u(r) \ = \ C \sinh\left(\sqrt{a\ R^{2-\alpha} - \lambda R^2}\ \frac{r}{R}\right).$$

Consider

$$\begin{aligned} u(R) \ &= \ C \sinh \sqrt{a\ R^{2-\alpha} - \lambda R^2}\,, \\ u'(R) \ &= \ C \cdot \sqrt{a\ R^{-\alpha} - \lambda}\ \cosh \sqrt{a\ R^{2-\alpha} - \lambda R^2}\,. \end{aligned}$$

$\left(\text{Observe: } u'(R) \ := \ \frac{du(r)}{dr} \mid_{r \uparrow R} \right)$. For $a > 0$ and $R \downarrow 0$ they stay finite only for an R-dependent choice of C, $C = \tilde{C}\left(\sqrt{aR^{-\alpha} - \lambda}\right)^{-1}$, such that in the limit we get (2) only with the boundary condition $u(0) = 0$, $u'(0) \neq 0$.

For $a < 0$ the situation is different. The real solution has for $r \uparrow R$ now the values

$$\begin{aligned} u(R) &= C \sin \sqrt{\mid a \mid R^{2-\alpha} + \lambda R^2}\,, \\ u'(R) &= C\sqrt{\mid a \mid R^{2-\alpha} + \lambda R^2}\,\frac{1}{R} \cos \sqrt{\mid a \mid R^{2-\alpha} + \lambda R^2}. \end{aligned}$$

For $\alpha = 3$ and $R \downarrow 0$ no limit is realizable up to $u(0) = u'(0) = 0$ allowing only the trivial solution $u(r) = 0$ of (2). (It is known from elementary quantum mechanics that the Schrödinger equation

$$-u'' + V_R^0(r)u = \lambda\, u\ ,\ r \geq 0,\ u(0) = 0 \tag{3}$$

now has an increasing number of bound states of increasing binding energy).

For $\alpha = 2$ the situation changes: For $0 > a > -\left(\frac{\pi}{2}\right)^2$ by choosing suitable $C(R)$ the boundary values $u(0) = 0$, $u'(0) \neq 0$ of the unperturbed Hamiltonian are obtained. For $a = -\left(\frac{\pi}{2}\right)^2$ however, $u'(R)$ vanishes for $R \downarrow 0$ for fixed C . (It is known that for $0 > a > -\left(\frac{\pi}{2}\right)^2$ there is no bound state of (3); for $-\left(\frac{\pi}{2}\right)^2 > a$ there is at least one). Consider now for $\alpha = 2$ a slightly changed

$$V_R^0 = -\left(\left(\frac{\pi}{2}\right)^2 \frac{1}{R^2} + \frac{D}{R}\right)\chi\left(\frac{r}{R}\right)$$

(compare [7]). Then

$$u(r) = C \sin \left(\sqrt{\left(\frac{\pi}{2}\right)^2 + DR + \lambda R^2} \cdot \frac{r}{R} \right)\ ,\ r \leq R$$

and

$$\begin{aligned} &u(R) \to C \text{ for } R \downarrow 0 \\ &u'(R) = C\,\frac{1}{R}\,\sqrt{\left(\frac{\pi}{2}\right)^2 + DR + \lambda R^2}\ \cos\ \sqrt{\left(\frac{\pi}{2}\right)^2 + DR + \lambda R^2} \to -\frac{CD}{2}, \end{aligned}$$

i.e., we now get the Schrödinger equation (2) with any boundary condition we want at $r = 0$. This shows that Friedman's theorem has an extension to case (a) and that the critical value of the coupling constant for suitably normalized perturbations is not zero but $-(\frac{\pi}{2})^2$.

Consider next case (b) relevant for the situation in atoms. For $\alpha = 3$ and for $\alpha = 2, a > -(\frac{\pi}{2})^2$ the discussion above may be repeated. Consider $\alpha = 2$ and replace V_R^0 by

$$V_R^0 = -\frac{1}{R^2}\left(\left(\frac{\pi}{2}\right)^2 - 2\,\frac{Rw'(R)}{w(R)}\right) \chi\left(\frac{r}{R}\right),$$

$\left(\frac{R \cdot w'(R)}{w(R)} = \frac{B R \log R + AR}{-B/q} + \cdots \right)$. For $0 < r < R$ the solution of (1) vanishing at the origin is

$$u(r) = C \sin\left(\sqrt{(\frac{\pi}{2})^2 - 2 \frac{Rw'(R)}{w(R)} + \lambda R^2} \frac{r}{R} \right).$$

Hence for small R

$$u(R) \to C$$

$$\frac{du}{dr}(R) \to C \cdot \frac{w'(R)}{w(R)} = B \log R + AR + \cdots .$$

Hence we can smoothly continue to an arbitrary solution of (2) ! This shows that Friedman's theorem has an extension to case (b) and that attractive perturbations are again possible.

Case (c) is not of physical interest (in contrast to the 2 dimensional case when a flux thread is present). Therefore I skip it here.

3. A (non-) application.

Let us consider one of the contact terms mentioned at the beginning, the hyperfine interaction term

$$\frac{2}{3} \vec{\mu} \cdot \vec{\mathcal{M}} \cdot 4\pi \cdot \delta^3(\vec{x}) \tag{4}$$

for the s-channel of a hydrogen-like atom. The δ-function here approximates the charge distribution of the nucleus, i.e. roughly

$$\frac{Z}{\frac{4\pi}{3} R_0^3} \chi(\frac{r}{R_0}), \quad R_0 = \text{nuclear radius} \approx 10^{-13}\text{cm}.$$

Looked upon from an orbiting electron, the difference may seem small and one might get the (wrong) idea that in view of Friedman's result there should be not a level splitting as calculated by perturbation calculus but an unsymmetric effect, a vanishing perturbation for positive values of $\vec{\mu} \cdot \vec{\mathcal{M}}$ and a large perturbation for negative values. Let us therefore look more closely to the numerical values.

Inserting the relevant number for the proton as nucleus (cf. [10]), (4) reads

$$2 \mu_0^2 \frac{1}{1836} 2 \vec{s} \cdot \vec{i} \frac{1}{R_0^3} \chi(\frac{r}{R_0})$$

where $\mu_0 = {}^{e\hbar}/2mc$, the Bohr's magneton, $\vec{i}$ = spin of proton, $\vec{s}$ = spin of electron, $2\vec{i} \cdot \vec{s} = -\frac{3}{2}$, $(+\frac{1}{2})$ for antiparallel (parallel) spin of proton and electron.

The radial Schrödinger equation for s-states then is

$$-u'' + (\hat{a} \chi(\frac{r}{R_0}) - q \frac{1}{R}) u = \lambda u$$

with

$$q = -\frac{2m}{\hbar^2}\, e^2$$
$$\hat{a} = \frac{2m}{\hbar^2}\, 2\,\mu_0^2\, \frac{1}{1836}\, (2\,\vec{s}\cdot\vec{i})\, \frac{1}{{R_0}^3}$$
$$\approx 1.5\cdot 10^{23} cm^{-2}\cdot(2\,\vec{s}\cdot\vec{i})\,,\ \mid \frac{q}{R_0} \mid \approx 4\cdot 10^{21} cm^{-2}$$
$$\lambda = \frac{2m}{\hbar^2}\, E \approx -3.6\cdot 10^{20} cm^{-2} \text{ for } E = E_0 = -13.6\ eV$$

such that $\mid \lambda \mid \ll \hat{a}$. For $a = {R_0}^2\hat{a}$, however, we get $a \approx 1.5\cdot 10^{-3}(2\,\vec{s}\cdot\vec{i})$, tiny compared to $(\frac{\pi}{2})^2$!

Since Whittaker functions are not very convenient to handle, we look at the simpler situation for a particle inside a 3-dimensional well of radius R_1 with infinitely high walls. The radial Schrödinger equation

$$-u'' + \hat{a}\,\frac{1}{R^3}\,\chi(\frac{r}{R}) = \lambda u$$

with $u(0) = u(R_1)$ is solved for $R < r < R_1$ by $\alpha\cdot\sin(\sqrt{\lambda}r + \eta)$, $\eta = 0$ for vanishing perturbation, with lowest eigenvalue $\lambda_0 = (\pi_1/_{R_1})^2$. By elementary calculation the following can be shown:

i) Case $\hat{\hat{a}} > 0$. The phase shift in the limit of small R is

$$\eta = -\sqrt{\lambda}R + \sqrt{\frac{\lambda}{\hat{\hat{a}}}R^{3/2}} + \cdots\,,$$

vanishing for $R \downarrow 0$ according to Friedman's theorem, and the shift of the smallest eigenvalue is

$$\Delta\lambda_0 = \frac{\pi^2}{R_1^2}\left(R - \frac{R^{3/2}}{\sqrt{\hat{\hat{a}}}}\right) + \cdots\,.$$

ii) Case $\hat{\hat{a}} > 0$, R fixed such that $1 \gg \hat{\hat{a}}/R \gg \lambda\ R^2$ (as in case of the hyperfine interaction; $a = R_0^2\hat{a} = \hat{a}/R_0$). Now one gets by elementary calculation

$$\eta = -\frac{1}{3}\,\sqrt{\lambda}\hat{\hat{a}} + \sqrt{\lambda}\cdot R\cdot 0\left[(\frac{\hat{\hat{a}}}{R})^2\right],$$
$$\Delta\lambda_0 = \frac{2\pi^2}{3}\,\frac{\hat{\hat{a}}}{R_1^3} + \cdots\,.$$

The latter coincides with the first order perturbation calculation using $\hat{\hat{a}}\cdot 4\pi\ \delta^3(\vec{x})$ as perturbation.

iii) Case $\hat{\hat{a}} < 0$. The case $R \downarrow 0$ for fixed $\hat{\hat{a}}$ need not be studied because we already know that there is no limit. For the range $0 < \lambda\ R^2 \ll\mid \hat{\hat{\alpha}}/R \mid\ll 1$ however, the result of elementary calculation now is

$$\eta = \frac{1}{3}\,\sqrt{\lambda}\hat{\hat{a}} + \cdots\,,$$

$$\Delta\lambda_0 = -\frac{2}{3}\,\pi^2\,\frac{\hat{\hat{a}}}{R_1^3} + \cdots$$

again coinciding with first order perturbation calculus using $\hat{\hat{a}} \cdot 4\pi\ \delta^3(\vec{x})$ as perturbation.

The conclusion of the present note therefore is twofold: There are generalizations of Friedman's theorem still to be proved. On the other side, one has to be careful with applications.

References

1. F.A. Berezin, L.D. Faddeev, Sov. Math. Dokl. **2** (1961), 372.
2. Ch. Friedman, J. Funct. Anal. **10** (1972), 346.
3. M. Reed, B. Simon, *Methods of Modern Mathematical Physics,* Vol. II, Academic Press (1975).
4. S. Albeverio, F. Gestesy, R. Høegh-Krohn, H. Holden, *Solvable Models in Quantum Mechanics*, Springer, Heidelberg (1988), and references quoted therein.
5. Th. Warzog, Diplomarbeit, Göttingen (1991).
6. B. Simon, *Functional Integration and Quantum Physics*, Academic Press, N.Y. (1979).
7. P. Gosdzinsky, R. Tarrach, Am.J.Phys. **59** (1991), 70.
8. J. Fernando Perez, F.A.B. Coutinho, Am.J.Phys. **59** (1991), 52.
9. P. Exner, to appear, Helv. Phys. Acta.
10. H.A. Bethe, E.E. Salpeter, *Quantum Mechanics of One- and Two-Electron Systems*, Handbuch der Physik XXV, S. Flügge ed., Springer, Berlin (1957).

INSTITUT FÜR THEORETISCHE PHYSIK, UNIVERSITÄT GÖTTINGEN, D-3400 GÖTTINGEN GERMANY

BEYOND COHERENT STATES: HIGHER ORDER REPRESENTATIONS

GEORG REENTS AND STEPHEN J. SUMMERS

ABSTRACT. A class of representations of the canonical commutation relations is presented which naturally subsumes coherent and symplectic (i.e. quasifree) representations. Necessary and sufficient conditions are presented for the unitary inequivalence of these representations with the Fock representation. Hence a new class of representations for bosonic systems with infinitely many degrees of freedom is opened up.

Dedicated to John Klauder on the occasion of his sixtieth birthday

1. Introduction

Since the canonical commutation relations (henceforth, the CCR) were introduced by Dirac in 1927 as a generalization of Heisenberg's commutation relations in order to discuss radiation theory [1], they have been the focus of a great deal of attention. Rigorous mathematical analysis of the CCR began in the 1950's, when it was also realized that in the case of infinitely many degrees of freedom, which is the circumstance of relevance in quantum statistical mechanics and quantum field theory, there are (uncountably) infinitely many unitarily inequivalent irreducible representations of the CCR [2] and that the choice of the proper representation is crucial in any physical application. Many different classes of representations of the CCR have been introduced and studied, and in this note we wish to present yet another class of representations. We shall introduce this new class in a setting that makes clear that it is a natural generalization of two classes that have been intensively studied earlier - the coherent representations and the symplectic representations (also called quasifree representations). We view the primary purpose of this note as pedagogical, hence here we shall not present proofs or the most general form of the results (these shall appear elsewhere [9]).

We first give a brief introduction to these three classes of representations in an elementary formulation and then develop some details in the body of the text. Let $\{q_k, p_k\}_{k=1}^{\infty}$ be a system of densely defined operators acting on the complex Hilbert space $\mathcal{K}_0$ and satisfying

$$[q_j, p_k] = i\delta_{jk} \cdot \mathbf{1}, \tag{1.1}$$

and for the standard annihilation and creation operators $a_k \equiv \frac{1}{\sqrt{2}}(q_k + ip_k)$ and $a_k^\dagger \equiv \frac{1}{\sqrt{2}}(q_k - ip_k)$ let there be a vector $\Omega \in \mathcal{K}_0$ such that $a_k\Omega = 0$ for all $k \in \mathbb{N}$. In other words, we consider the Fock representation of a bosonic system with infinitely many degrees of freedom. This is one, very well-understood representation of the CCR.

Another class of representations is that of the so-called coherent representations [6]. For any sequences $\{\alpha_k\}_{k=1}^\infty, \{\beta_k\}_{k=1}^\infty \subset \mathbb{R}$ we define

$$Q_k \equiv q_k + \alpha_k \cdot \mathbf{1}, \qquad P_k \equiv p_k + \beta_k \cdot \mathbf{1}, \tag{1.2}$$

or equivalently

$$A_k \equiv a_k + \lambda_k \cdot \mathbf{1}, \qquad A_k^\dagger \equiv a_k^\dagger + \overline{\lambda_k} \cdot \mathbf{1}, \tag{1.3}$$

with the complex numbers $\lambda_k \equiv \frac{1}{\sqrt{2}}(\alpha_k + i\beta_k)$. Of course, this new set of operators $\{Q_k, P_k\}_{k=1}^\infty$ also satisfies the CCR. It is known that this second representation is unitarily equivalent to the Fock representation if and only if $\sum_k |\lambda_k|^2 < \infty$; hence the transformation (1.2) will provide a new representation if and only if the sum $\sum_k |\lambda_k|^2$ diverges.

In this same elementary setting, symplectic representations (or quasifree representations) are obtained from the Fock representation by Bogoliubov transformations, which have the form

$$Q_k \equiv q_k, \qquad P_k \equiv p_k + \sum_l \lambda_{kl}\, q_l, \tag{1.4}$$

with λ_{kl} the entries of a certain real, symmetric matrix (see below). Here it is known that the new representation is unitarily equivalent to the Fock representation if and only if $\sum_{k,l} \lambda_{kl}^2 < \infty$, with the obvious resulting condition for the transformation (1.4) to yield something new.

The representations discussed so far are of the form†

$$\begin{aligned} (A) \quad & Q_k = q_k, \qquad P_k = p_k + \beta_k \cdot \mathbf{1}, \\ (B) \quad & Q_k = q_k, \qquad P_k = p_k + \sum_l \lambda_{kl}\, q_l. \end{aligned}$$

Both transformations have the general structure $Q_k = q_k, \quad P_k = p_k + F_k(q_1, q_2, \dots)$, where the $F_k(q_1, q_2, \dots)$ are real functions of the operators $q_1, q_2, \dots$. We shall examine such transformations formally, make some conjectures and then state some recently proven rigorous results. Note that the transformations above, leading to coherent, resp. symplectic representations, are generated by the choice $F_k \equiv \frac{\partial F}{\partial x_k}$, with

†In this elementary paper we simplify the presentation by considering primarily transformations changing only the momentum. Nonetheless, the transformations (1.2) with nonzero α_k are treated in Section 3.

$$(A) \quad F(x_1, x_2, \dots) = \sum_k \beta_k x_k,$$

$$(B) \quad F(x_1, x_2, \dots) = \frac{1}{2} \sum_{k,l} \lambda_{kl} : x_k x_l :\equiv \frac{1}{2} \sum_{k,l} \lambda_{kl}(x_k x_l - \frac{1}{2}\delta_{kl}).$$

Since the obvious next step would be to choose $F = \frac{1}{3} \sum_{k,l,m} \lambda_{klm} : x_k x_l x_m :$, which would give rise to the transformation

$$Q_k = q_k, \qquad P_k = p_k + \sum_{l,m} \lambda_{klm} : q_l q_m :,$$

we find it natural to call these new representations higher-order representations of the CCR. It has already been well-established that coherent and symplectic representations have many uses; it can be hoped that also these higher-order representations will find application. A natural possibility for such an application are the multi-wave mixers discussed by Klauder and co-workers in [17].

In this paper we shall always assume our linear spaces to be separable and infinite dimensional.

2. Notation

There are many different formulations of the CCR in the literature, each with its own advantages and disadvantages. We shall use a number of these; hence, in order to fix notation and to facilitate the passage between these formulations, we present a brief review. For us the basic starting point is a real (nondegenerate) symplectic space (H, σ) with an associated (regular) Weyl system $(\mathcal{K}, W(f))$ consisting of a complex Hilbert space $\mathcal{K}$ and a mapping $W : H \to \mathcal{U}(\mathcal{K})$ from H into the group $\mathcal{U}(\mathcal{K})$ of unitary operators on $\mathcal{K}$ that satisfies the following axioms [13][7]:

$$W(f)W(g) = e^{-i\sigma(f,g)/2} W(f+g), \quad \forall f, g \in H, \tag{2.1}$$

$$W(f)^* = W(-f), \quad \forall f \in H, \tag{2.2}$$

and

$$t \in \mathbb{R} \to W(tf) \in \mathcal{B}(\mathcal{K}) \text{ is weakly continuous for all } f \in H. \tag{2.3}$$

In this introductory paper (H, σ) will always arise from a complex Hilbert space $\mathcal{H}$ with its inner product $\langle \cdot, \cdot \rangle$, in which case H will be $\mathcal{H}$ understood as a real Hilbert space and $\sigma(f, g) = \mathrm{Im}\langle f, g \rangle$. Condition (2.3) entails that the map $t \in \mathbb{R} \to W(tf) \in \mathcal{B}(\mathcal{K})$ is actually strongly continuous, hence by Stone's Theorem one knows that for each $f \in H$ there exists a selfadjoint operator $\Phi(f)$ on $\mathcal{K}$ such that $W(tf) = e^{it\Phi(f)}$ for all $t \in \mathbb{R}$ and by (2.1) the map $f \to \Phi(f)$ is real linear. In fact, there exists a dense domain of vectors $D_W \subset \mathcal{K}$ that is a core for and left invariant by every $\Phi(f)$ [10][5]; it is on this domain that the linearity just mentioned can be verified and that all operator equations in this paper hold. On this domain one also verifies that the generators satisfy the CCR:

$$\Phi(f)\Phi(g) - \Phi(g)\Phi(f) = i\sigma(f,g) \cdot \mathbf{1}, \quad \forall f,g \in H. \tag{2.4}$$

We therefore call the Weyl system $(\mathcal{K}, W(f))$ and its associated generators as above a regular representation of the CCR over (H,σ).

We shall denote by $\mathcal{A}(H,\sigma)$ the C^*-algebra on $\mathcal{K}$ generated by the operators $\{W(f) \mid f \in H\}$. As the notation indicates, the algebra $\mathcal{A}(H,\sigma)$ does not depend on the choice of representation of the Weyl operators $\{W(f) \mid f \in H\}$ [16]. $\mathcal{A}(H,\sigma)$ is a simple, nonseparable C^*-algebra [16][7].

For any real linear map $j : H \to H$ satisfying $\sigma(jf,g) = -\sigma(f,jg)$, $-\sigma(jf,f) \geq 0$ and $j^2 = -1$ one can introduce a complex structure on H as follows: $(\alpha + i\beta)f \equiv \alpha f + \beta jf$, for all $f \in H$, and $\alpha, \beta \in \mathbb{R}$. Moreover, $\langle f,g\rangle \equiv -\sigma(jf,g) + i\sigma(f,g)$ defines a scalar product on H such that $(H, \langle\cdot,\cdot\rangle)$ is a complex preHilbert space. With a choice of such a 'σ-admissible' complex structure j on (H,σ), there is an important representation of $\mathcal{A}(H,\sigma)$ called the Fock representation. This is given as the GNS-representation space $\{\mathcal{K}, \pi, \Omega\}$ associated with the pure state [8] ω_j determined on $\mathcal{A}(H,\sigma)$ by

$$\omega_j(W(f)) \equiv e^{\sigma(jf,f)/4}. \tag{2.5}$$

Given such a representation, one can define the following 'annihilation' and 'creation' operators:

$$a(f) \equiv \frac{1}{\sqrt{2}}(\Phi(f) + i\Phi(jf)), \qquad a^\dagger(f) \equiv \frac{1}{\sqrt{2}}(\Phi(f) - i\Phi(jf)), \tag{2.6}$$

where $\pi(W(tf)) = e^{it\Phi(f)}$. One has then $a(f)\Omega = 0$ for all $f \in H$. In the notation for the harmonic oscillator, we define

$$Q(f) \equiv \frac{1}{\sqrt{2}}(a(f) + a^\dagger(f)) = \Phi(f) \quad \text{and} \quad P(f) \equiv \frac{1}{i\sqrt{2}}(a(f) - a^\dagger(f)) = \Phi(jf).$$

These relations might clarify for the reader the physical significance of the choice of complex structure.

For pedagogical reasons in the introduction but also for our convenience, determined by the method of proof we utilize, we often make calculations in the so-called basis-dependent approach: one effectively introduces a basis $\{f_k\}_{k\in\mathbb{N}} \subset \mathcal{H}$ and with $a_k \equiv a(f_k)$ and $a_k^\dagger \equiv a^\dagger(f_k)$ defines the following operators in any regular representation of the CCR:

$$q_k \equiv \frac{1}{\sqrt{2}}(a_k + a_k^\dagger), \qquad p_k \equiv \frac{1}{i\sqrt{2}}(a_k - a_k^\dagger).$$

These operators satisfy the CCR $[q_k, p_l] = i\delta_{kl} \cdot \mathbf{1}$ on the invariant domain D_W mentioned above. As shown by Hegerfeldt [4], every representation of the basis-independent approach in a separable Hilbert space can be obtained from a suitable basis-dependent representation by extending the latter in a natural metric introduced by Hegerfeldt and Klauder in [3] (inducing a topology on H which is the weakest vector topology for which $f \to W(f)$ is strongly continuous) and identifying isomorphic test function spaces. This reduces the problem of classifying all representations of the CCR for all possible (separable) test function spaces to the study of basis-dependent representations.

3. Coherent and Symplectic Representations

Coherent representations can be obtained most readily in the following manner. If $\{\mathcal{K}, W(f)\}$ is a representation of $\mathcal{A}(H,\sigma)$ and $\Gamma : H \to \mathbb{R}$ is a linear map, then

$$\hat{W}(f) \equiv e^{i\Gamma(f)} W(f), \quad f \in H, \tag{3.1}$$

determines a representation of $\mathcal{A}(H,\sigma)$ generally called a coherent representation if $\{\mathcal{K}, W(f)\}$ is a Fock representation [11]. This leads to the following relation between the generators of the representations: $\hat{\Phi}(f) = \Phi(f) + \Gamma(f) \cdot \mathbf{1}$, $f \in H$. Given a σ-admissible complex structure j on (H,σ), one then has the operators $a(f) = \frac{1}{\sqrt{2}}(\Phi(f) + i\Phi(jf))$ (and $a^\dagger(f) = \frac{1}{\sqrt{2}}(\Phi(f) - i\Phi(jf))$). And introducing a basis $\{f_k\}_{k\in\mathbb{N}}$ in the complexification (see Section 2) $\mathcal{H} = (H, \langle\cdot,\cdot\rangle)$ one has

$$\hat{q}_k = q_k + \alpha_k \cdot \mathbf{1} \quad , \quad \hat{p}_k = p_k + \beta_k \cdot \mathbf{1},$$

where $\alpha_k = \Gamma(f_k)$, $\beta_k = \Gamma(jf_k)$.

One may equivalently start with a Fock state ω_0 on $\mathcal{A}(H,\sigma)$ with associated representation $\{\mathcal{K}_0, W_0(f)\}$ and define a coherent state ω_Γ by

$$\omega_\Gamma(f) \equiv \omega_0(f) e^{i\Gamma(f)}, \quad f \in H.$$

Of course, the GNS representation of $\mathcal{A}(H,\sigma)$ corresponding to ω_Γ is given on $\mathcal{K}_0$ by

$$W_\Gamma(f) \equiv e^{i\Gamma(f)} W_0(f), \quad f \in H. \tag{3.2}$$

Note that with such a σ-admissible complex structure j on (H,σ) and with $\mathcal{H} = (H, \langle\cdot,\cdot\rangle)$ the associated complex Hilbert space, for every real linear map $\Gamma : H \to \mathbb{R}$, there is a uniqe complex linear functional $\Lambda = \Lambda_\Gamma$ on $\mathcal{H}$, namely $\Lambda(f) \equiv \Gamma(f) - i\Gamma(jf)$, such that $\Gamma(f) = \mathrm{Re}(\Lambda(f))$ for all $f \in H$. Conversely, an arbitrary linear functional Λ on $\mathcal{H}$ determines via $\Gamma(f) = \mathrm{Re}(\Lambda(f))$ a real linear map $\Gamma = \Gamma_\Lambda$ on H.

We recast the transformations (1.2) and (1.3) in this more general setting. The sequence $\{\lambda_k\}_{k=1}^\infty = \{\alpha_k + i\beta_k\}_{k=1}^\infty$ defines a linear functional on $\ell_2(\mathbb{C})$: in fact, for $\hat{c} \equiv (c_1, c_2, \dots) \in \ell_2$, let $\tilde{\Lambda}(\hat{c}) = \sum_k \overline{\lambda}_k c_k$. On the other hand, a given functional $\tilde{\Lambda} : \ell_2(\mathbb{C}) \to \mathbb{C}$ with a basis $\{\hat{f}_k\}_{k=1}^\infty$ in its domain determines a sequence $\{\lambda_k\}_{k=1}^\infty \subset \mathbb{C}$ by $\overline{\lambda}_k = \tilde{\Lambda}(\hat{f}_k)$.

More generally, if Λ is a linear functional on $\mathcal{H}$ we can choose a basis $\{f_k\}_{k=1}^{\infty}$ in the domain of definition of Λ and map $\mathcal{H}$ isometrically onto $\ell_2(\mathbb{C})$ by $Sf = S\sum_k c_k f_k \equiv \hat{c} \equiv (c_1, c_2, \dots)$. Then defining $\tilde{\Lambda}$ by $\tilde{\Lambda}(\hat{c}) = \Lambda(S^{-1}\hat{c})$, we see that $\{\hat{f}_k\}_{k=1}^{\infty} \equiv \{Sf_k\}_{k=1}^{\infty}$ is a basis contained in the domain of definition of $\tilde{\Lambda}$. Henceforth, wherever confusion will not arise, we shall drop the tildes.

We therefore see that the coherent representations of the CCR algebra $\mathcal{A}(\mathcal{H})$ over a complex Hilbert space $\mathcal{H}$ are in one-to-one correspondence with linear functionals Λ on $\mathcal{H}$. Moreover, we have the following result.

Theorem 3.1. *Let $\{q_k, p_k\}_{k=1}^{\infty}$ be a (regular) representation of the CCR in the sense of Section 2. Then the representation $\{Q_k, P_k\}_{k=1}^{\infty}$ defined by equation (3.1) (or (3.2)) is unitarily equivalent to $\{q_k, p_k\}_{k=1}^{\infty}$ if and only if the corresponding linear functional Λ as above is bounded.*

Proof. Based on the work of Gårding and Wightman [2] and Segal [12], Shelupsky [14] showed that for transformations of the form (1.3), the resulting representation is unitarily equivalent to the original Fock representation if and only if $\sum\limits_k |\lambda_k|^2 < \infty$. In the discussion above one sees that $\Lambda : \ell_2 \to \mathbb{C}$ (equivalently, $\tilde{\Lambda} : \mathcal{H} \to \mathbb{C}$) is bounded if and only if $\sum\limits_k |\lambda_k|^2 < \infty$. □

Hence an unbounded map Λ will generate a new representation.

Symplectic representations [12][13] are generally obtained from a given (Fock) representation of the CCR-algebra $\mathcal{A}(H, \sigma)$ by a Bogoliubov transformation

$$\Phi_T(f) \equiv \Phi(Tf) \qquad (\text{or equivalently } W_T(f) \equiv W(Tf)) \tag{3.3}$$

using a symplectic operator T, i.e. one leaving the symplectic form invariant: $\sigma(Tf, Tg) = \sigma(f, g)$, all $f, g \in H$. Every bounded surjective symplectic operator T on a complex Hilbert space $\mathcal{H}$ (so $\mathrm{Im}\langle Tf, Tg\rangle = \mathrm{Im}\langle f, g\rangle$ for all $f, g \in \mathcal{H}$) is of the form

$$T = U(\mathbf{1} - \frac{i}{2}\Lambda(\mathbf{1} - C)),$$

where U is unitary, Λ bounded and selfadjoint, and C is a conjugation which commutes with Λ. Since the unitary operator plays no role in determining the unitary equivalence of the representation defined by T, there is in every such equivalence class a symplectic operator T of the form $T = \mathbf{1} - \frac{i}{2}\Lambda(\mathbf{1} - C)$. With a suitable basis $\{f_k\}_{k \in \mathbb{N}}$, such that $Cf_k = f_k$ and $Cif_k = -if_k$, one has

$$\begin{aligned} Q_k &\equiv \Phi(Tf_k) = \Phi(f_k) = q_k, \\ P_k &\equiv \Phi(Tif_k) = p_k + \Phi(\Lambda f_k) = p_k + \sum_j \lambda_{kj}\, q_j, \end{aligned} \tag{3.4}$$

where the numbers $\lambda_{kj} \equiv \langle f_k, \Lambda f_j\rangle$ are the entries of a real and symmetric matrix. In this formulation we recall the following necessary and sufficient conditions for the unitary implementability of this canonical transformation.

Theorem 3.2. *Let $\{q_k, p_k\}_{k=1}^{\infty}$ be a (regular) representation of the CCR in the sense of Section 2. Then the representation $\{Q_k, P_k\}_{k=1}^{\infty}$ defined by equation (3.3) (or (3.4)) is unitarily equivalent to $\{q_k, p_k\}_{k=1}^{\infty}$ if and only if the corresponding selfadjoint operator Λ as above is Hilbert-Schmidt, i.e. $\sum_{k,l} \lambda_{kl}^2 < \infty$.*

Proof. This follows essentially from Shale's characterization [13] of symplectic representations that are unitarily equivalent to the original Fock representation. For details, see [9]. □

In the next section we discuss higher order representations of the CCR.

4. Higher Order Representations

As indicated in the introduction, we are interested in studying canonical transformations of the form

$$Q_k = q_k, \qquad P_k = p_k + F_k(q_1, q_2, \ldots), \tag{4.1}$$

where the $F_k(q_1, q_2, \ldots)$ are real functions of the operators $q_1, q_2, \ldots$. We find it very convenient to use the realization of the Fock space representation of the CCR introduced by Segal (see [15] for an introduction) that is called the Q-space representation. In this representation the operators $F_k(q_1, q_2, \ldots)$ act as multiplication by the functions $F_k(x_1, x_2, \ldots)$.[1] It is then easily seen that in order for the operators P_k to mutually commute these functions must satisfy

$$\frac{\partial F_k}{\partial x_j}(x_1, \ldots) = \frac{\partial F_j}{\partial x_k}(x_1, \ldots), \qquad \text{for all} \qquad j, k \in \mathbb{N}. \tag{4.2}$$

Hence we require the existence of a suitably smooth function $F(x_1, x_2, \ldots)$ such that $F_k = \partial F / \partial x_k$. One must also keep a watchful eye on domain problems, but in this paper we shall be limiting ourselves to quadratic representations, which are generated by the function $F = \frac{1}{3} \sum_{k,l,m} \lambda_{klm} : x_k x_l x_m :$, giving rise to the transformation

$$Q_k = q_k, \qquad P_k = p_k + \sum_{l,m} \lambda_{klm} : q_l q_m : . \tag{4.3}$$

For such transformations (indeed for any involving only Wick ordered products of fields) the set $\mathcal{F}_0$ of finite particle vectors provides a common, dense, invariant core for these operators. The condition (4.2) requires here that λ_{klm} be symmetric with respect to permutations of the indices k, l, m.

We have found it convenient to express all transformations in terms of the (Segal) field $\Phi(f)$, instead of Q and P. We begin with a complex Hilbert space $\mathcal{H}$ with $H = \text{Re}\mathcal{H} \oplus \text{Re}\mathcal{H}$, where $\text{Re}\mathcal{H} = \{f \in \mathcal{H} \mid Cf = f\}$ for some specified conjugation C. With j a specified σ-admissible complex structure on (H, σ) and $\{\tilde{f}_k\}_{k \in \mathbb{N}}$ an

[1] Indeed it is in the Q-space representation where mathematical sense is made of the phrase 'the $F_k(q_1, q_2, \ldots)$ are real functions of the operators $q_1, q_2, \ldots$'.

orthonormal basis in $\mathcal{H}$, then the vectors $e_k \equiv \frac{1}{2}(\tilde{f}_k + C\tilde{f}_k) \oplus \frac{1}{2i}(\tilde{f}_k - C\tilde{f}_k)$ and $e'_k \equiv je_k$ in H form a symplectic orthonormal basis $\{e_k, e'_k\}_{k\in\mathbb{N}}$ in (H,σ). Then $q_k = \Phi(e_k)$ and $p_k = \Phi(e'_k)$, and one sees there is no loss of generality.

For $F \in H \otimes H$, $F = \sum_{i=1}^{n} \alpha_i(f_i \otimes g_i)$, $\alpha_i \in \mathbb{R}$, $f_i, g_i \in H$ we define the operator $\Psi(F)$ by

$$\Psi(F) \equiv \sum_{i=1}^{n} \alpha_i : \Phi(f_i)\Phi(g_i) : .$$

(It can be shown that this operator is independent of the particular linear combination chosen to express F.) One sees that since the map $H \otimes H \ni F \to \Psi(F)$ is strongly continuous on $\mathcal{F}_0$, it can be extended continuously to all of $H \otimes H$. In addition, for every $F \in H \otimes H$ the operator $\Psi(F)$ is essentially selfadjoint on $\mathcal{F}_0$.

For any linear, densely defined map $\Lambda : H \to H \otimes H$ we use the above quadratic expression to define a transformation on the field $\Phi(\cdot)$ as follows: $\Phi(f) \to \Phi_\Lambda(f) = \Phi(f) + \Psi(\Lambda f)$. In fact, [9] $\Phi_\Lambda(f)$ is essentially selfadjoint on $\mathcal{F}_0$ for every f in the domain of Λ. In order for this transformation to be canonical, i.e. in order for the CCR $[\Phi_\Lambda(f), \Phi_\Lambda(g)] = i\sigma(f,g)\cdot \mathbf{1}$ to hold, one verifies that the following conditions must obtain.

Proposition 4.1 [9]. *For $f \in H$ and $F, G \in H \otimes H$ let $\sigma(F,G) \in \mathbb{C}$, $b(f,F) \in H$ and $B(F,G) \in H \otimes H$ be defined implicitly by*

$$[\Phi(f), \Psi(F)] = 2i\Phi(b(f,F)), \quad \text{and} \quad [\Psi(F), \Psi(G)] = 2i\{2\Psi(B(F,G)) + \sigma(F,G)\cdot \mathbf{1}\}$$

The transformation $\Phi(f) \to \Phi_\Lambda(f) = \Phi(f) + \Psi(\Lambda f)$ is canonical if and only if $b(f, \Lambda g) = b(g, \Lambda f)$ and $B(\Lambda f, \Lambda g) = 0$ for all f, g in the domain of Λ (which, in fact, entails that $[\Phi(f), \Psi(\Lambda g)] = [\Phi(g), \Psi(\Lambda f)]$ and $[\Psi(\Lambda f), \Psi(\Lambda g)] = 0$).

Proof. This is a simple calculation; see [9] for details. □

These conditions are not immediately transparent, but it has been shown [9] that for any such transformation there exists a symplectic orthonormal basis $\{e_k, e'_k\}_{k\in\mathbb{N}}$ in (H,σ) and real numbers λ_{ijk}, $i,j,k = 1,2,\ldots$, which are totally symmetric in the indices and fulfill the condition $\sum_{i,j} \lambda_{ijk}^2 < \infty$ for every k, such that

$$\Lambda f = \sum_{i,j}\sum_{k} \lambda_{ijk}\, \sigma(e_k, f) e_i \otimes e_j, \tag{4.4}$$

for all f in the domain of Λ.

Necessary and sufficient conditions for such a canonical transformation to be unitarily implementable from the Fock representation have also been established.

Theorem 4.2 [9]. *The canonical transformation $\Phi(f) \to \Phi_\Lambda(f) = \Phi(f) + \Psi(\Lambda f)$ is unitarily implementable in the Fock representation if and only if the map $\Lambda : H \to H \otimes H$ is Hilbert-Schmidt; in terms of the characterization (4.4), this means $\sum_{i,j,k} \lambda_{ijk}^2 < \infty$.*

Proof. The characterization (4.4) is used crucially in this argument [9]. □

Hence, for any map $\Lambda : H \to H \otimes H$ of the form (4.4) which is not Hilbert-Schmidt, the canonical transformation $\Phi(f) \to \Phi_\Lambda(f) = \Phi(f) + \Psi(\Lambda f)$ yields a new representation of the CCR. We believe that this representation can be exponentiated to yield a new representation of the C^*-algebra $\mathcal{A}(H, \sigma)$, but the proof is not yet completed.

We have presented in this section the case of quadratic transformations of representations of the CCR, but clearly the same questions can be posed for yet higher order transformations. And similar results can be expected. These will be presented in a forthcoming manuscript [9]. Another direction for generalization which will be left for the future is the following. Throughout this paper the reference representation has been a Fock representation. However, if $\{Q_k, P_k, \mathcal{K}\}$ is any (regular) representation of the CCR, the transformation (4.3), and more generally any tranformation of the form (4.1) satisfying (4.2), generates another representation of the CCR. It would be interesting to know under what set of circumstances this transformation would result in a representation that was not (quasi-) equivalent to the original representation. To make this managable, one would restrict the class of the original representation - for example, to the class of quasifree representations.

Acknowledgement: One of us (G.R.) wishes to thank John Klauder and the University of Florida for their hospitality and the Max Kade Foundation for a grant that enabled him to spend the academic year 1989-1990 in Gainesville, where this collaboration was begun. Both of us also thank Martin Proksch for allowing us to report on some results obtained with his participation.

REFERENCES

1. P.A.M. Dirac, *The quantum theory of the emission and absorption of radiation*, Proc. Roy. Soc. London **114** (1927), 243-265.
2. L. Gårding and A.S. Wightman, *Representations of the commutation relations*, Proc. Nat. Acad. Sci. **40** (1954), 622-626.
3. G.C. Hegerfeldt and J.R. Klauder, *Metrics on test function spaces for canonical field operators*, Commun. Math. Phys. **16** (1970), 329-346.
4. G.C. Hegerfeldt, *Equivalence of basis-dependent and basis-independent approach to canonical field operators*, J. Math. Phys. **12** (1971), 167-172.
5. G.C. Hegerfeldt, *Gårding domains and analytic vectors for quantum fields*, J. Math. Phys. **13** (1972), 821-827.
6. J.R. Klauder and B.-S. Skagerstam, *Coherent States: Applications in Physics and Mathematical Physics*, World Scientific, 1985.
7. J. Manuceau, *C^*-algèbre de relations de commutation*, Ann. Inst. Henri Poincaré **8** (1968), 139-161.
8. J. Manuceau and A. Verbeure, *Quasi-free states of the CCR-algebra and Bogoliubov transformations*, Commun. Math. Phys. **9** (1968), 293-302.
9. M. Proksch, G. Reents and S.J. Summers, *A New class of representations of the canonical commutation relations* (to appear).

10. M. Reed, *A Gårding domain for quantum fields*, Commun. Math. Phys. **14** (1969), 336-346.
11. G. Roepstorff, *Coherent photon states and spectral condition*, Commun. Math. Phys. **19** (1970), 301–314.
12. I.E. Segal, *Foundations of the theory of dynamical systems of infinitely many degrees of freedom, I*, Mat. Fys. Medd. Danske Vid. Selsk. **31** (1959), # 12.
13. D. Shale, *Linear symmetries of the free boson field*, Trans. A.M.S. **103** (1962), 149-167.
14. D. Shelupsky, *Translations of the discrete Bose-Einstein operators*, J. Math. Phys. **7** (1966), 163-166.
15. B. Simon, *Functional Integration and Quantum Physics*, Academic Press, 1979.
16. J. Slawny, *On factor representations and the C^*-algebra of canonical commutation relations*, Commun. Math. Phys. **24** (1971), 151-170.
17. B. Yurke, S.L. McCall and J.R. Klauder, *SU(2) and SU(1,1) interferometers*, Phys. Rev. A **33** (1986), 4033-4054.

PHYSIKALISCHES INSTITUT, UNIVERSITÄT WÜRZBURG, D-8700 WÜRZBURG, GERMANY AND DEPARTMENT OF MATHEMATICS, UNIVERSITY OF FLORIDA, GAINESVILLE, FL 32611, USA

YET ANOTHER PROOF OF BLATTNER'S THEOREM

P. L. Robinson

Abstract. The inner Bogoliubov automorphisms of the von Neumann Clifford algebra were characterized by Blattner; we offer a new proof of his result, elaborating upon an idea of Carey.

For John Klauder on his sixtieth birthday

1. Introduction

In his construction of outer automorphic group representations, Blattner [4] completely characterized the inner Bogoliubov automorphisms of the von Neumann algebra generated by the trace representation of the complex Clifford algebra associated to any real Hilbert space. A new proof of this characterization was presented by de la Harpe and Plymen [7] alongside a new proof of the Shale-Stinespring [10] characterization of all inner Bogoliubov automorphisms of the C* Clifford algebra itself.

Our aim here is to offer yet another proof of Blattner's theorem on inner Bogoliubov automorphisms of the von Neumann Clifford algebra, after reviewing relevant material on Clifford algebras and their trace and Fock representations. In principle, our proof uses a variant of the standard doubling procedure supplemented by information on half-spin representations in Fock space; it originates in ideas of Carey [6] and takes advantage of work by Araki [2].

2. Background material

Let V be an infinite dimensional real inner product space with inner product denoted by $(\cdot|\cdot)$ and corresponding norm denoted by $\|\cdot\|$. For our purposes, we may with impunity suppose V to be complete but need not assume V to be separable. We begin by recalling briefly elementary facts pertaining to the complex Clifford algebra of V and its C* algebra completion; further detail on these matters may be found in [2][5][9].

The complex Clifford algebra $C(V)$ is a unital associative complex algebra with involution containing V as a real subspace of self-adjoint elements and having the following universal property: if B is a unital associative complex algebra with involution and $f : V \to B$ is a real-linear map into the self-adjoint part of B such that $f(v)^2 = \|v\|^2\,\mathbf{1}$ whenever

$v \in V$, then there exists a unique involution-preserving algebra map $F : C(V) \to B$ such that $F|V = f$.

As a routine consequence of this universal property, it follows that each orthogonal transformation $g \in O(V)$ of V extends to an automorphism $\theta(g)$ of $C(V)$; we shall refer to $\theta(g)$ as a Bogoliubov automorphism. In particular, the Bogoliubov automorphism $\theta(-I)$ will be denoted by γ and referred to as the grading automorphism of $C(V)$: its fixed algebra is the even Clifford algebra $C_+(V)$ whilst the complementary subspace on which γ acts as minus the identity is denoted by $C_-(V)$.

Every nonzero involution-preserving algebra map from $C(V)$ into a C* algebra induces the same norm on $C(V)$. The completion of $C(V)$ in this unique norm is a simple C* algebra which we call the C* Clifford algebra $C[V]$ of V. This C* algebra has the following universal property: if B is a unital complex C* algebra and $f : V \to B$ is a real-linear map into the self-adjoint part of B such that $f(v)^2 = \|v\|^2\,\mathbf{1}$ whenever $v \in V$, then there exists a unique C* algebra map $F : C[V] \to B$ such that $F|V = f$.

Of course, this universal property has the consequence that each $g \in O(V)$ also induces a Bogoliubov automorphism $\theta(g)$ of the C* Clifford algebra $C[V]$. Again, we refer to $\gamma = \theta(-I)$ as the grading automorphism of $C[V]$ and note that it engenders a decomposition $C[V] = C_+[V] \oplus C_-[V]$ in which γ acts as $+I$ on the even C* Clifford algebra $C_+[V]$ and as $-I$ on the complementary subspace $C_-[V]$.

Now, the C* Clifford algebra $C[V]$ has a unique tracial state τ: the corresponding cyclic representation furnished by the GNS construction is called the trace representation; the von Neumann algebra that it generates is a factor which we call the vN Clifford algebra $\mathcal{A}[V]$ of V. Uniqueness of the tracial state ensures that each $g \in O(V)$ gives rise to a Bogoliubov automorphism $\theta(g)$ of the vN Clifford algebra $\mathcal{A}[V]$. In these terms, Blattner's theorem provides necessary and sufficient conditions on g in order that $\theta(g)$ be inner as an automorphism of $\mathcal{A}[V]$. It is convenient to record certain facts regarding the trace representation; for amplification we refer to [4][7][9].

Write H_τ for the complex vector space $C(V)$ endowed with the inner product given by $(\phi|\psi) = \tau(\psi^*\phi)$ for all $\phi, \psi \in C(V)$ and write $\mathbb{H}_\tau$ for its Hilbert space completion. Note that the operation of taking adjoints on $C(V) = H_\tau$ extends to define an anti-involution Σ on the Hilbert space $\mathbb{H}_\tau$. The left regular representation λ of $C(V)$ on H_τ given by

$$a \in C(V),\ \phi \in H_\tau \implies \lambda(a)\phi = a\phi$$

extends to define a representation

$$\lambda : C[V] \longrightarrow B(\mathbb{H}_\tau)$$

where $B(\mathbb{H}_\tau)$ signifies the C* algebra of all bounded linear operators on $\mathbb{H}_\tau$. In like manner, the right regular antirepresentation ρ of $C(V)$ on H_τ given by

$$a \in C(V),\ \phi \in H_\tau \implies \rho(a)\phi = \phi a$$

extends to an antirepresentation

$$\rho : C[V] \longrightarrow B(\mathbb{H}_\tau).$$

Note that $\Omega := \mathbf{1} \in \mathbb{H}_\tau$ is a cyclic unit vector for both λ and ρ.

Let us denote by $\mathcal{A}_\lambda$ the von Neumann algebra $\lambda[C(V)]''$ generated by λ in $B(\mathbb{H}_\tau)$; likewise, let us put $\mathcal{A}_\rho = \rho[C(V)]''$. Now, if $a \in C(V)$ then $\lambda(a)$ and $\rho(a^*)$ are intertwined by Σ; thus $\Sigma \cdot \mathcal{A}_\rho \cdot \Sigma = \mathcal{A}_\lambda$. Also, multiplications on left and right commute; thus $\mathcal{A}_\lambda$ is contained in the commutant $\mathcal{A}'_\rho$ of $\mathcal{A}_\rho$. Lastly, if $A \in \mathcal{A}'_\rho$ then $\Sigma(A \cdot \Omega) = A^* \cdot \Omega$, so the usual arguments involving the double commutant theorem yield $\Sigma \cdot \mathcal{A}'_\rho \cdot \Sigma \subset \mathcal{A}_\rho$ and therefore $\mathcal{A}'_\rho \subset \Sigma \cdot \mathcal{A}_\rho \cdot \Sigma$. In summary, we have the following result.

Theorem 1. $\mathcal{A}'_\rho = \Sigma \cdot \mathcal{A}_\rho \cdot \Sigma = \mathcal{A}_\lambda$.

□

Of course, this theorem embodies a specific instance of the Tomita-Takesaki modular theory in which Σ is the corresponding modular conjugation operator. Needless to say, the theorem remains valid upon interchange of λ and ρ.

The von Neumann algebra $\mathcal{A}_\lambda \subset B(\mathbb{H}_\tau)$ generated by the trace representation λ is what we have called the vN Clifford algebra $\mathcal{A}[V]$ of V. The grading automorphism γ extends to $\mathcal{A}[V]$ and as usual induces a decomposition $\mathcal{A}[V] = \mathcal{A}_+[V] \oplus \mathcal{A}_-[V]$ with $\mathcal{A}_+[V]$ the even vN Clifford algebra. In addition, the automorphism γ of $C(V) = H_\tau$ extends to a symmetry Γ on $\mathbb{H}_\tau$ which implements γ in the trace representation, so that

$$a \in C(V) \implies \lambda(\gamma(a)) = \Gamma\lambda(a)\Gamma$$

and indeed

$$A \in \mathcal{A}[V] \implies \gamma(A) = \Gamma A \Gamma.$$

In our study of inner Bogoliubov automorphisms of the vN Clifford algebra, we shall require information on Fock representations. We shall be content here to present a brief summary of the essentials relevant to our needs; for extended accounts of these matters, see [2][3][5].

A Fock representation of $C[V]$ is determined by the choice of a unitary structure on V: that is, a complex structure J with the property that the prescription

$$x, y \in V \implies < x|y >= (x|y) + i(x|Jy)$$

defines a complex inner product on V. The Fock state σ_J is now the unique state of $C[V]$ with the property that if $v \in V$ and $z = v + iJv$ then $\sigma_J(z^*z) = 0$. The corresponding cyclic representation of $C[V]$ is the Fock representation π_J on Fock space $\mathbb{H}_J$ with cyclic unit vector Ω_J. We remark that the property singling out the Fock state σ_J corresponds to the following Fock vacuum property of the cyclic vector Ω_J under π_J: that if $v \in V$ then $\pi_J(v + iJv)\Omega_J = 0$.

As a consequence of the vacuum property, the Fock state σ_J is pure and the Fock representation π_J is irreducible. Moreover, the grading automorphism γ of $C[V]$ is implemented in π_J by a symmetry Γ_J on $\mathbb{H}_J$ so that

$$a \in C(V) \Longrightarrow \pi_J(\gamma(a)) = \Gamma_J \pi_J(a) \Gamma_J .$$

Conjugation by Γ_J now defines an extension of γ to all of $B(\mathbb{H}_J) = \pi_J[C(V)]''$. Accordingly, we are entitled to say that $A \in B(\mathbb{H}_J)$ is even iff $\Gamma_J A \Gamma_J = A$ and odd iff $\Gamma_J A \Gamma_J = -A$.

We should perhaps remark that the Fock representation π_J of $C[V]$ is usually constructed explicitly in terms of creators and annihilators on the completed space of alternating tensors over V made complex Hilbert via J; in this context, Γ_J is taken to be the operator acting as $+I$ on the even tensors and as $-I$ on the odd tensors. However, for our purposes it will be enough to recognize π_J by means of its vacuum property.

It is natural to ask when the Bogoliubov automorphism $\theta(g)$ induced by the orthogonal transformation $g \in O(V)$ is implemented in the Fock representation π_J of $C[V]$. The definitive answer to this question was provided by Shale and Stinespring [10] with supplementary considerations of parity decided by Araki [2]; it may be stated as follows.

Theorem 2. *$\theta(g)$ is implemented in π_J iff the commutator $[J, g]$ is Hilbert-Schmidt; its implementers have the same parity as the complex dimension of $ker\,(g - JgJ)$.*

□

Thus: if the complex dimension of $ker\,(g - JgJ)$ is even then the unitary implementers U of $\theta(g)$ in π_J satisfy $\Gamma_J U \Gamma_J = U$ whilst if $ker\,(g - JgJ)$ is odd-dimensional then the implementers U satisfy $\Gamma_J U \Gamma_J = -U$. Notice here that $g - JgJ$ commutes with J and so has J-stable kernel, whence it makes sense to speak of the complex dimension of $ker\,(g - JgJ)$.

3. Theorem and Proof

In order to state Blattner's theorem, we denote by G_+ the group of all orthogonal transformations $g \in O(V)$ such that $g - I$ is Hilbert-Schmidt and $g + I$ has even-dimensional kernel, by G_- the set of all $g \in O(V)$ such that $g + I$ is Hilbert-Schmidt and $g - I$ has odd-dimensional kernel, and by G the group $G_+ \cup G_-$.

Theorem. *(Blattner): The Bogoliubov automorphism $\theta(g)$ of the vN Clifford algebra $\mathcal{A}[V]$ is inner iff $g \in G$.*

The presentation of our proof of Blatttner's theorem will be facilitated by the use of a simple notational device: we introduce the parity symbol ε to denote either $+$ or $-$. We shall speak of elements of $\mathcal{A}_\varepsilon[V]$ as having parity ε, applying analogous terminology to operators in Fock representations, and shall refer to a vector space as having parity $+$ or $-$ according to whether its dimension is even or odd. In these terms, $G = G_+ \cup G_-$ where G_ε comprises all $g \in O(V)$ such that $g - \varepsilon I$ is Hilbert-Schmidt and $ker\,(g + \varepsilon I)$ has parity ε.

As preparation for our proof of Blattner's theorem, we reconsider the trace representation. Let $W = V \oplus V$ be the orthogonal direct sum of two copies of the inner product space V. Define a real-linear map π from W to $B(\mathbb{H}_\tau)$ by

$$\pi(x \oplus y) = \lambda(x) + i\rho(y)\Gamma$$

so that

$$\pi(x \oplus y)\phi = x\phi + i\Gamma(\phi)y$$

for $x, y \in V$ and $\phi \in H_\tau$. It is readily verified that if $x, y \in V$ then $\pi(x \oplus y)$ is self-adjoint and satisfies

$$\pi(x \oplus y)^2 = (\|x\|^2 + \|y\|^2)\, I = \|x \oplus y\|^2\, I.$$

The universal property of the C* Clifford algebra now guarantees that π extends to a representation

$$\pi : C[W] \longrightarrow B(\mathbb{H}_\tau)$$

with cyclic unit vector Ω.

Let J be the unitary structure on W given by $J(x \oplus y) = (-y) \oplus x$ for $x, y \in V$ so that

$$J = \begin{bmatrix} O & -I \\ I & O \end{bmatrix}$$

in block form relative to the orthogonal decomposition $W = V \oplus V$. A direct calculation reveals that $\pi(w + iJw)\,\Omega = 0$ whenever $w \in W$. This cyclic vacuum property implies that π is the Fock representation of $C[W]$ determined by J. Notice that we may take the parity operator Γ_J to be Γ itself, since Γ anticommutes with $\pi(w)$ whenever $w \in W$.

Given $g \in O(V)$ we define $g_\varepsilon \in O(W)$ in block form by

$$g_\varepsilon = \begin{bmatrix} g & O \\ O & \varepsilon I \end{bmatrix}.$$

Block form calculations show that

$$[J, g_\varepsilon] = \begin{bmatrix} O & g - \varepsilon I \\ g - \varepsilon I & O \end{bmatrix}$$

and

$$g_\varepsilon - Jg_\varepsilon J = \begin{bmatrix} g + \varepsilon I & O \\ O & g + \varepsilon I \end{bmatrix}.$$

In particular, it follows that $[J, g_\varepsilon]$ is Hilbert-Schmidt on W iff $g - \varepsilon I$ is Hilbert-Schmidt on V and that the (complex) parity of $ker\,(g_\varepsilon - Jg_\varepsilon J)$ and (real) parity of $ker\,(g + \varepsilon I)$ coincide.

We have now assembled all the necessary ingredients and proceed directly to our proof of Blattner's theorem.

Suppose first that $g \in G$; indeed, let $g \in G_\varepsilon$. By definition, $g - \varepsilon I$ is Hilbert-Schmidt and $ker\,(g + \varepsilon I)$ has parity ε; it follows that $[J, g_\varepsilon]$ is Hilbert-Schmidt and $ker\,(g_\varepsilon - Jg_\varepsilon J)$ has parity ε. According to Theorem 2, the Bogoliubov automorphism $\theta(g_\varepsilon)$ of $C[W]$ is implemented in π by a unitary operator $U = \varepsilon \Gamma U \Gamma$ of parity ε on $\mathbb{H}_\tau$. It now follows that if $x, y \in V$ then

$$U\pi(x \oplus y) = \pi(gx \oplus \varepsilon y)U$$

so that in particular if $v \in V$ then

$$\begin{aligned} U\pi(0 \oplus v) &= \varepsilon\pi(0 \oplus v)U \\ \therefore \quad U\rho(v)\Gamma &= \varepsilon\rho(v)\Gamma U = \rho(v)U\Gamma \\ \therefore \quad U\rho(v) &= \rho(v)U. \end{aligned}$$

As a consequence, U actually lies in $\mathcal{A}'_\rho$ and therefore in $\mathcal{A}_\lambda = \mathcal{A}[V]$ in accordance with Theorem 1. Thus: if $g \in G_\varepsilon$ then $\theta(g)$ is inner, implemented by a unitary $U \in \mathcal{A}_\varepsilon[V]$.

Conversely, let $g \in O(V)$ and suppose $\theta(g)$ to be inner, implemented by the unitary $U \in \mathcal{A}[V]$. Since $\gamma^2 = I$ and $\mathcal{A}[V]$ is a factor, it follows that $\gamma(U) = \pm U$; we shall suppose that $U \in \mathcal{A}_\varepsilon[V]$. Now, let $x, y \in V$. Since U implements $\theta(g)$ we have $U\lambda(x) = \lambda(gx)U$; in view of Theorem 1 we also have $U\rho(y) = \rho(y)U$. Consequently,

$$\begin{aligned} U\pi(x \oplus y) &= U(\lambda(x) + i\rho(y)\Gamma) \\ &= \lambda(gx)U + i\rho(y)U\Gamma \\ &= \lambda(gx)U + i\varepsilon\rho(y)\Gamma U \\ &= \pi(gx \oplus \varepsilon y)U. \end{aligned}$$

It follows from this that $\theta(g_\varepsilon)$ is implemented in π by a unitary operator of parity ε. Theorem 2 now implies that $[J, g_\varepsilon]$ is Hilbert-Schmidt and $ker\,(g_\varepsilon - Jg_\varepsilon J)$ has parity ε; as a result, $g - \varepsilon I$ is Hilbert-Schmidt and $ker\,(g + \varepsilon I)$ has parity ε. Thus: if $\theta(g)$ is implemented by a unitary $U \in \mathcal{A}_\varepsilon[V]$ then $g \in G_\varepsilon$.

This concludes our proof.

Acknowledgement: The author is grateful for partial financial support from the National Science Foundation.

References

1. H. Araki, *On quasifree states of CAR and Bogoliubov automorphisms*, Publ. RIMS Kyoto Univ. **6** (1970/71), 385–442.
2. H. Araki, *Bogoliubov automorphisms and Fock representations of canonical anticommutation relations*, AMS Contemporary Mathematics **62** (1987), 23–141.
3. E. Balslev, J. Manuceau and A. Verbeure, *Representations of anticommutation relations and Bogoliubov transformations*, Comm. Math. Phys. **8** (1968), 315–326.
4. R. J. Blattner, *Automorphic group representations*, Pacific J. Math. **8** (1958), 665–677.
5. O. Bratteli and D.W. Robinson, *Operator Algebras and Quantum Statistical Mechanics II*, Springer-Verlag, 1979.

6. A.L. Carey, *Inner automorphisms of hyperfinite factors and Bogoliubov transformations*, Ann. Inst. Henri Poincaré, Phys. théor. **40** (1984), 141–149.
7. P. de la Harpe and R.J. Plymen, *Automorphic group representations: a new proof of Blattner's theorem*, J. London Math. Soc. **19** (1979), 509–522.
8. S.N.M. Ruijsenaars, *On Bogoliubov transformations, II: The general case*, Ann. Phys. **116** (1978), 105–134.
9. D. Shale and W.F. Stinespring, *States of the Clifford algebra*, Ann. Math. **80** (1964), 365–381.
10. D. Shale and W.F. Stinespring, *Spinor representations of infinite orthogonal groups*, J. Math. Mech. **14** (1965), 315–322.

DEPARTMENT OF MATHEMATICS, UNIVERSITY OF FLORIDA, GAINESVILLE FL 32611, USA

HOW TO GENERATE THERMAL PHOTONS – ON THE COMPUTER

Manfred R. Schroeder

Abstract. Although photons are usually associated with light, their arrival times at a given target can be simulated on a computer for demonstration purposes. The statistics depends of course on the "preparation" of the light beam. For incoherent photons with a geometric (Bose–Einstein) number distribution, the random time intervals t_1 between adjacent photons can be generated from a uniform random variable u_1 in the interval $(0, 1]$ by setting $t_1 = u_1^{-1} - 1$. The arrival times of the nth photon are generated by setting $t_n = u_1^{-1} u_2^{-1/2} \ldots u_n^{-1/n} - 1$, where the u_k are independent uniform random variables.

For John Klauder on his 60th Birthday

Coherent Photons

The number of photons n in a finite interval of a coherent beam of single–mode, non-squeezed laser light is given by the Poisson distribution

$$\nu(n) = \frac{m^n}{n!} e^{-m}, \quad n = 0, 1, 2, \ldots \tag{1}$$

where m, the average number of photons, is proportional to the beam's intensity. Such a process can be simulated on a computer either directly using equation (1) or by means of the corresponding distribution of the time intervals t between adjacent photons:

$$\begin{aligned} p_1(t) &= m\, e^{-mt} && \text{for } t \geq 0 \\ p_1(t) &= 0 && \text{for } t < 0 \end{aligned} \tag{2}$$

The simplest method of numerically simulating an exponential distribution, such as that given by equation (2), is to consider the integral of the distribution

$$P_1(t) := \int_t^\infty p_1(t')\, dt' = 1 - e^{-mt} \tag{3}$$

and to set it (or $1 - P(t)$) equal to a uniform variable u in the interval $(0,1]$. Solving for t then yields the time intervals

$$t = -\frac{1}{m} \ln u \tag{4}$$

Successive intervals in a Poisson process are independent. Thus, independent values of u will generate a proper sequence of photon intervals.

The arrival time t_n of the nth photon is obtained by $(n-1)$–fold convolution of the interval distribution given by equation (2). This gives

$$\begin{aligned} p_n(t_n) &= \frac{m^n}{(n-1)!}\, t_n^{n-1}\, e^{-mt_n} \quad \text{for } t_n \geq 0 \\ p_n(t_n) &= 0 \quad \text{for } t_n < 0 \end{aligned} \tag{2a}$$

Thermal Photons

For certain kinds of incoherent light, the number distribution, instead of being Poisson, is the geometric distribution also called Bose–Einstein distribution (see Appendix). For a detailed discussion of thermal light, see the book by Klauder and Sudarshan [3]:

$$w_t(n) = \frac{t^n}{(1+t)^{n+1}}\,,\ n = 0, 1, 2, \ldots \tag{5a}$$

Here, to unclutter subsequent equations, we have chosen our unit of time to make the expected number of photons equal to t. As is well known, the variance σ^2 of a geometrically distributed variable equals the mean squared *plus* the mean:

$$\sigma^2 = m^2 + m \tag{5b}$$

The additional term, m, reflects the *granularity* of electromagnetic radiation. It was from this additional contribution to the fluctuation of black–body radiation that Einstein originally inferred the existence of discrete photons.

The corresponding time interval distribution between adjacent photons is found with the help of the following observation: The probabiliaty $P_1(t)$ that the interval *exceeds* the value t equals the probability $w_t(0)$ that zero photons fall into an interval t:

$$P_1(t) = w_t(0) \tag{6}$$

or

$$p_1(t) = -\frac{d}{dt}\, w_t(0)$$

This relation yields a simple *power law* for the distribution of intervals between adjacent photons

$$p_1(t) = \frac{1}{(1+t)^2}\,,\ t \geq 0 \tag{7}$$

More generally, the interval distribution $p_n(t)$ between one photon and the nth photon following it is obtained as follows. The probability $P_n(t)$ that the nth photon arrives after a time interval exceeding t equals the probability $W_t(n-1)$ that the number of photons in the interval t is at most $n-1$:

$$P_n(t) = W_t(n-1) \tag{8}$$

where

$$W_t(n-1) := \sum_{k=0}^{n-1} w_t(k) = 1 - \left[\frac{t}{1+t}\right]^n \tag{9}$$

is the cumulative probability that no more that $n-1$ photons fall into the time interval t. Differentiation yields

$$\begin{aligned} p_n(t_n) &= \frac{n\, t_n^{n-1}}{(1+t_n)^{n+1}} \quad && \text{for } t_n \geq 0 \\ p_n(t_n) &= 0 && \text{for } t_n < 0 \end{aligned} \tag{10}$$

Such power–law distributions [5] have no mean and are best characterized by a median $(t_n)_{med}$ (and other percentiles). From equation (9), we obtain easily

$$(t_n)_{med} = \frac{1}{2^{1/n}-1} \approx 1.44\, n$$

for large n. Another unfailing "mean" to characterize such meanless distributions is the logarithmic mean $\overline{\ln t_n}$ or, equivalently, the *geometric mean* $(t_n)_{geom}$ defined as

$$(t_n)_{geom} := \exp(\overline{\ln t_n}) \tag{11}$$

Its experimental determination does not require sorting data as percentiles do. With equation (10), one obtains [2]

$$(t_1)_{geom} = 1$$

and for $n > 1$

$$(t_n)_{geom} = \exp\left[1 + \frac{1}{2} + \ldots \frac{1}{n-1}\right] \approx 1.78\,(n-1) \tag{12}$$

for large n.

The ratio $(t_n)_{geom}/(t_n)_{med}$ approaches the numerical constant $\ln 2e^C \approx 1.2345$. Here $C = 0.5772\ldots$ is Euler's constant.

The Time Intervals Between Thermal Photons

In contrast to coherent light, for thermal light the memory is not lost after arrival of a photon and therefore the time intervals between successive photons are not independent ("photon bunching"). (The same is true for squeezed states which show photon "anti-bunching" [6].)

The time of occurence of the nth photon after $t_0 = 0$ depends on that of the $(n-1)$th photon as follows (see Appendix):

$$p_n(t_n|t_{n-1}) = \frac{n(1+t_{n-1})^n}{(1+t_n)^{n+1}}\,, \; t_n \geq t_{n-1} \tag{13}$$

To generate the photon arrival times on a computer, we need the integral

$$P_n(t_n|t_{n-1}) := \int_{t_n}^{\infty} p_n(t'_n|t_{n-1})\,dt'_n\,, \; t_n \geq t_{n-1} \tag{14}$$

With equation (13), we obtain the simple power law

$$P_n(t_n|t_{n-1}) = \left[\frac{1+t_{n-1}}{1+t_n}\right]^n \tag{15}$$

Setting P_n equal to a uniform variable u_n in $(0,1]$ and solving for t_n, yields the random variable

$$t_n = (1+t_{n-1})\,u_n^{-1/n} - 1 \tag{16}$$

With $t_0 = 0$ and $t_1 = u_1^{-1} - 1$ we obtain, by induction on n,

$$t_n = u_1^{-1} u_2^{-1/2} u_3^{-1/3} \ldots u_n^{-1/n} - 1 \tag{17}$$

For the time interval $\Delta_n := t_n - t_{n-1}$, we find

$$\Delta_n = u_1^{-1} u_2^{-1/2} u_3^{-1/3} \ldots u_{n-1}^{-1/(n-1)} (u_n^{-1/n} - 1) \tag{19}$$

which is thus seen to depend on the arrival times of prior photons.

Interestingly, the geometric mean of Δ_n does not depend on n. In fact, we have, for $n > 1$,

$$\overline{\ln \Delta_n} = -\left[1 + \frac{1}{2} + \cdots + \frac{1}{n-1}\right]\overline{\ln u} + \overline{\ln(u_n^{-1/n} - 1)}$$

Here $\overline{\ln u}$ equals -1. The second average is equal to

$$-1 - \frac{1}{2} - \cdots - \frac{1}{n-1}$$

Thus,

$$\overline{\ln \Delta_n} = 0$$

and consequently, independently of n,

$$(\Delta_n)_{geom} = \exp(\overline{\ln \Delta_n}) = 1 \tag{20}$$

In actual tests of equation (19) on a personal (notebook) computer, the photon bunching of thermal photons can be readily observed using the BEEP command to signal the "arrival" of photons.

APPENDIX

We assume our photon statistics to be describable by a modulated Poisson process. The number distribution $\nu(n)$ is then given by the *Poisson transform* [4]

$$\nu(n) = \int_0^\infty \frac{m^n}{n!} e^{-m} q(m)\, dm, \tag{A1}$$

where $q(m)$ is the distribution of the average number m of photons. If there is only one value of m, i.e. if $q(m)$ is a Dirac–delta function, then the $\nu(n)$ is the Poisson distribution, see equation (1).

For an exponential distribution of m with a mean value of μ

$$\begin{aligned} q(m) &= \frac{1}{\mu} e^{-m/\mu} && \text{for } m \geq 0 \\ q(m) &= 0 && \text{for } m < 0 \end{aligned} \tag{A2}$$

the Poisson transform $(A1)$ yields the geometric distribution, also called Bose–Einstein distribution by physicists:

$$\nu(n) = \frac{\mu^n}{(1+\mu)^{n+1}} \tag{A3}$$

For the unit of time we have chosen, the parameter μ equals the observation time t. Thus,

$$\nu_t(n) = \frac{t^n}{(1+t)^{n+1}} \tag{A4}$$

which is our equation (5).

While for the Poisson process time intervals between subsequent photons are independent, this is not the case for the modulated Poisson process. The conditional probability density $p_{\Delta_n}(\Delta_n|t_{n-1})$ for the nth interval $\Delta_n := t_n - t_{n-1}$, given t_{n-1}, can be calculated as follows from the conditional probabilities involving m:

$$p_{\Delta_n}(\Delta_n|t_{n-1}) = \int_0^\infty p_1(\Delta_n|m)\, w(m|t_{n-1})\, dm \tag{A5}$$

With, see equation (2),

$$p_1(\Delta_n|m) = m\, e^{-m\Delta_n} \quad \text{for } \Delta_n \geq 0$$

and with Bayes' theorem [1]

$$w(m|t_{n-1}) = \frac{p_{n-1}(t_{n-1}|m)\, q(m)}{p_{n-1}(t_{n-1})}$$

where, see equation (2a),

$$p_{n-1}(t_{n-1}|m) = \frac{m^{n-1}}{(n-2)!}(t_{n-1})^{n-2}\, e^{-mt_n}$$

together with equation (A2), with μ equal to 1,

$$q(m) = e^{-m}$$

one obtains from (A5)

$$p_{\Delta_n}(\Delta_n|t_{n-1}) = \frac{n(1+t_{n-1})^n}{(1+t_{n-1}+\Delta_n)^{n+1}}\,,\ \Delta_n \geq 0 \tag{A6}$$

or, equivalently,

$$p_n(t_n|t_{n-1}) = \frac{n(1+t_{n-1})^n}{(1+t_n)^{n+1}}\,,\ t_n \geq t_{n-1} \tag{A7}$$

which is our equation (13).

References

1. H. Cramér, *Mathematical Methods of Statistics*, Chapter 34, Princenton University Press, Princeton, 1971.
2. I. S. Gradshteyn and I. M. Ryzhik, *Tables of Integrals, Series, and Products*, Formula 2.727, Academic Press, New York, 1972.
3. J. R. Klauder and E.C.G. Sudarshan, *Fundamentals of Quantum Optics*, W.A. Benjamin, New York, 1968.
4. B.E.A. Saleh, *Photonelectron Statistics*, Springer, New York, 1978.
5. M. Schroeder, *Fractals, Chaos, Power Laws: Minutes from an Infinite Paradise*, W. H. Freeman, New York, 1991.
6. M. C. Teich and B.E.A. Saleh, *Squeezed and Antibunched Light*, Physics Today (June 1990), 26-34.

DRITTTES PHYSIKALISSCHES INSTITUT, UNIVERSITÄT GÖTTINGEN, GERMANY

A MODEL FOR PHASE TRANSITIONS AKIN TO THE ISING MODEL

LARRY SHEPP

Contribution in honor of the 60th birthday of John R. Klauder

John will certainly have an opinion on the central thesis of this paper — that the Ising model is not engraved in the sky, and that another model for phase transition should be considered. Indeed, my joint work with John was based on a physics model close in spirit to the Ising model. Thus I may have to avoid John until he forgets about this paper!

1. The Model

Imagine a collection of atoms, $\{1, 2, 3, \dots\}$, with bonds between some of the pairs. At low temperatures, enough bonds may be present, and the collection has the properties of a solid. As the temperature goes up, some bonds break and the collection moves into a disconnected state, say liquid, or gas, or other.

To treat this phenomenon mathematically, one might use the Ising model introducing a Gibbs energy or Gibbs random field, [7] (see also [1], [4]) but there seems to be no fundamental reason to use such a model, and the problem of justifying one or another Gibbs field via a direct observation of the Gibbs potential energy seems far from being realizable in the laboratory. In the Gibbs-Ising model, the probability of any state, $\lambda = \lambda(i)$, $i = 1, 2, 3, \dots$ of the system is often given in terms of a Gibbs energy, E:

$$\text{Prob}(\lambda) = \frac{1}{Z} e^{-cE}$$

where $c = \beta/T$, with T = temperature, and β a constant. Z is the normalizing partition function, and the Gibbs energy E is defined by a kernel φ :

$$E = \sum(\varphi(\lambda_i, \lambda_j))$$

where the sum is over neighboring atoms i and j. λ may be ± 1 valued, or more general. In the usual Ising model there is a critical temperature $T = T_0$ such that some macroscopic observable develops a singularity as T crosses over T_0. As Dobrushin and Griffiths showed [7], one can state this in a meaningful mathematical way in terms of the so-called Gibbs-Markov property. They showed that for certain $\varphi(x, y)$'s and neighborhood structures, the above critical value of $T = T_0$, can be defined in a different way without looking for singularities in macroscopic observables — i.e. for $T > T_0$ there is only one random field which has the Gibbs-Markov property, i.e. for which the conditional probability of the value of λ at an atom i, given the values of λ at the other atoms only depends on the immediate neighbors and is given in terms of the Gibbs potential $\varphi(x, y)$. But for $T < T_0$, there is more than one such Gibbs-Markov random field.

The Gibbs-Ising model is accepted by most physicists as being quite a physically pleasing model despite several limitations: Although elegant, it is rather complicated, far from the domain of verifiability, rather hypothetical, and difficult to get explicit answers with, or even bounds for checking theory against experiments.

A different model was suggested by Lester Dubins, actually during a talk on the Ising model, where the speaker was merely able to prove the existence of a phase transition and remarked about the difficulty in actually calculating the critical value of the temperature. Dubins remarked politely that one should ask an easier question before working on this hard one, and then went on to pose one.

Dubins suggested placing a connecting bond between each pair i and j of atoms randomly (and independently for each pair) with a probability, $p(i, j, T)$, depending on the locations of i and j and on the temperature, T. One thinks of the atoms as the vertices of a graph, and the connections as the edges, with each edge either present or absent — present with a probability $p = p(i, j, T)$, $0 \le p \le 1$, and absent with probability $1 - p$. If the graph is connected with high probability, then the object comprising the atom is likely in a solid state, else it falls apart, and is in another state.

Dubins probably saw that the most natural model, where the graph is infinite and $p(i, j) = c$ (where c is a fixed constant), would have no phase transition, and this is indeed the case, since for every value of c it is connected with probability one, as is not hard to see. So he invented another case — where $p(i, j) = c/\max(i, j)$, and conjectured that this would have a phase transition in the value of c, which plays the role of an inverse temperature. It turned out, that he was right.

The good news about the Dubins model is that it is possible to find explicit values for the phase transition constant, T_0. Namely it is 1/4. Thus, one can show [8] that if in the graph on $\{1, 2, 3, \dots\}$ one places an edge between nodes i and j with probability $p(i, j)$ and one omits this edge with the complementary probability, where

$$p(i, j) = \frac{c}{\max(i, j)}$$

independently for each $1 \le i < j < \infty$, then the graph is connected with probability one if $c > c_0 = 1/T_0 = .25$ and is connected with probability zero if $c \le .25$. A complete proof is given in [8]; here we just discuss the model and give a rough sketch of the basic ideas in the proof which go back to Erdös and Renyi [3], and also rely on ideas of Kalikow and Weiss [5] and on later work of Durrett and Kesten [2]. Indeed Durrett and Kesten proved that if $p(i, j)$ is replaced by:

$$p(i, j) = \frac{c}{i + j}$$

then the critical $c_0 = 1/\pi$, where $\pi = 3.14159\dots$. In each of the 2 cases above, $p(i, j)$ is of the form

$$p(i, j) = c\, h(i, j)$$

where $h(x,y)$ is continuous, positive, and homogeneous of order -1, i.e.

$$h(cx,cy) = h(x,y)/c$$

In this general case, Durrett and Kesten were able to show that the critical constant can be given explicitly: if $c \leq c_0(h)$ given by

$$c_0 = c_0(h) = 1\bigg/\int_0^\infty h(x,1)/\sqrt{x}\,dx \tag{*}$$

then the probability of connectedness is 0, while if $c > c_0$ then the graph is connected with probability one. In the case $h(x,y) = 1/\max(x,y)$ or $h(x,y) = 1/(x+y)$, the integral becomes 4 or π, respectively.

So much for the good news. The bad news is that no one wants to regard the model as physics, even the mathematicians say it is a "toy" model, whatever that means. Isn't the Ising model also a toy model? Isn't it really also far from experiment and in the realm of fantasy? The role of $h(x,y)$ seems exactly analogous to the role of the potential kernel $\varphi(x,y)$ in the Gibbs model. Inverse temperature comes into both models and the c in the Dubins model can be written with an official constant β,

$$c = \beta/T$$

if one is in a decorative frame of mind.

Some younger and perhaps naive mathematicians strongly objected, saying that the Dubins model is "obviously not physics", that the Ising model has the honor of being confirmed by experimental evidence and that no method of verification seems possible for the $p(i,j)$ of the Dubins model. I would argue that this is also an objection for the $\varphi(i,j)$ of the Ising model. I am not aware of any experimental evidence (which ought to be the only way to put down such a model) against Dubins' picture. This naive mathematician always thought there should be some experiments if one rejects a model as non-physical.

Some physicists, perhaps with a vested interest in the Ising model, raised the objection that in the Ising model, $\varphi(x,y)$ is a function of the difference,

$$\varphi(x,y) = \varphi(x-y)$$

whereas in the Dubins model, if $p(i,j) = c\,h(i,j)$, where $h(x,y) = h(x-y)$, it is a consequence of a theorem of Michael Keane [6] that there can be no phase transition. This is a harder objection to deal with, and does seem to be a stumbling block for the Dubins model — i.e. why shouldn't the model treat all atoms in the same homogeneous way? I must of course admit I have no real answer to this, but perhaps there is, in higher dimensional arrangements of the atoms $1,2,3,\ldots,$ (we considered only one-dimensional arrangements above) with full or with more homogeneity, which could be introduced and where the phase transition phenomenon could be retained. This perhaps deserves further study. Also it is possible that some physical problem will appear where homogeneity is not wanted or not required. It is generally believed that the phenomenon of a phase transition can't occur with short range interaction in one dimension. Of course the connectedness model involves long-range interaction and does not contradict this belief (theorem). The fact that the graph is infinite in our case plays no essential role and is analogous the use of infinite lattices in the Ising case.

2. Sketch of ideas underlying the proofs

The zero-one law of probability theory says that any event which depends only on the tail field of a sequence of independent random variables has probability either 0 or 1. Here an event, A, is such a tail event if whether it occurs or not is decidable only by looking at the random variables in any tail of the sequence, i.e. any finite number of them can be ignored. This is because it is approximable by some event defined by only a finite number of the variables but is independent of this event because it is a tail event. But then it is independent of itself and so $P(A) = P(A \cap A) = P(A)^2$, i.e. $P(A) = 0$ or 1.

An example of a tail event is the event that infinitely many of events A_n occur, where A_n depends only on the nth random variable. Clearly no finite number of the variables can influence the outcome. The Borel-Cantelli lemma says that infinitely many of a sequence of independent events occur with probability one if and only if the sum of the probabilities is ∞.

The event that the graph is connected is not a tail event, but under the condition of Steven Kalikow and Benjy Weiss [5] it has probability 0 or 1. The condition of Kalikow and Weiss is that

$$\sum_{i \in A} \sum_{j \in A^c} p(i,j) \equiv \infty \tag{K-W}$$

for all proper sets A contained in $\{1, 2, \dots\}$ where the sum is over $i \in A$, $j \in A^c$, where A and A^c (which is the complement of A) are not empty. This resembles the series of the Borel-Cantelli lemma, except that there is such an event for every A, and there are uncountably many subsets of the integers, in the sense of Cantor, i.e. there is no $1-1$ correspondence between subsets of the integers and the integers themselves. For each fixed A, the condition says that there is a link between A and A^c with probability one. But then why isn't it true that the graph is connected with probability one? After all, 1 is connected with probability 1 to some other element (take $A = \{1\}$). Then those two elements are connected to a third, etc. in this way the connected set grows and grows, and can never die because the condition says that every set has a link to its complementary set. Where is this argument wrong ? Why isn't it always true that the graph is always connected under the (K-W) condition ? It must be wrong, if the earlier results about a phase transition are right because for any c, it is easy to see that (because of the divergence of $\sum(1/n)$) that if

$$p(i,j) = c/\max(i,j) \ ,$$

for example, then (K-W) holds for all $c > 0$.

The error in the above argument is that there are uncountably many sets A, so that even though it is true that each fixed set A has probability 0 of being isolated (no links to A^c), there may be some set A which cannot be named in advance. It is like choosing a number at random from [0,1]. Any fixed x has probability 0 to be chosen, but some number is chosen. Probably I am fussing too much over mathematical "trivialities", but the reason I do so is to indicate where the ideas are not.

Indeed the proofs of the theorems are finitistic in the sense that they deal with the connectedness of finite subsets of $\{1, 2, 3, \dots\}$. In this way and only in this way, paradoxes of the angels-on-heads-of-pins as indicated above can be avoided.

There is an argument that goes back to Peierls which gives one-half of the result we want, that is, it proves that for $c \leq c_0(h)$, see $(*)$, the graph is disconnected almost surely. The idea is to calculate the expected number of simple paths connecting 1 and 2. A simple path is non-reentrant, i.e. self-avoiding. It can be shown (Durrett-Kesten [2]) that if $c \leq$ the maximum eigenvalue of the infinite matrix $h(i,j)$, then the expected number of simple paths connecting 1 and 2 is finite.

Suppose the expected number were actually < 1. Then the probability that 1 and 2 are connected by any path at all must be less than the expected number of simple paths and so is also < 1. But by the (K-W) condition, it must be zero. If a finite number of probabilities are changed, then the expected number of simple paths can be made < 1, and so again, the K-W theorem says that the connectedness probability must be 0. So the Peierls argument gives the right answer. This was very surprising to the experts in percolation theory. Usually this simple argument gives only a sufficient condition, but next we indicate that in this case it is also necessary for connectedness. For more details on the first half, see [2], where a proof is given that avoids using the K-W $0-1$ law. The paper [9] gives a special argument for the case $h(x,y) = 1/\max(x,y)$ which does not generalize to arbitrary homogeneous h of order -1; this was done in [2].

To prove the graph is connected for $c > c_0(h)$, given in $(*)$, we follow an idea of [5]. We can show that for $c > c_0(h)$ there is, in the subgraph G_n of vertices

$$G_n = \{i : a^n < i < a^{(n+1)}\},$$

a "giant" component of size $> b\, a^n$, with probability at least d, where $a > 1$, and b and d are uniformly positive and do not depend on n. In G_n, only connections within G_n are considered as edges. If this is true, let us look at the chance that 1 and 2 are not joined by a path. One way for this to happen is to have a single edge from 1 and a single edge from 2 both leading to the same or different elements of the giant component of G_n. But then 1 and 2 are connected, since the giant component is a connected set. The chance that 1 is not connected to one of the bn elements of the giant component is $1-(1-c/n)^{bn}$ and so is < 1. The same argument shows that 2 is connected to some element of the giant component with probability $1 > d > 0$, where d doesn't depend on n. But the G_n subgraphs are all disjoint and each one affords 1 and 2 to be connected with a fixed probability > 0. Hence the probability that 1 and 2 are connected is 1. Since the argument can be repeated for 1 and 3, 1 and 4, etc., and since there are only a countable number of vertices, we see that the full graph is connected if there is positive probability of a giant component in G_n.

But why is there a giant component in G_n? This is an argument based on a martingale-like branching process, where the idea discussed above of starting with a few elements in G_n, connected to each other, and then looking at all the elements (again of G_n) they are directly linked to not already in the clique, then adding again all vertices not already connected to them which are connected by a direct link to one just connected, etc. It is a branching process because each node begets "children" until the race of nodes dies and the full component of G_n is generated. This argument generalizes an earlier one for the case when $p(i,j) = p$, independently of i and j, which was first done by Erdös and Renyi [3]. For more details see [5] and [8].

3. A further remark and extension

Here is a problem which, if answered, would shed light on the "order of connectedness", which might be used to distinguish the liquid state from the solid state of the collection of atoms. This is all in the spirit of the probabilistical physics generated by the Dubins model. Suppose the temperature is low enough that $c > c_0$ and connectedness occurs. Of course the probability that the subgraph $\{1, 2, \dots, n\}$ is connected is $<< 1$ for large n because the probability that atom n is isolated is greater than

$$(1 - \text{constant}/n)^n$$

for some constant, which depends on h. Since the latter is uniformly > 0, and there may be other isolated atoms, or groups of atoms, $\{1, 2, \dots\}$ is not connected with a high probability. But, because $c > c_0(h)$, $\{1, 2, \dots, n\}$ must be connected within $\{1, 2, \dots, N(n)\}$ with probability approaching 1 as $n \to \infty$ if $N(n) \to \infty$ fast enough. So how fast is fast enough? Paul Zhang [10] has shown that $N(n)$ need increase no faster than a polynomial in n, which was surprising to me. But he was unable to find the exact degree of the polynomial as a function of $c > c_0$. No doubt as c decreases to c_0, this degree must necessarily tend to ∞, since the graph is not connected at $c = c_0$.

References

1. C. Domb and M.S. Green, *Phase Transitions and Critical Phenomena, Vol. 1*, Acad. Press, London, New York, 1972.
2. R. Durrett and H. Kesten, *The critical parameter for connectedness of some random graphs*, in Papers dedicated to Paul Erdös (Baker, Bollobas, and Hajnal, eds.), Cambridge U. Press, Cambridge, 1990.
3. P. Erdös and A. Renyi, *On the evolution of random graphs*, Magyar Tud Akad Mat Kutato Int Kozl **5** (1960), 17–61.
4. R. Israel, *Convexity and the Theory of Lattice Gases*, Princeton U. Press, Princeton,NJ, 1978.
5. S. Kalikow and B. Weiss, *When are random graphs connected*, Isr. J. Math. (1990).
6. M. Keane, G.R. Grimmett and J.M. Marstrand, *On the connectedness of a random graph*, Proc. Camb. Phil. Soc. **96** (1984), 151–166.
7. R. Kindermann and J.L. Snell, *Markov Random Fields and their Applications*, AMS, Providence, 1980.
8. L. Shepp, *Connectedness of certain random graphs*, Isr. J. Math. **67** (1989), 23–33.
9. L. Shepp, *On the evolution of islands*, Isr. J. Math. **67** (1989), 34–42.
10. P. Zhang, manuscript (in preparation).

AT&T Bell Laboratories, Murray Hill, NJ 07974, USA

LOCALIZATION OF MASSLESS SPINNING PARTICLES AND THE BERRY PHASE

B.-S. SKAGERSTAM

ABSTRACT. The components of the position operator, at a fixed time, for a massless and spinning particle with given helicity λ described in terms of bosonic degrees of freedom have an anomalous commutator proportional to λ. The position operator generates translations in momentum space. We show that a ray-representation for these translations emerges due to the non-commuting components of the position operators and relate this to the Berry-phase for photons. The Tomita-Chiao experiment then gives support for this relativistic and quantum mechanical description of photons in terms of non-commuting position operators.

Contribution in honor of the 60th birthday of John R. Klauder

1. Introduction

The concept of localization of relativistic elementary systems has a long and intriguing history in physics (see e.g. [1,2,3,4]). Observations of physical phenomena take place in space and time. The notion of *localizability* of particles, elementary or not, then refers to the empirical fact that particles, at a given instant of time, appear to be localizable in the physical space.

In the realm of non-relativistic quantum mechanics the concept of localizability of particles is built into the theory at a very fundamental level and is expressed in terms of the canonical commutation relation between a position operator and the corresponding generator of translations, i.e. the canonical momentum of a particle. In relativistic theories the concept of localizability of physical systems is deeply connected to our notion of space-time, the arena of physical phenomena, as a 4-dimensional continuum. In the context of the general theory of relativity the localization of light rays in space-time is a fundamental ingredient. In fact, it has been argued [5] that the Riemannian metric is basically determined by the properties of light propagation.

In a fundamental paper by Newton and Wigner [1] it was argued that in the context of relativistic quantum mechanics a notion of point-like localization of a single particle can be uniquely determined by kinematics. Wightman [2] extended this notion to localization to finite domains of space and it was rigourously shown that massive particles are always localizable if they are elementary, i.e. if they are described in terms of irreducible representations of the Poincaré group [6]. Massless elementary systems with non-zero helicity, like a gluon, graviton, neutrino or a photon, are not localizable in the sense of Wightman. The axioms used by Wightman can, of course, be weakened. It was actually shown by Jauch, Piron and Amrein [3] that in such a sense the photon is *weakly localizable.* The notion of weak localizability essentially corresponds to allowing for non-commuting observables in order to characterize the localization of massless and spinning particles in general.

Localization of relativistic particles, at a fixed time, as alluded to above, has been shown to be incompatible with a natural notion of (Einstein-) causality [7]. If a relativistic elementary system has an exponentially small tail outside a finite domain of localization at $t = 0$, then, according to the hypothesis of a weaker form of causality, this should remain true at later times, i.e. the tail should only be shifted further out to infinity. As was shown by Hegerfeldt [8], even this notion of causality is incompatible with the notion of a positive and bounded observable whose expectation value gives the probability to find a particle inside a finite domain of space at a given instant of time.

In the present paper we will reconsider some of these questions related to the concept of localizability in terms of a quantum mechanical description of a massless particle with given helicity λ [9,10,11]. We will show how one can extend this description to include both positive and negative helicities. We will then be able, for instance, to describe the motion of a linearly polarized photon in the framework of relativistic quantum mechanics.

2. Position Operators for Massless Particles

It is easy to show that the components of the position operators for a massless particle must be non-commuting[1] if the helicity $\lambda \neq 0$. If J_k are the generators of rotations and p_k the diagonal momentum, $k = 1,2,3$, then $\mathbf{J} \cdot \mathbf{p} = \pm\lambda$; here $\mathbf{J} = (J_1, J_2, J_3)$ and $\mathbf{p} = (p_1, p_2, p_3)$; with $\hbar = 1$:

$$[J_k, p_l] = i\epsilon_{klm}p_m \quad . \tag{2.1}$$

If a canonical position operator $\mathbf{x}$ exists with components x_k such that

$$[x_k, x_l] = 0 \quad , \tag{2.2}$$

$$[x_k, p_l] = i\delta_{kl} \quad , \tag{2.3}$$

$$[J_k, x_l] = i\epsilon_{klm}x_m \quad , \tag{2.4}$$

then we can define generators of orbital angular momentum in the conventional way, i.e.

$$L_k = \epsilon_{klm}x_l p_m \quad . \tag{2.5}$$

"Spin"- generators are then defined by

$$S_k = J_k - L_k \quad . \tag{2.6}$$

They fulfill the algebra

$$[S_k, S_l] = i\epsilon_{klm}S_m \quad , \tag{2.7}$$

and they, furthermore, commute with $\mathbf{x}$ and $\mathbf{p}$. Then, however, the spectrum of $\mathbf{S} \cdot \mathbf{p}$ is $\lambda, \lambda - 1, ..., -\lambda$, which contradicts the requirement $\mathbf{J} \cdot \mathbf{p} = \pm\lambda$ since, by construction, $\mathbf{J} \cdot \mathbf{p} = \mathbf{S} \cdot \mathbf{p}$.

---

[1] This argument has, as far as we know, first been suggested by N. Mukunda.

As has been discussed in detail in the literature, the non-zero commutator of two components of the position operator for a massless particle primarily emerges due to the non-trivial topology of the momentum space [9,10,11]. The irreducible representations of the Poincaré group for massless particles [6] can be constructed from a knowledge of the little group G of a light-like momentum four-vector $p = (p^0, \mathbf{p})$. This group is the Euclidean group $E(2)$. Physically, we are interested in possible finite-dimensional representations of the covering of the little group. We therefore restrict ourselves to the compact subgroup, i.e. we represent the $E(2)$-translations trivially and consider $G = SO(2) = U(1)$. Since the origin in the momentum space is excluded for massless particles one is therefore led to consider G-bundles over S^2 since the energy of the particle can be kept fixed. Such G-bundles are classified by mappings from the equator to G, i.e. by the first homotopy group $\Pi_1(U(1))=\mathbf{Z}$, where it turns out that each integer corresponds to twice the helicity of the particle. A massless particle with helicity λ and *sharp momentum* is thus described in terms of a non-trivial line bundle characterized by $\Pi_1(U(1)) = \{2\lambda\}$ [12].

This consideration can easily be extended to higher space-time dimensions. If D is the number of space-time dimensions, the corresponding G-bundles are classified by the homotopy groups $\Pi_{D-3}(Spin(D-2))$. These homotopy groups are in general non-trivial. It is a remarkable fact that the only trivial homotopy groups of this form in higher space-time dimensions correspond to $D = 5$ and $D = 9$ due to the existence of the quaternions and the Cayley numbers (see e.g. [13]). In these space-time dimensions, and for $D = 3$, it then turns out that one can explicitly construct canonical *and* commuting position operators for massless particles [11]. The mathematical fact that the spheres S^1, S^3 and S^7 are parallelizable can then be expressed in terms of the existence of canonical *and* commuting position operators for massless spinning particles in $D = 3$, $D = 5$ and $D = 9$ space-time dimensions.

In terms of the canonical momenta p_i and coordinates x_j satisfying the canonical commutation relation (2.3) we can easily derive the commutator of two components of the position operator $\mathbf{x}$ by making use of a simple consistency argument as follows. If the massless particle has a given helicity λ, then the generators of angular momentum are given by:

$$J_k = \epsilon_{klm} x_l p_m + \lambda \frac{p_k}{|\mathbf{p}|} \quad . \tag{2.8}$$

The canonical momentum then transforms as a vector under rotations, i.e.

$$[J_k, p_l] = i\epsilon_{klm} p_m \quad , \tag{2.9}$$

without any condition on the commutator of two components of the position operator $\mathbf{x}$. The position operator will, however, not transform like a vector unless the following commutator is postulated

$$i[x_k, x_l] = \lambda \epsilon_{klm} \frac{p_m}{|\mathbf{p}|^3} \quad , \tag{2.10}$$

where we notice that this commutator formally corresponds to a pointlike Dirac magnetic monopole [14] localized at the origin in momentum space with strength $4\pi\lambda$. The energy p^0 of the massless particle is, of course, given by $\omega = |\mathbf{p}|$. In terms of a singular $U(1)$

connection $\mathcal{A}_l \equiv \mathcal{A}_l(\mathbf{p})$ we can write

$$x_k = i\partial_k - \mathcal{A}_k \quad , \tag{2.11}$$

where $\partial_k = \partial/\partial p_k$ and

$$\partial_k \mathcal{A}_l - \partial_l \mathcal{A}_k = \lambda \epsilon_{klm} \frac{p_m}{|\mathbf{p}|^3} \quad . \tag{2.12}$$

Out of the observables x_k and the energy ω one can easily construct the generators (at time $t = 0$) of Lorentz boosts, i.e.

$$K_m = (x_m \omega + \omega x_m)/2 \quad , \tag{2.13}$$

and verify that J_l and K_m lead to a realization of the Lie algebra of the Lorentz group, i.e.

$$[J_k, J_l] = \; i\epsilon_{klm} J_m \quad , \tag{2.14}$$

$$[J_k, K_l] = \; i\epsilon_{klm} K_m \quad , \tag{2.15}$$

$$[K_k, K_l] = -i\epsilon_{klm} J_m \quad . \tag{2.16}$$

The components of the Pauli-Lubanski operator W_μ are given by

$$W^\mu = (W^0, \mathbf{W}) = (\mathbf{J} \cdot \mathbf{p}\, , \mathbf{J}p^0 + \mathbf{K} \times \mathbf{p}) = \lambda p^\mu \; , \tag{2.17}$$

i.e. we also obtain an irreducible representation of the Poincaré group. The additional non-zero commutators are

$$[K_k, \omega] = i p_k \quad , \tag{2.18}$$

$$[K_k, p_l] = i\delta_{kl} \omega \quad . \tag{2.19}$$

At $t \equiv x^0(\tau) \neq 0$ the Lorentz boost generators K_m as given by (2.13) are extended to

$$K_m = (x_m \omega + \omega x_m)/2 - t p_m \quad . \tag{2.20}$$

In the Heisenberg picture, the quantum equation of motion of an observable $\mathcal{O}(t)$ is obtained by using

$$\frac{d\mathcal{O}(t)}{dt} = \frac{\partial \mathcal{O}(t)}{\partial t} + i[H, \mathcal{O}(t)] \quad , \tag{2.21}$$

where the Hamiltonian H is given by ω. One then finds that all generators of the Poincaré group are conserved as they should. The equation of motion for $\mathbf{x}(t)$ is

$$\frac{d}{dt}\mathbf{x}(t) = \frac{\mathbf{p}}{\omega} \quad , \tag{2.22}$$

which is an expected equation of motion for a massless particle.

The non-commuting components x_k of the position operator $\mathbf{x}$ transform as the components of a vector under spatial rotations. Under Lorentz boosts we find in addition that

$$i[K_k, x_l] = \frac{1}{2}\left(x_k \frac{p_l}{\omega} + \frac{p_l}{\omega} x_k\right) - t\delta_{kl} + \lambda \epsilon_{klm} \frac{p_m}{|\mathbf{p}|^2} \quad . \tag{2.23}$$

The first two terms in (2.23) correspond to the correct limit for $\lambda = 0$ since the proper-time condition $x^0(\tau) \approx \tau$ is not Lorentz invariant [15]. The last term in (2.23) is due to the non-zero commutator (2.10). This anomalous term can be dealt with by introducing an appropriate two-cocycle for finite transformations consisting of translations generated by the position operator $\mathbf{x}$, rotations generated by $\mathbf{J}$ and Lorentz boosts generated by $\mathbf{K}$. For pure translations this two-cocycle will be explicitly constructed in Section 3.

The algebra discussed above can be extended in a rather straightforward manner to incorporate both positive and negative helicities needed in order to describe e.g. linearly polarized light. As we now will see this extension corresponds to a replacement of the Dirac monopole in momentum space with a $SU(2)$ Wu-Yang [16] monopole. The procedure below follows a rather standard method of imbedding the singular $U(1)$ connection $\mathcal{A}_l$ into a *regular* $SU(2)$ connection. Let us specifically consider a massless, spin-one particle. The Hilbert space, $\mathcal{H}$, of one-particle transverse wave-functions $\phi_\alpha(\mathbf{p}), \alpha = 1, 2, 3$ is defined in terms of a scalar product

$$(\phi, \psi) = \int d^3p \phi^*_\alpha(\mathbf{p}) \psi_\alpha(\mathbf{p}) \quad , \tag{2.24}$$

where $\phi^*_\alpha(\mathbf{p})$ denotes the complex conjugated $\phi_\alpha(\mathbf{p})$. In terms of a Wu-Yang connection $\mathcal{A}^a_k \equiv \mathcal{A}^a_k(\mathbf{p})$, i.e.

$$\mathcal{A}^a_k(\mathbf{p}) = \epsilon_{alk} \frac{p_l}{|\mathbf{p}|^2} \quad . \tag{2.25}$$

Equation (2.11) is extended to

$$x_k = i\partial_k - \mathcal{A}^a_k(\mathbf{p}) S_a \quad , \tag{2.26}$$

where

$$(S_a)_{kl} = -i\epsilon_{akl} \tag{2.27}$$

are the spin-one generators. By means of a singular gauge-transformation the Wu-Yang connection can be transformed into the singular $U(1)$-connection $\mathcal{A}_l$ times the third component of the spin generators S_3. The position operator defined by (2.26) is compatible with the transversality condition on the one-particle wave-functions, i.e. $x_k \phi_\alpha(\mathbf{p})$ is transverse. With suitable conditions on the one-particle wave-functions the position operator $\mathbf{x}$ therefore has a well–defined action on $\mathcal{H}$. Furthermore,

$$i[x_k, x_l] = \mathcal{F}^a_{kl} S^a = \epsilon_{klm} \frac{p_m}{|\mathbf{p}|^3} \hat{\mathbf{p}} \cdot \mathbf{S} \quad , \tag{2.28}$$

where

$$\mathcal{F}^a_{kl} = \partial_k \mathcal{A}^a_l - \partial_l \mathcal{A}^a_k - \epsilon_{abc} \mathcal{A}^b_k \mathcal{A}^c_l = \epsilon_{klm} \frac{p_m p_a}{|\mathbf{p}|^4} \quad , \tag{2.29}$$

is the non-Abelian $SU(2)$ field strength tensor and $\hat{\mathbf{p}}$ is a unit vector in the direction of the particle momentum $\mathbf{p}$. The generators of angular momentum are now defined as follows

$$J_k = \epsilon_{klm} x_l p_m + \frac{p_k}{|\mathbf{p}|} \hat{\mathbf{p}} \cdot \mathbf{S} \quad . \tag{2.30}$$

The helicity operator $\Sigma \equiv \hat{\mathbf{p}} \cdot \mathbf{S}$ is covariantly constant, i.e.

$$\partial_k \Sigma + i\,[A_k, \Sigma] = 0 \quad , \tag{2.31}$$

where $A_k \equiv \mathcal{A}_k^a(\mathbf{p}) S_a$. The position operator $\mathbf{x}$ therefore commutes with $\hat{\mathbf{p}} \cdot \mathbf{S}$. One can therefore verify in a straightforward manner that the observables p_k, ω, J_l and $K_m = (x_m \omega + \omega x_m)/2$ close to the Poincaré group. At $t \neq 0$ the Lorentz boost generators K_m are defined as in (2.20), and (2.23) is extended to

$$i[K_k, x_l] = \frac{1}{2}\left(x_k \frac{p_l}{\omega} + \frac{p_l}{\omega} x_k\right) - t\delta_{kl} + i\omega [x_k, x_l] \quad . \tag{3.32}$$

For helicities $\hat{\mathbf{p}} \cdot \mathbf{S} = \pm\lambda$ one extends the previous considerations by considering $\mathbf{S}$ in the spin $|\lambda|$-representation. (2.28), (2.30) and (2.32) are then valid in general. A reducible representation for the generators of the Poincaré group for an arbitrary spin has therefore been constructed for a massless particle. We observe that the helicity operator Σ can be interpreted as a generalized "magnetic charge", and since Σ is covariantly conserved one can use the general theory of topological quantum numbers [17] and derive the quantization condition

$$\exp(i4\pi\Sigma) = 1 \quad , \tag{2.33}$$

i.e. the helicity is properly quantized.

3. Topological Spin

Coadjoint orbits on a group G admit a symplectic two-form (see e.g. [18]) which can be used to construct topological Lagrangians, i.e. Lagrangians constructed by means of Wess-Zumino terms [19] (for a general account, see e.g. [20]). Let us illustrate the basic ideas for a non-relativistic spin and $G = SU(2)$. Let $\mathcal{K}$ be an element of the Lie algebra $\mathcal{G}$ of G in the fundamental representation. Without loss of generality we can write $\mathcal{K} = \lambda_\alpha \sigma_\alpha = \lambda \sigma_3$, where σ_α, $\alpha = 1, 2, 3$ denote the three Pauli spin matrices. Let H be the little group of $\mathcal{K}$. Then the coset space G/H is isomorphic to S^2 and defines an adjoint orbit (for semi-simple Lie groups adjoint and coadjoint representations are equivalent due to the existence of the non-degenerate Cartan-Killing form). The action for the spin degrees of freedom is then expressed in terms of the group G itself, i.e.

$$S_P = -i \int \langle \mathcal{K}, g^{-1}(\tau) dg(\tau)/d\tau \rangle \, d\tau \quad , \tag{3.1}$$

where $\langle A, B \rangle$ denotes the trace-operation of two Lie-algebra elements A and B in $\mathcal{G}$ and where

$$g(\tau) = \exp(i\sigma_\alpha \xi_\alpha(\tau)) \tag{3.2}$$

defines the (proper-)time dependent dynamical group element. We observe that S_P has a gauge-invariance, i.e. the transformation

$$g(\tau) \longrightarrow g(\tau)\exp(i\theta(\tau)\sigma_3) \tag{3.3}$$

only changes the Lagrangian density $\langle\mathcal{K}, g^{-1}(\tau)dg(\tau)/d\tau\rangle$ by a total time derivative. The gauge-invariant components of spin, $S_k(\tau)$, are defined in terms of $\mathcal{K}$ by the relation

$$S(\tau) \equiv S_k(\tau)\sigma_k = \lambda g(\tau)\sigma_3 g^{-1}(\tau) \quad , \tag{3.4}$$

such that

$$S^2 \equiv S_k(\tau)S_k(\tau) = \lambda^2 \quad . \tag{3.5}$$

By adding a non-relativistic particle kinetic term as well as a conventional magnetic moment interaction term to the action S_P, one can verify that the components $S_k(\tau)$ obey the correct classical equations of motion for spin precession [9,20].

Let $M = \{\sigma,\tau|\sigma \in [0,1]\}$ and $(\sigma,\tau) \to g(\sigma,\tau)$ parametrize $\tau-$dependent paths in G such that $g(0,\tau) = g_0$ is an arbitrary reference element and $g(1,\tau) = g(\tau)$. The Wess-Zumino term in this case is given by

$$\omega_{WZ} = -id\langle\mathcal{K}, g^{-1}(\sigma,\tau)dg(\sigma,\tau)\rangle = i\langle\mathcal{K}, (g^{-1}(\sigma,\tau)dg(\sigma,\tau))^2\rangle \quad , \tag{3.6}$$

where d denotes exterior differentiation and where now

$$g(\sigma,\tau) = \exp(i\sigma_\alpha\xi_\alpha(\sigma,\tau)) \quad . \tag{3.7}$$

Apart from boundary terms which do not contribute to the equations of motion, we then have that

$$S_P = S_{WZ} \equiv \int_M \omega_{WZ} = -i\int_{\partial M}\langle\mathcal{K}, g^{-1}(\tau)dg(\tau)\rangle \quad , \tag{3.8}$$

where the one-dimensional boundary ∂M of M , parametrized by τ, can play the role of (proper-) time. ω_{WZ} is now gauge-invariant under a larger $U(1)$ symmetry, i.e. (3.3) is now extended to

$$g(\sigma,\tau) \longrightarrow g(\sigma,\tau)\exp(i\theta(\sigma,\tau)\sigma_3) \quad . \tag{3.9}$$

ω_{WZ} is therefore a two-form defined on the coset space G/H. A canonical analysis then shows that there are no gauge-invariant dynamical degrees of freedom in the interior of M. The Wess-Zumino action (3.8) is the topological action for spin degrees of freedom.

As for the quantization of the theory described by the action (3.8), one may use methods from geometrical quantization and especially the Borel-Weil-Bott theory of representations of compact Lie groups [18,20]. One then finds that λ is half an integer, i.e. $|\lambda|$ corresponds to the spin. This quantization of λ also naturally emerges by demanding that the action (3.8) is well defined in quantum mechanics for periodic motion as was recently discussed by e.g. Klauder [21], i.e.

$$4\pi\lambda = \int_{S^2}\omega_{WZ} = 2\pi n \quad , \tag{3.10}$$

where n is an integer. The symplectic two-form ω_{WZ} must then belong to an integer class cohomology. This geometrical approach is in principle straightforward, but it requires explicit coordinates on G/H. An alternative approach, as used in [9,20], is a canonical Dirac analysis and quantization [15]. This procedure leads to the condition $\lambda^2 = s(s+1)$, where s is half an integer. The fact that one can arrive at different answers for λ illustrates a certain lack of uniqueness in the quantization procedure of the action (3.8). The quantum theories obtained describes, however, the same physical system namely one irreducible representation of the group G.

The action (3.8) was first proposed in [22]. The action can be derived quite naturally in terms of a coherent state path integral (for a review see e.g. [23]) using spin coherent states. It is interesting to notice that the structure of the action (3.8) actually appears in such a language already in a paper by Klauder on continuous representation theory [24].

A classical action which after quantization leads to a description of a massless particle in terms of an irreducible representation of the Poincaré group can be constructed in a similar fashion [9]. Since the Poincaré group is non-compact the geometrical analysis referred to above for non-relativistic spin must be extended and one should consider coadjoint orbits instead of adjoint orbits (D=3 appears to be an exceptional case due to the existence of a non-degenerate bilinear form on the D=3 Poincaré group Lie algebra [25]. In this case there is a topological action for irreducible representations of the form (3.8) [26]). The action then takes the form

$$S = \int d\tau \left(p_\mu(\tau)\dot{x}^\mu(\tau) + \frac{i}{2}\mathrm{Tr}[\mathcal{K}\Lambda^{-1}(\tau)\frac{d}{d\tau}\Lambda(\tau)] \right) \quad . \tag{3.11}$$

Here $[\sigma_{\alpha\beta}]_{\mu\nu} = -i(\eta_{\alpha\mu}\eta_{\beta\nu} - \eta_{\alpha\nu}\eta_{\beta\mu})$ are the Lorentz group generators in the spin one representation and $\eta_{\mu\nu} = (-1,1,1,1)$ is the Minkowski metric. The Lorentz group Lie-algebra element $\mathcal{K}$ is chosen to be $\lambda\sigma_{12}$. The τ-dependence of the Lorentz group element $\Lambda_{\mu\nu}(\tau)$ is defined by

$$\Lambda_{\mu\nu}(\tau) = \left[\exp\left(i\sigma_{\alpha\beta}\xi^{\alpha\beta}(\tau)\right)\right]_{\mu\nu} \quad . \tag{3.12}$$

The momentum variable $p_\mu(\tau)$ is defined by

$$p_\mu(\tau) = \Lambda_{\mu\nu}(\tau)k^\nu \quad , \tag{3.13}$$

where the constant reference momentum k^ν is given by

$$k^\nu = (\omega, 0, 0, |\mathbf{k}|) \quad , \tag{3.14}$$

where $\omega = |\mathbf{k}|$. The momentum $p_\mu(\tau)$ is light-like by construction. The action (3.11) leads to the equations of motion

$$\frac{d}{d\tau}p_\mu(\tau) = 0 \quad , \tag{3.15}$$

and

$$\frac{d}{d\tau}\left\{x_\mu(\tau)p_\nu(\tau) - x_\nu(\tau)p_\mu(\tau) + S_{\mu\nu}(\tau)\right\} = 0 \quad . \tag{3.16}$$

Here we have defined gauge-invariant spin degrees of freedom $S_{\mu\nu}(\tau)$ by

$$S_{\mu\nu}(\tau) = \frac{1}{2}\mathrm{Tr}[\Lambda(\tau)\mathcal{K}\Lambda^{-1}(\tau)\sigma_{\mu\nu}] \tag{3.17}$$

in analogy with (3.4). These spin degrees of freedom satisfy the relations

$$p_\mu(\tau)S^{\mu\nu}(\tau) = 0 \quad , \tag{3.18}$$

and

$$\frac{1}{2}S_{\mu\nu}(\tau)S^{\mu\nu}(\tau) = \lambda^2 \ . \tag{3.19}$$

Inclusion of external electromagnetic and gravitational fields leads to the classical Bargmann-Michel-Telegdi and Papapetrou equations of motion respectively [9]. Since the equations derived are expressed in terms of *bosonic* variables these equations of motion admit a straightforward classical interpretation. (An alternative *bosonic* treatment of internal degrees of freedom can be found in [27].)

Canonical quantization of the system described by *bosonic* degrees of freedom and the action (3.11) leads to a realization of the Poincaré Lie algebra with generators p_μ and $J_{\mu\nu}$ where

$$J_{\mu\nu} = x_\mu p_\nu - x_\nu p_\mu + S_{\mu\nu} \quad . \tag{3.20}$$

The four–vectors x_μ and p_ν commute with the spin generators $S_{\mu\nu}$ and are canonical, i.e.

$$[x_\mu, x_\nu] = [p_\mu, p_\nu] = 0 \quad , \tag{3.21}$$

$$[x_\mu, p_\nu] = i\eta_{\mu\nu} \quad . \tag{3.22}$$

The spin generators $S_{\mu\nu}$ fulfill the conventional algebra

$$[S_{\mu\nu}, S_{\lambda\rho}] = i(\eta_{\mu\lambda}S_{\nu\rho} + \eta_{\nu\rho}S_{\mu\lambda} - \eta_{\mu\rho}S_{\nu\lambda} - \eta_{\nu\lambda}S_{\mu\rho}) \quad . \tag{3.23}$$

The mass-shell condition $p^2 = 0$ as well as the constraints (3.18) and (3.19) are all first-class constraints [15]. In the proper-time gauge $x^0(\tau) \approx \tau$ one obtains the system described in Section 2, i.e. we obtain an irreducible representation of the Poincaré group with helicity λ [9]. For half-integer helicity, i.e. for fermions, one can verify in a straightforward manner that the wave-functions obtained change with a minus-sign under a 2π rotation [9,11,20] as they should.

4. The Berry Phase for Photons

We have constructed a set of $O(3)$-covariant position operators of massless particles corresponding to a reducible representation of the Poincaré group corresponding to a combination of positive and negative helicity. It is interesting to notice that the construction above leads to observable effects. Let us specifically consider photons and the motion of photons along an optical fibre. Berry has argued [28] that a spin in an adiabatically changing magnetic field leads to the appearance of an observable phase factor, called the Berry phase. It was suggested [29] that a similar geometric phase could appear for photons. We will now, within the framework of *relativistic quantum mechanics*, provide a derivation of this geometrical phase in terms of the operator realization of the Poincaré group discussed above. The Berry phase for photons can then be obtained as follows. We consider the motion of a photon with fixed energy moving in an optical fibre. We assume that as the photon moves in the fibre, the momentum vector traces out a closed loop in momentum space on the constant energy surface, i.e. on a two-sphere S^2. We therefore consider wave-functions $|\mathbf{p}\rangle$ which are diagonal in momentum. We also define the translation operators $U(\mathbf{a}) = \exp(i\mathbf{a}\cdot\mathbf{x})$. It is straightforward to show using (2.10) that

$$U(\mathbf{a})U(\mathbf{b})|\mathbf{b}\rangle = \exp(i\gamma[\mathbf{a},\mathbf{b};\mathbf{p}])|\mathbf{p}+\mathbf{a}+\mathbf{b}\rangle \quad , \tag{4.1}$$

where the two-cocycle phase $\gamma[\mathbf{a},\mathbf{b};\mathbf{p}]$ is equal to the flux of the magnetic monopole in momentum space through the simplex spanned by the vectors $\mathbf{a}$ and $\mathbf{b}$ localized at the point $\mathbf{p}$, i.e.

$$\gamma[\mathbf{a},\mathbf{b};\mathbf{p}] = \lambda \int_0^1 \int_0^1 d\xi_1 d\xi_2 a_k b_l \epsilon_{lkm} B_m(\mathbf{p}+\xi_1\mathbf{a}+\xi_2\mathbf{b}) \quad , \tag{4.2}$$

where $B_m(\mathbf{p}) = p_m/|\mathbf{p}|^3$. The non-trivial phase appears because the second de Rham cohomology group of S^2 is non-trivial. The two-cocycle phase $\gamma[\mathbf{a},\mathbf{b};\mathbf{p}]$ is therefore not a coboundary and hence it cannot be removed by a redefinition of $U(\mathbf{a})$. This result has a close analogy in the theory of magnetic monopoles [30]. The anomalous commutator (2.10) therefore leads to a ray-representation of the translation in momentum space.

A closed loop in momentum space, starting and ending at $\mathbf{p}$, can then be obtained by using a sequence of infinitesimal translations $U(\delta\mathbf{a})|\mathbf{p}\rangle = |\mathbf{p}+\delta\mathbf{a}\rangle$ such that $\delta\mathbf{a}$ is orthogonal to the argument of the wave-function on which it acts (this defines the adiabatic transport of the system). The momentum vector $\mathbf{p}$ then traces out a closed curve on the constant energy surface S^2 in momentum space. The total phase of these translations then gives a phase γ which is λ times the solid angle of the closed curve the momentum vector traces out on the constant energy surface. This phase does not depend on the Planck constant. This is precisely the Berry phase for the photon with a given helicity λ. In the experiment by Tomita and Chiao [31] one considers a linearly polarized photon. The same line of arguments above but making use (2.28) instead of (2.10) leads to the desired change of polarization as the photon moves along the optical fibre. An alternative derivation of the Berry phase for photons is based on the observation that the covariantly conserved helicity operator Σ can be interpreted as a generalized "magnetic charge". Let Γ denote a closed path in momentum space parametrized by $\sigma \in [0,1]$ such that $\mathbf{p}(\sigma=0) = \mathbf{p}(\sigma=1) = \mathbf{p}_0$ is fixed. The parallel transport of a one-particle state $\phi_\alpha(\mathbf{p})$ along the path Γ is then determined by a path-ordered exponential, i.e.

$$\phi_\alpha(\mathbf{p}_0) \longrightarrow \left[P\exp\left(i\int_\Gamma A_k(\mathbf{p}(\sigma))\frac{dp_k(\sigma)}{d\sigma}d\sigma\right)\right]_{\alpha\beta}\phi_\beta(\mathbf{p}_0) \quad , \tag{4.3}$$

where $A_k(\mathbf{p}(\sigma)) \equiv \mathcal{A}_k^a(\mathbf{p}(\sigma))S_a$. By making use of a non-Abelian version of Stokes theorem [17] one can then show that

$$P\exp\left(i\int_\Gamma A_k(\mathbf{p}(\sigma))\frac{dp_k(\sigma)}{d\sigma}d\sigma\right) = \exp\left(i\Sigma\Omega[\Gamma]\right) \quad , \tag{4.4}$$

where $\Omega[\Gamma]$ is the solid angle subtended by the path Γ on the two-sphere S^2. This result leads again to the desired change of linear polarization as the photon moves along the path described by Γ. This derivation does not require that $|\mathbf{p}(\sigma)|$ is constant along the path.

In the experiments so far considered the photon flux is large. In order to strictly apply our results under such conditions one can consider a second quantized version of the theory we have presented following e.g. the discussion of Amrein [3]. By making use of coherent states of the electromagnetic field in a standard and straightforward manner (see e.g. [23]) one then realizes that our considerations survive. This is so since the coherent states are parametrized in terms of the one-particle states. By construction the coherent states then inherit the transformation properties of the one-particle states discussed above.

5. Final Remarks and Conclusions

In the analysis of Wightman, corresponding to commuting position variables, the natural mathematical tool turned out to be systems of imprimitivity for the representations of the three-dimensional Euclidean group. In the case of non-commuting position operators we have also seen that notions from differential geometry are important. It is interesting to see that such a broad range of mathematical methods enters into the study of the notion of localizability of physical systems. We have in particular argued that Abelian as well as non-Abelian magnetic monopole field configurations reveal themselves in a description of localizability of massless spinning particles. Concerning the physical existence of magnetic monopoles Dirac remarked in 1981 [32] that "*I am inclined now to believe that monopoles do not exist. So many years have gone by without any encouragement from the experimental side*". The "monopoles" we are considering appear, however, as mathematical objects in the momentum space of the massless particles. Their existence, we have argued, is then only indirectly revealed to us by the properties of e.g. the photons moving along optical fibres.

Localized states of massless particles will necessarily develop non-exponential tails in space as a consequence of the Hegerfeldt theorem [8]. Various number operators representing the number of massless, spinning particles localized in a finite volume V at time t has been discussed in the literature. The non-commuting position observables we have discussed for photons correspond to the pointlike limit of the weak localizability of Jauch, Piron and Amrein [3]. This is so since our construction, as we have seen in Section 2, corresponds to an explicit enforcement of the transversality condition of the one-particle wave-functions.

In a finite volume, photon number operators appropriate for weak localization [3] do not agree with the photon number operator introduced by Mandel [33] for sufficiently small wavelengths as compared to the linear dimension of the localization volume. It would be interesting to see if there are measurable differences. A necessary ingredient in answering such a question would be the experimental realization of a localized one-photon state. It is interesting to notice that such states can be generated in the laboratory [34].

In concluding we find it appealing that the quantization of the system describing a photon, in general linearly polarized, is "*geometry, after all*" [21].

Acknowledgement

This research was supported by the Swedish National Research Council under contract no. 8244-103. We are greatful for a fruitful collaboration with A. P. Balachandran, G. Marmo and A. Stern on projects related to the subject in question. We also acknowledge discussions with M. Berry and D. Kastler. I am also very greatful to John R. Klauder for sharing with me his deep insights in theoretical physics in a spirit of friendship and search for truth.

References

1. T. Newton and E. P. Wigner, *Localized States for Elementary Systems*, Rev. Mod Phys. **21** (1949), 400.
2. A. S. Wightman, *On the Localizability of Quantum Mechanical Systems*, Rev. Mod. Phys. **34** (1962), 845.
3. J. M. Jauch and C. Piron, *Generalized Localizability*, Helv. Phys. Acta **40** (1967), 559; W. O. Amrein, *Localizability for Particles of Mass Zero*, Helv. Phys. Acta **42** (1969), 149.
4. T. F. Jordan and N. Mukunda, *Lorentz-Covariant Position Operators for Spinning Particles*, Phys. Rev. **132** (1963), 1842; A. L. Licht, *Local States*, J. Math. Phys. **7** (1966), 1656; K. Krauss, *Position Observables of the Photon*, in: The Uncertainty Principle and Foundations of Quantum Mechanics (W.C. Price and S.S. Chissick, eds.), John Wiley & Sons, New York (1977); M. I. Shirokov, *Strictly Localized States and Particles*, Theor. Math. Phys. **42** (1980), 134; S. N. M. Ruijsenaars, *On Newton-Wigner Localization and Superluminal Propagation Speeds*, Ann. Phys. (N.Y.) **137** (1981), 33; E. Prugovecki, *Stochastic Quantum Mechanics and Quantum Spacetime*, Kluwer Academic, Hingham, Mass., 1984.
5. J. Ehlers, A. E. Pirani and A. Schild, *The Geometry of Free-Fall and Light Propagation*, in: General Relativity. Papers in Honor of J. L. Synge (L. O'Raifeartaigh, ed.), Oxford University Press, London, 1972.
6. E. Wigner, *Unitary Representations of the Inhomogeneous Lorentz Group*, Ann. Math. **40** (1939), 149.
7. G. C. Hegerfeldt, *Remark on the Causality and Particle Localization*, Phys. Rev. **D10** (1974), 3320; B.-S. Skagerstam, *Some Remarks Concerning the Question of Localization of Elementary Particles*, Int. J. Theor. Phys. **15** (1976), 213; J. F. Perez and I. F. Wilde, *Localization and Causality in Relativistic Quantum Mechanics*, Phys. Rev. **D16** (1977), 315; G. C. Hegerfeldt and S. N. M. Ruijsenaar, *Remarks on Causality, Localization and Spreading of Wave Packets*, Phys. Rev. **D22** (1980), 377.
8. G. C. Hegerfeldt, *Violation of Causality in Relativistic Quantum Theory ?*, Phys. Rev. Lett. **54** (1985), 2395.
9. A. P. Balachandran, G. Marmo, B.-S. Skagerstam and A. Stern, *Spinning Particles in General Relativity*, Phys. Lett. **89B** (1980), 199; Gauge Symmetries and Fibre Bundles: Applications to Particle Dynamics, Lecture Notes in Physics, Vol. **188**, Springer Verlag, Berlin (1983); B.-S. Skagerstam and A. Stern, *Lagrangian Description of Classical Charged Particles with Spin*, Physica Scripta **24** (1981), 493.
10. M. Atre, A. P. Balachandran and T. V. Govindarajan, *Massless Spinning Particles in All Dimensions and Novel Magnetic Monopoles*, Int. J. Mod. Phys. **A2** (1987), 453.
11. B.-S. Skagerstam and A. Stern, *Light-Cone Gauge Versus Proper-Time Gauge for Massless Spinning Particles*, Nucl. Phys. **B294** (1987), 636.
12. B. Ek and B. Nagel, *Differentiable Vectors and Sharp Momentum States of Helicity Representations of the Poincaré Group*, J. Math. Phys. **25** (1984), 1662.
13. G. W. Whitehead, *Elements of Homotopy Theory*, Springer Verlag, Berlin, 1978.
14. P. A. M. Dirac, *Quantized Singularities in the Electromagnetic Field*, Proc. Roy. Soc. **A133** (1931), 60; *The Theory of Magnetic Monopoles*, Phys. Rev. **74** (1948), 817.
15. A. J. Hansson, T. Regge and C. Teitelboim, *Constrained Hamiltonian Systems*, Accademia Nazionale dei Lincei, Roma (1976).

16. T. T. Wu and C. N. Yang, *Some Remarks About Unquantized Non-Abelian Gauge Fields*, Phys. Rev. **D12** (1975), 3843.
17. P. Goddard, J. Nuyts and D. Olive, *Gauge Theories and Magnetic Charge*, Nucl. Phys. **B125** (1977), 1.
18. A. A. Kirillov, Elements of the Theory of Representations, A Series of Comprehensive Studies in Mathematics **220**, Springer Verlag, Berlin (1976); E. Witten, *Coadjoint Orbits of the Virasoro Group*, Commun. Math. Phys. **114** (1988), 1; A. Aleksev, L. Fadeev and S. Shatashvili, *Quantization of the Symplectic Orbits of the Compact Lie Groups by Means of the Functional Integral*, J. Geom. Phys. **1** (1989), 3; A. Aleksev and S. Shatashvili, *Path Integral Quantization of the Coadjoint Orbits of the Virasoro Group and 2D Gravity*, Nucl. Phys. **B323** (1989), 719; O. Alvarez, I. M. Singer and P. Windey, *Quantum Mechanics and the Geometry of the Weyl Character Formula*, Nucl. Phys. **B337** (1990), 467; B. Rai and V. G. J. Rodgers, *From Coadjoint Orbits to Scale Invariant WZNW Type Actions and 2-D Quantum Gravity Action*, Nucl. Phys. **B341** (1990), 119; G. W. Delius, P. van Nieuwenhuizen and V. G. J. Rodgers, *The Method of Coadjoint Orbits: An Algorithm for the Construction of Invariant Actions*, Int. J. Mod. Phys. **A5** (1990), 3943.
19. J. Wess and B. Zumino, *Consequences of Anomalous Ward Identities*, Phys. Lett. **37B** (1971), 95.
20. A. P. Balachandran, G. Marmo, B.-S. Skagerstam and A. Stern, *Classical Topology and Quantum States*, World Scientific, Singapore, 1991.
21. J. R. Klauder, *Quantization Is Geometry, After All*, Ann. Phys. (N.Y.) **188** (1988), 120.
22. A. P. Balachandran, S. Borchardt and A. Stern, *Lagrangian and Hamiltonian Descriptions of Yang-Mills Particles*, Phys. Rev. **D11** (1978), 3247.
23. J. R. Klauder and B.-S. Skagerstam, *Coherent States - Applications in Physics and Mathematical Physics*, World Scientific, Singapore, 1985; Beijing, 1988.
24. J. R. Klauder, *Continuous-Representation Theory.II. Generalized Relation Between Quantum and Classical Dynamics*, J. Math. Phys. **4** (1963), 1058.
25. E. Witten, *2 + 1 Dimensional Gravity As An Exactly Solvable System*, Nucl. Phys. **B311** (1988), 46.
26. B.-S. Skagerstam and A. Stern, *Topological Quantum Mechanics in 2+1 Dimensions*, Int. J. Mod. Phys. **A5** (1990), 1575.
27. A. P. Balachandran, P. Salomonson, B.-S. Skagerstam and J.-O. Winnberg, *Classical Description of a Particle Interacting With a Non-Abelian Gauge Field*, Phys. Rev. **D15** (1977), 2308.
28. M. V. Berry, *Quantal Phase Factors Accompanying Adiabatic Changes*, Proc. Roy. Soc. (London) **392** (1984), 45; B. Simon, *Holonomy, the Quantum Adiabatic Theorem, and Berry's Phase*, Phys. Rev. Lett. **51** (1983), 2167; I. J. R. Aitchison, *Berry Phases, Magnetic Monopoles, and Wess-Zumino Terms or How the Skyrmion Got Its Spin*, Acta. Phys. Polon. **B18** (1987), 207; *Berry's Topological Phase in Quantum Mechanics and Quantum Field Theory*, Physica Scripta **T23** (1988), 12; M. V. Berry, *Quantum Adiabatic Anholonomy*, in: Anomalies, Phases, Defects..., (M. Bregola, G. Marmo and G. Morandi, eds.), Bibliopolis, Napoli (1990); S. I. Vinitskiǐ, V. L. Derbov, V. N. Dubovik, B. L. Markovski and Yu. P. Stepanovskiǐ, *Topological Phases in Quantum Mechanics and Polarization Optics*, Sov. Phys. Usp. **33** (1990), 403.
29. R. Y. Chiao and Y.-S. Wu, *Manifestations of Berry's Topological Phase for the Photon*, Phys. Rev. Lett. **57** (1986), 933.
30. R. Jackiw, *Three-Cocycle in Mathematics and Physics*, Phys. Rev. Lett. **54** (1985), 159; *Magnetic Sources and 3-Cocycles (Comment)*, Phys. Lett. **154B** (1985), 303; B. Grossman, *A 3-Cocycle in Quantum Mechanics*, Phys. Lett. **152B** (1985), 93; *Three-Cocycle in Quantum Mechanics II*, Phys. Rev. **D33** (1986), 2922; Y. S. Wu and A. Zee, *Cocycles and Magnetic Monopole*, Phys. Lett. **152B** (1985), 98; J. Mickelsson, *Comment on "Three-Cocycle in Mathematics and Physics"*, Phys. Rev. Lett. **54** (1985), 2379; D. Boulware, S. Deser and B. Zumino, *Absence of 3-Cocycles in the Dirac Monopole Problem*, Phys. Lett. **153B** (1985), 307; Bo-Yu Hou, Bo-Yuan Hou and Pei Wang, *How to Eliminate the Dilemma in 3-Cocycle*, Ann. Phys. (N.Y.) **171** (1986), 172.
31. A. Tomita and R. Y. Chiao, *Observation of Berry's Topological Phase by Use of an Optical Fiber*, Phys. Rev. Lett. **57** (1986), 937.
32. N. S. Craige, P. Goddard and W. Nahm (eds.), *Monopoles in Quantum Field Theory*, World Scientific, Singapore, 1982.

33. L. Mandel, *Configuration-Space Photon Number Operators in Quantum Optics*, Phys. Rev. **144** (1966), 1071; *Photon Interference and Correlation Effects Produced by Independent Quantum Sources*, Phys. Rev. **A28** (1983), 929.
34. C. K. Hong and L. Mandel, *Experimental Realization of a Localized One-Photon State*, Phys. Rev. Lett. **56** (1986), 58.

INSTITUTE OF THEORETICAL PHYSICS, CHALMERS UNIVERSITY OF TECHNOLOGY, S-412 96 GÖTEBORG, SWEDEN ; EMAIL ADDRESS: TFEBSS@SECTHF51.BITNET OR TFEBSS@FY.CHALMERS.SE

A LOCAL QUANTUM THEORY WITHOUT POSITIVE ENERGY REPRESENTATIONS

R. F. STREATER

ABSTRACT.

We construct a C^*−algebra, associated with the free Dirac equation, which has a local structure obeying the Haag–Kastler axioms. We show that the algebra possesses no representations having positive energy. This provides a simple proof that the positive–energy axiom is independent of the other axioms, a result first proved by Doplicher, Regge and Singer in 1968.

Dedicated to John Klauder at the occasion of his sixtieth birthday

1. Introduction

Haag and Kastler have given [1] a set of axioms that ought to be satisfied by any reasonable relativistic theory of local observables. In a modern version, the axioms are

1. To each bounded open set $\mathcal{O}$ of space–time, $\mathbb{R}^4$, is assigned a C^*−algebra $\mathcal{A}(\mathcal{O})$ containing an identity.
2. If $\mathcal{O}_1 \subseteq \mathcal{O}_2$ then $\mathcal{A}(\mathcal{O}_1) \subseteq \mathcal{A}(\mathcal{O}_2)$; the inclusion provides an injection of C^*−algebras. This axiom allows us to define the C^*−algebra $\mathcal{A}$ of quasi–local observables as the inductive limit: $\mathcal{A} = ind \lim_{\mathcal{O}\uparrow\mathbb{R}^4} \mathcal{A}(\mathcal{O})$.
3. Let $L = (a, \Lambda)$ denote an element of the Poincaré group $P_+^\uparrow$ of proper, orthochronous, inhomogeneous Lorentz transformations $L : \mathbb{R}^4 \to \mathbb{R}^4$ thus: $x \to Lx = \Lambda x + a$.
   For each $L \in P_+^\uparrow$ is given a $*$−automorphism τ_L of $\mathcal{A}$ which is an isomorphism from $\mathcal{A}(\mathcal{O})$ onto $\mathcal{A}(\mathcal{O}_L)$, where $\mathcal{O}_L = \{x \in \mathbb{R}^4 : L^{-1}x \in \mathcal{O}\}$, for each $\mathcal{O}$. Moreover, the map $L \to \tau_L$ from $P_+^\uparrow$ into Aut $\mathcal{A}$ is a group homomorphism.
4. Local commutativity: if $\mathcal{O}_1$ and $\mathcal{O}_2$ are space–like, then $\mathcal{A}(\mathcal{O}_1)$ and $\mathcal{A}(\mathcal{O}_2)$ commute.

Any infinite–dimensional C^*−algebra $\mathcal{A}$ has many inequivalent representations i.e. *-homomorphism $\pi : \mathcal{A} \to \mathcal{B}(\mathcal{H}_\pi)$, where $\mathcal{H}_\pi$ is a Hilbert space and $\mathcal{B}(\mathcal{H}_\pi) = C^*$−algebra of all bounded operators on $\mathcal{H}_\pi$. A representation π of the quasi–local algebra $\mathcal{A}$, obeying 1) – 4), is said to obey the spectral condition if

(a) the automorphisms $\tau_{a,1} = \tau_a$ are spatial in π i.e. there exist unitary operators $U(a) \in \mathcal{B}(\mathcal{H}_\pi)$ such that $\pi(\tau_a A) = U(a)\pi(A)U(a)^{-1}$ for all $A \in \mathcal{A}$, $a \in \mathbb{R}^4$

(b) $a \to U(a)$ is strongly continuous, and the joint spectrum p^μ of the generators P^μ of $U(a)(\mu = 0,1,2,3)$ can be chosen to be in the forward light cone.

The quasi–local algebra $\mathcal{A}$ is said to obey the spectral condition if it possesses at least one representation π obeying a) and b).

We can now formulate Axiom 5

5. The algebra $\mathcal{A}$ obeys the spectral condition.

In 1968, Doplicher, Regge and Singer constructed a quasi–local algebra $\mathcal{A}$ and proved that it possessed no representation π obeying the spectral condition [3]. They thereby showed that axiom 5 is independent of the other axioms. In this paper we give a simpler example of a quasilocal algebra showing the same thing.

2. The quasilocal algebra

The Hilbert space $\mathcal{H} = L^2(\mathbb{R}^3; \mathbb{C}^4, d^3x)$ carries the representation $[m, \frac{1}{2}]_+ \oplus [m, \frac{1}{2}]_-$ of the covering group $IS(2,\mathbb{C}) = \tilde{P}_+^\uparrow$ of the Poincaré group $P_+^\uparrow$. The action $U(a, A)$ of $\widetilde{P}_+^\uparrow$ on $\mathcal{H}$ does not obey the spectral condition, because $[m, \frac{1}{2}]_-$ is the negative–energy representation of mass $m \geq 0$ and spin $\frac{1}{2}$. The action of $U(a, A)$ is constructed as follows. Each $\psi \in \mathcal{H}$ can be identified with the time–zero value (up to sets of measure zero) of a square–integrable solution $\Psi(\vec{x}, t)$ of the Dirac equation [2, (1-41)]

$$(\gamma^\mu \frac{\partial}{\partial_{x^\mu}} + m)\Psi(\vec{x}, t) = 0 \ ; \ \Psi(\vec{x}, t)|_{t=0} = \psi(\vec{x}) \ a.e.$$

To each solution Ψ we have another, Ψ_L, $L \in \widetilde{P}_+^\uparrow$ given by $\Psi_L(x) = S(A)\Psi(L^{-1}x)$, $x \in \mathbb{R}^4$, $L = (a, A)$. Then the action of $U(L)$ on $\mathcal{H}$ is $(U(a,\Lambda)\psi(\vec{x}) = \Psi_L(\vec{x}, t)|_{t=0}$. Here, $S(A)$ is Dirac's spin matrix [2, 1-44]. To each bounded open set $\triangle \subseteq \mathbb{R}^3$ let $\mathcal{A}(\triangle)$ be the C^*–algebra generated by $1_\mathcal{H}$ and all operators on $\mathcal{H}$ of the form $M_f \otimes 1$ and $M_f \otimes \alpha^k$, $k = 1,2,3$. Hence, M_f is the multiplication operator by $f \in C_0(\triangle) =$ continuous functions with support in $\triangle$, and $\alpha^k = \gamma^0\gamma^k$ are the Dirac matrices. We note that $\mathcal{A}(\triangle)$ is elementwise invariant under the operators $\exp(iM_f)$, which can be regarded as gauge tranformations of the second kind. Naturally, $\exp(iM_f)$ does not commute with $U(L)$ and in particular, the Hamiltonian (of the free Dirac particle) is not gauge invariant. To each bounded open set $\triangle \subseteq \mathbb{R}^3$, let $\mathcal{O}(\triangle)$ be the double cone subtended by $\triangle$, that is $\mathcal{O}(\triangle) = \triangle_0''$ where, for any set S in $\mathbb{R}^4$,

$$S' = \{x \in \mathbb{R}^4 : x - y \text{ is space–like for all } y \in S\}$$

and $\triangle_0 = \{(\vec{x}, t) \in \mathbb{R}^4 : \vec{x} \in \triangle \text{ and } t = 0\}$. By acting with $L \in P_+^\uparrow$ we transform $\mathcal{O}(\triangle)$ to $\mathcal{O}(\triangle_L) = \triangle_L''$, where $\triangle_L = \{x \in \mathbb{R}^4 : L^{-1}x \in \triangle_0\}$. Thus $\mathcal{O}(\triangle_L)$ is a double cone based on $\triangle_L$. We can now define the C^*–algebra $\mathcal{A}(\mathcal{O})$ for any open bounded $\mathcal{O} \subseteq \mathbb{R}^4$, to be the C^*–algebra generated by operators of the form $B = U_L A U_{L^{-1}}$ where $A \in \mathcal{A}(\triangle)$ and $\mathcal{O}(\triangle_L) \subseteq \mathcal{O}$, for some $L \in P_+^\uparrow$ and some bounded open $\triangle \subseteq \mathbb{R}^3$. Axioms 1,2 are trivially true for the system $\{\mathcal{A}(\mathcal{O})\}$. Moreover, we may for $B \in \mathcal{A}(\mathcal{O})$ define $\tau_L B = U_L B U_{L^{-1}}$.

This maps an element of the form $U_{L_1}AU_{L^{-1}}$, $A \in \mathcal{O}(\Delta)$ to $U_{LL_1}AU_{(LL_1)^{-1}}$ $A \in \mathcal{O}(\Delta))$ and so maps $\mathcal{A}(\mathcal{O})$ into $\mathcal{A}(\mathcal{O}_L)$. By the same argument, $\tau_{L^{-1}}$ maps $\mathcal{A}(\mathcal{O}_L)$ into $\mathcal{A}(\mathcal{O})$ and therefore, as $\tau_{L^{-1}} = \tau_L^{-1}$, we see that τ_L is an isomorphism of $\mathcal{A}(\mathcal{O})$ onto $\mathcal{A}(\mathcal{O}_L)$. Hence τ_L can be extended uniquely to an automorphism of $\mathcal{A}$, the inductive limit of all the $\mathcal{A}(\mathcal{O})$. This proves axiom 3. To prove axiom 4, we first verify local commutativity for $A \in \mathcal{A}(\Delta)$ and $B \in \mathcal{A}(\mathcal{O}(\Delta_{1L}))$ when Δ_0 and Δ_{1L} are space–like. Try $A = M_f \otimes \gamma^\nu, supp f \subseteq \Delta$. Since B is a sum (convergent in norm) of operators of the form $U_{L_i}A_iU_{L_i}{}^{-1}$ with $A_i \in \mathcal{A}(\Delta_i)$, it is enough to verify local commutativity for each A, A_i. Thus if $A_1 = \sum_\mu M_{g_\mu} \otimes \gamma^\mu$ and $B = U_LA_1U_L^{-1}$ we see that, if $supp\, g_\mu = \Delta_1$ then $A_1U_L^{-1}\psi$ has support in $\Delta_1 \subseteq \mathbb{R}^3, \psi \in \mathcal{H}$. Thus $A_1U_L^{-1}\psi$ defines (by the Dirac equation) a solution $\Psi_1(\vec{x},t)$ zero in Δ_1', by the hyperbolic nature of the Dirac equation. Thus $B\psi = U_L(A_1U_L^{-1}\psi)$ defines a solution Ψ_2 which is zero in $L\Delta_1'$. But $L\Delta_1'$ contains $\mathcal{O}(\Delta)$ since by assumption $L\Delta_1$ i.e. Δ_{1L} is space–like to Δ_O. Hence $AB\psi = 0$, since $\Delta_O \subseteq \mathcal{O}(\Delta)$ and $supp f \subseteq \Delta$. Similarly $BA\psi = 0$. Thus $AB = BA$. For a general pair of algebras $\mathcal{A}(\mathcal{O}_1)$, $\mathcal{A}(\mathcal{O}_2)$ we pick a pair of elements, each of which is the norm limit of a sum $\sum U_{L_i}A_iU_{L_i}^{-1}$. For the ith such term we reduce the problem to the previous case by conjugating with $U_{L_i}^{-1}$. This proves local commutativity, axiom 4.

3. Conclusion

We have shown that for any two operators, A and B in $\mathcal{A}$, that are localized in space–like separated regions, we have $AB = 0$. This property was found also in ref [3] for their construction. As in ref [3] we note that this violates the Reeh–Schlieder theorem, which holds for any local theory in a representation π obeying the spectral condition: therefore, no such π exists.

The operators $M_f \otimes 1$ and $M_g \otimes \alpha^k$ are related to the smeared charge operator $\rho(f) = \int \overline{\hat{\psi}(\vec{x},0)}\gamma^0\hat{\psi}(\vec{x},0)\, f(\vec{x})\, d^3x$ and current $\vec{j}(g) = \int \overline{\hat{\psi}(\vec{x},0)}\vec{\gamma}\hat{\psi}(\vec{x},0)\, g(\vec{x})\, d^3x$ respectively, in the (unphysical) Fock representation $\hat{\psi}$ of the second quantized Dirac field; these operators act on $\mathbb{C}\oplus\mathcal{H}\oplus(\mathcal{H}\otimes\mathcal{H})_{\mathcal{A}}\oplus,\ldots$ the antisymmetric Fock space over $\mathcal{H}$, and leave $\mathcal{H}$ invariant (there is no vacuum polarization in this representation). Then M_f is $\rho(f)|_{\mathcal{H}}$ and $M_g \otimes \alpha^k$ is $j^k(g)|\mathcal{H}$. In the physical representation of the Dirac field, ρ and $\vec{j}$ are not even operators (at sharp time). This shows that the algebraic properties of currents, unlike those of the CCR or the CAR, are strongly representation dependent.

Doplicher [4] has given an algebraic criterion for the existence of a positive–energy representation, in terms of the existence of non–trivial, proper "positive–energy" ideal $\mathcal{F}$ inside $\mathcal{A}$. The definition of the ideal depends on the continuity in norm of the action of $\mathbb{R}^4$ on $\mathcal{A}$ (otherwise the Bochner integrals used in the definition of $\mathcal{F}$ cannot be defined). But in our model, the continuity of the action of $\mathbb{R}^4$ on $\mathcal{A}$ can be established. It is enough to establish continuity on a dense set, and hence on local operators $M_f \otimes \gamma^\mu$, $supp f \subseteq \Delta$, f continuous. But f is uniformly continuous on $\Delta \subseteq \mathbb{R}^3$, and hence $\|A - U_{\vec{a}}AU_{\vec{a}}^{-1}\| = \sup_{\vec{x}\in\Delta}|f(\vec{x}) - f(\vec{x} - \vec{a})| \to 0$ as $\vec{a} \to 0$. So $\mathbb{R}^3$ acts continuously. For any time t write

$A_t = U(\vec{b})^{-1}U(\Lambda)\, B_{\vec{a}}\, U(\Lambda)^{-1}U(\vec{b})^{-1}$ where $B = U(\Lambda)^{-1}\, A\, U(\Lambda)$, $(t,\vec{b}) = \Lambda(0,\vec{a})$ and $B_{\vec{a}} = U(\vec{a})BU(\vec{a})^{-1}$. Keep Λ fixed and let $\vec{a} \to 0$ (so $t \to 0, \vec{b} \to 0$). Then $\|A_t - A\| =$

$\|U(\Lambda)B_{\vec{a}}U^{-1}(\Lambda)-A_{\vec{b}}\| \le \|U(\Lambda)B_{\vec{a}}U^{-1}(\Lambda)-A\|+\|A-A_{\vec{b}}\| = \|B_{\vec{a}}-B\|+\|A-A_{\vec{b}}\| \to 0$ as $\vec{a}\to 0$. Hence $\mathbb{R}^4$ acts continuously. Doplicher's ideal is generated by elements of $\mathcal{A}$ with energy–momentum in the forward–light cone. Only multiples of the identity can have this property (with zero–energy momentum), at least among the multiplication operators, and 1 generates $\mathcal{A}$, which is not a proper ideal. The absence of positive–energy representations shows that Doplicher's ideal does not exist.

References

1. R. Haag and D. Kastler, J. Math. Phys. **5** (1964), 848-861.
2. R. F. Streater and A. S. Wightman, *PCT Spin and Statistics, and All That*, 3rd Ed., Addison Wesley (1989).
3. S. Doplicher, T. Regge and I. M. Singer, Commun. Math. Phys. **11** (1968), 51-54.
4. S. Doplicher, *An Algebraic Spectrum Condition*, Commun. Math. Phys. **1** (1965), 1-5.

Center for Transport Theory and Mathematical Physics, Virginia Tech, Blacksburg, VA 24061, USA
Permanent address: Department of Mathematics, King's College, Strand, London WC2R2LS, England

STABILIZATION OF BOTTOMLESS SYSTEMS BASED ON STOCHASTIC QUANTIZATION WITH A FIELD-DEPENDENT KERNELED LANGEVIN EQUATION

SATOSHI TANAKA AND MIKIO NAMIKI

ABSTRACT. We propose a new method for stochastically quantizing systems with bottomless action functionals, based on a field-dependent kerneled Langevin equation. We prove that our kerneled Langevin equation describes a substitute stochastic process with a thermal equilibrium state. The proof makes clear the relation between our method and the conventional path-integral quantization. We also demonstrate numerically that dynamical evolution leads to the desired equilibrium state. We check that our method gives the same perturbative results as in the conventional path-integral quantization. We also show that our method is applicable to numerical calculation of nonperturbative expectation values that are consistent with the perturbative results in the case of a small coupling constant.

Contribution[1] in honor of the 60th birthday of John R. Klauder

1. Introduction

Since the stochastic quantization method (SQM) was first proposed [1] many authors have pointed out that this method extends the territory of quantum theory. For example, Greensite and Halpern proposed a scheme to apply the SQM to bottomless systems [2]. Their scheme is based on the $(d+1)$-dimensional path-integral representation of the SQM [3–6], and it was recently applied to 2-dimensional gravity [7]. This scheme is interesting but unsatisfactory, in that the relation to the conventional d-dimensional path-integral quantization method is unclear. The key problem is that the stochastic process corresponding to the $(d+1)$-dimensional path-integral representation does not have an equilibrium state, reflecting the unboundedness of the action functional. As a result the Feynman path-integral measure is not given as an equilibrium Fokker-Planck distribution.

In this paper, we propose a new stochastic quantization scheme to quantize bottomless systems based on a kerneled Langevin equation [8]. (This paper is based on [9] and [10].) Our kerneled Langevin equation describes a stochastic process which has an equilibrium state, and the equilibrium Fokker-Planck distribution makes clear the relation between our

[1] Work in collaboration with I.Ohba, M.Mizutani, N.Komoike and M.Kanenaga.

method and the conventional d-dimensional path-integral quantization method. The presence of an equilibrium state makes it straightforward to apply our scheme to 2-dimensional gravity, for example.

Let us briefly review the SQM based on a kerneled Langevin equation. A scalar field ϕ characterized by an action functional $S[\phi]$ is quantized by setting up the following kerneled Langevin equation,

$$\begin{aligned}\frac{\partial}{\partial t}\phi(x,t) = &-\int d^dx' K(x,x';\phi)\frac{\delta S}{\delta\phi(x',t)} \\ &+\int d^dx' \frac{\delta K(x,x';\phi)}{\delta\phi(x',t)} \\ &+\int d^dx' G(x,x';\phi)\,\eta(x',t),\end{aligned} \tag{1.1a}$$

where t is the fictitious-time variable, and the kernel $K(x,x';\phi)$ is a real functional depending only on ϕ at fictitious time t. It is assumed that $K(x,x';\phi)$ is factorizable as

$$K(x,x';\phi) = \int d^dx'' G(x,x'';\phi)G(x',x'';\phi), \tag{1.1b}$$

with a real functional $G(x,x';\phi)$. The Gaussian white noise $\eta(x,t)$ is characterized by the statistical properties

$$<\eta(x,t)>=0\ , \qquad <\eta(x,t)\eta(x',t')>=2\delta^d(x-x')\delta(t-t')\ . \tag{1.1c}$$

The Fokker-Planck equation corresponding to (1.1) is

$$\frac{\partial}{\partial t}P[\phi,t] = \int d^dx \int d^dx' \frac{\delta}{\delta\phi(x)}K(x,x';\phi)\Big[\frac{\delta}{\delta\phi(x')}+\frac{\delta S}{\delta\phi(x')}\Big]P[\phi,t]\ . \tag{1.2}$$

It is clear that the Feynman path-integral measure, $\exp[-S]$, is a stationary solution of (1.2). When the action $S[\phi]$ is positive definite (has a bottom), it is easily proved that this stationary solution is realized in the equilibrium limit, and that we can calculate correct path-integral expectation values by making use of (1.1) and by taking the limit $t\to\infty$. What we propose in this paper is to apply (1.1) to quantize systems in which the action functional $S[\phi]$ is bottomless, i.e. the system is dynamically unstable.

2. General prescription to quantize bottomless systems

In this section, we formulate a general prescription to quantize a system with a bottomless action based on the kerneled Langevin equation described above. Let us consider a scalar field ϕ characterized by a bottomless action $S[\phi]$. Following the naive prescription of the SQM, we set up a Langevin equation (1.1) with a kernel $K(x,x';\phi)=\delta^d(x-x')$. But solutions of this naive Langevin equation run away to plus or minus infinity, so that an equilibrium state does not exist. In such a case, we cannot calculate expectation values of physical quantities in the limit $t\to\infty$. This means that the naive prescription of the SQM fails. Therefore we quantize our bottomless system using the kerneled Langevin equation (1.1), instead of the Langevin equation with the naive kernel. Let us choose a field-dependent kernel of the following form:

$$K(x,x';\phi)=\delta^d(x-x')\overline{K}[\phi]\ , \tag{2.1a}$$

where $\overline{K}[\phi]$ is assumed to be a real and positive functional of ϕ. When K has the form (2.1a), $G(x,x',;\phi)$ defined by (1.1b) is given by

$$G(x,x';\phi)=\delta^d(x-x')\overline{K}^{1/2}[\phi]\ . \tag{2.1b}$$

Then the kerneled Langevin equation (1.1a) takes the following form:

$$\frac{\partial}{\partial t}\phi(x,t)=-\overline{K}[\phi]\frac{\delta S[\phi]}{\delta\phi(x,t)}+\frac{\delta\overline{K}[\phi]}{\delta\phi(x,t)}+\overline{K}^{1/2}[\phi]\ \eta(x,t)\ . \tag{2.2}$$

Here we introduce a functional $S_-[\phi]$ defined by

$$S_-[\phi]\equiv S[\phi]-\ln\overline{K}[\phi]\ . \tag{2.3}$$

By using $S_-[\phi]$, we rewrite (2.2) as

$$\frac{\partial}{\partial t}\phi(x,t)=-\overline{K}[\phi]\frac{\delta S_-[\phi]}{\delta\phi(x,t)}+\overline{K}^{1/2}[\phi]\ \eta(x,t)\ . \tag{2.4}$$

Since we assume $\overline{K}[\phi]$ is positive, the drift term $-\overline{K}[\phi]\cdot\delta S_-[\phi]/\delta\phi(x,t)$ becomes a damping term *if we properly choose $\overline{K}[\phi]$ so as to have a positive-definite functional $S_-[\phi]$ (keeping a bottom).* In this case, solutions of (2.4) are confined to a finite region, and therefore we expect that an equilibrium state exists.

As an example, we consider the following bottomless ϕ^4-theory:

$$S[\phi]=S_2[\phi]-S_4[\phi]\ , \tag{2.5a}$$

$$S_2[\phi]=\int d^dx\ \left[\frac{1}{2}(\partial_\mu\phi)^2+\frac{1}{2}m^2\phi^2\right]\ , \tag{2.5b}$$

$$S_4[\phi]=\int d^dx\ \left[\frac{1}{4}\lambda\phi^4\right]\ ,\qquad \lambda>0\ . \tag{2.5c}$$

For the model (2.5), let us choose $\overline{K}[\phi]$ as

$$\overline{K}[\phi]=\exp[-S_4[\phi]]\ , \tag{2.6}$$

then $S_-[\phi]$ becomes

$$S_-[\phi]=S[\phi]+S_4[\phi]=S_2[\phi]\ , \tag{2.7}$$

which is now positive definite. Thus the Langevin equation (2.4) turns to

$$\frac{\partial}{\partial t}\phi(x,t) = -\exp[-S_4]\frac{\delta S_2}{\delta\phi(x,t)} + \exp[-S_4/2]\,\eta(x,t)\ . \tag{2.8}$$

The drift term of (2.8) has clearly a damping effect, so that the solution is confined to a finite region if a finite initial value is chosen. Note that in the limit $\lambda \to 0$, (2.8) reduces correctly to the Langevin equation for a free field in the naive SQM prescription, where $K(x,x';\phi) = \delta^d(x-x')$.

Returning back to the general case, we prove that the stochastic process (2.4) has an equilibrium state if we choose a proper $\overline{K}[\phi]$ such that $S_-[\phi]$ is positive definite and the drift term of (2.4) becomes a damping term. Note that the conventional arguments leading to the equilibrium distribution proportional to $\exp[-S]$ (see [4], for example) cannot be applied to this problem, since normalizability of $\exp[-S]$ is assumed in these arguments.

When the drift term of (2.4) is a damping term, we know that the Fokker-Planck distribution $P[\phi,t]$ is subject to the following boundary condition:

$$P[\phi,t] \underset{|\phi|\to\infty}{\longrightarrow} 0\ . \tag{2.9}$$

Recall that we need (2.9) to derive the Fokker-Planck equation equivalent to (2.4) or (2.2), since surface terms should vanish in performing the functional integration by parts arising in the usual way of derivation. The derived Fokker-Planck equation is

$$\frac{\partial}{\partial t}P[\phi,t] = \mathcal{H}\,P[\phi,t]\ , \tag{2.10a}$$

$$\mathcal{H} = \int d^dx\ \frac{\delta}{\delta\phi(x)}\overline{K}[\phi]\Big[\frac{\delta}{\delta\phi(x)} + \frac{\delta S[\phi]}{\delta\phi(x)}\Big]\ . \tag{2.10b}$$

Using (2.1), we can easily show that this Fokker-Planck equation is nothing but (1.2) with (2.1).

By making use of $S_-[\phi]$ given by (2.3), we can show that the Fokker-Planck operator $\mathcal{H}$ of (2.10b) becomes

$$\mathcal{H} = \int d^dx\ \frac{\delta}{\delta\phi(x)}\Big[\frac{\delta}{\delta\phi(x)} + \frac{\delta S_-[\phi]}{\delta\phi(x)}\Big]\overline{K}[\phi]\ . \tag{2.11}$$

Here let us introduce $\overline{\mathcal{H}}$ and $\overline{P}$ defined by

$$\overline{\mathcal{H}} = \mathcal{H}\overline{K}^{-1}[\phi]\ , \tag{2.12a}$$

$$\overline{P}[\phi,\tau(t)] = \overline{K}[\phi]P[\phi,t]\ , \tag{2.12b}$$

with

$$\tau(t) \equiv \overline{K}[\phi]t\ . \tag{2.12c}$$

By means of (2.12), the Fokker-Planck equation (2.10a) with (2.11) is transformed to

$$\frac{\partial}{\partial \tau}\overline{P}[\phi,\tau] = \overline{\mathcal{H}}\,\overline{P}[\phi,\tau]\,, \tag{2.13a}$$

$$\overline{\mathcal{H}} = \int d^d x\, \frac{\delta}{\delta\phi(x)}\left[\frac{\delta}{\delta\phi(x)} + \frac{\delta S_-[\phi]}{\delta\phi(x)}\right]. \tag{2.13b}$$

Remember that we have performed not only transformations of the Fokker-Planck operator and the distribution functional ($\mathcal{H} \to \overline{\mathcal{H}}$, $P \to \overline{P}$) but also a scale transformation of the fictitious time ($t \to \tau$): note that $\overline{K}[\phi]$ is positive.

Equation (2.13) has exactly the same form as the Fokker-Planck equation of the naive SQM prescription, where its action functional is $S_-[\phi]$ and $K(x,x';\phi) = \delta^d(x-x')$. We know that $\overline{\mathcal{H}}$ is a negative semi-definite operator having negative eigenvalues. Therefore, when $S_-[\phi]$ is positive definite, the asymptotic behavior of $\overline{P}[\phi,\tau]$ is

$$\overline{P}_{\text{eq}}[\phi] \equiv \lim_{\tau\to\infty} \overline{P}[\phi,\tau] = \text{const}\cdot\exp[-S_-[\phi]]\,. \tag{2.14}$$

Since $\exp[-S_-]$ is a normalizable functional, following the conventional arguments (see [4], for example) concerned with a Fokker-Planck equation allows us to prove (2.14) in this case. Rewriting (2.14) by means of (2.12b) and (2.3), we can obtain the asymptotic behavior of the Fokker-Planck distribution itself as

$$P_{\text{eq}}[\phi] \equiv \overline{K}^{-1}[\phi]\overline{P}_{\text{eq}}[\phi] = \text{const}\cdot\exp[-S[\phi]]\,. \tag{2.15}$$

Thus we have shown that our stochastic process (2.4) has the equilibrium state expressed by the Feynman path-integral measure $\exp[-S]$, which is clearly a solution of the Fokker-Planck equation (2.10), as far as $\overline{K}$ is regular. Note that, in principle, the equilibrium state should be supplemented with the boundary condition (2.9).

3. Perturbation theory

Let us perform perturbative calculations based on our method and see its consistency with the conventional path-integral quantization method which is meaningful only at the perturbative level.

We apply the above general scheme to a zero-dimensional ϕ^3-model, whose action is given by

$$S[\phi] = S_2[\phi] - S_3[\phi]\,, \tag{3.1a}$$

$$S_2[\phi] = \frac{1}{2}m^2\phi^2\,, \tag{3.1b}$$

$$S_3[\phi] = \frac{1}{3}g\phi^3\,. \tag{3.1c}$$

(Extension of the following arguments to higher dimensional cases is quite straightforward.) For the model (3.1), we choose the following $\overline{K}[\phi]$:

$$\overline{K}[\phi] = \exp[-S_3[\phi]]\,. \tag{3.2}$$

For the choice (3.2), the Langevin equation (2.4) becomes

$$\frac{\partial}{\partial t}\phi(t) = -m^2 \exp[-S_3]\phi(t) + \exp[-S_3/2]\ \eta(t)\ . \tag{3.3}$$

In the limit $g \to 0$, (3.3) reduces correctly to the Langevin equation for the free field theory (Gaussian model) in the naive stochastic quantization prescription, where $\overline{K}[\phi] = 1$.

Let us perform the Langevin perturbation [1] for the model (3.1) based on the Langevin equation (3.3). Expanding the exponential factors in the drift and random terms of (3.3) up to the second order in g, we obtain

$$\frac{\partial}{\partial t}\phi(t) = -m^2\phi(t) + \eta(t) + \frac{m^2}{3}g\phi^4(t) - \frac{1}{6}g\eta(t)\phi^3(t) - \frac{m^2}{18}g^2\phi^7(t) + \frac{1}{72}g^2\eta(t)\phi^6(t)\ . \tag{3.4}$$

The stationary solution of (3.4) for a very large t is

$$\phi(t) = \int_{-\infty}^{\infty} ds\ G(t-s)\Big[\eta(s) + \frac{m^2}{3}g\phi^4(s) - \frac{1}{6}g\eta(s)\phi^3(s) - \frac{m^2}{18}g^2\phi^7(s) + \frac{1}{72}g^2\eta(s)\phi^6(s)\Big]\ , \tag{3.5}$$

where $G(t-s)$ is the Green's function,

$$G(t-s) \equiv \theta(t-s)\exp\left[-m^2(t-s)\right]\ , \tag{3.6}$$

with the Heaviside step function defined by

$$\theta(t \geq 0) = 1\ , \qquad \theta(t < 0) = 0\ . \tag{3.7}$$

Note that we are using the convention, $\theta(0) = 0$, based on the Ito calculus. Validity of this convention in perturbative calculation is explained in references [5], [11] (see also [6] and [12]).

Equation (3.5) can be solved iteratively. We introduce the diagramatical expression summarized in Table 1 for the perturbative solution: $G(t-s)$ is expressed by the directed line and η by the cross. Then the solution up to the order of g^2 is expressed as Fig.1.

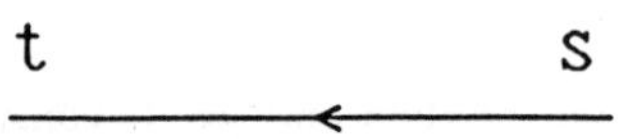

$\equiv G(t-s)$

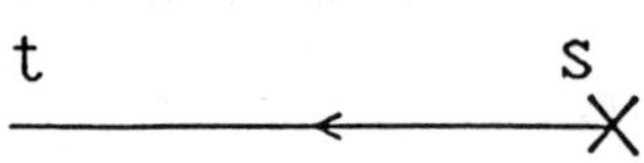

$\equiv \int ds\ G(t-s)\eta(s)$

$\equiv \eta$

TABLE 1: RULES FOR DIAGRAMATICAL EXPRESSION OF A PERTURBATIVE SOLUTION OF THE LANGEVIN EQUATION.

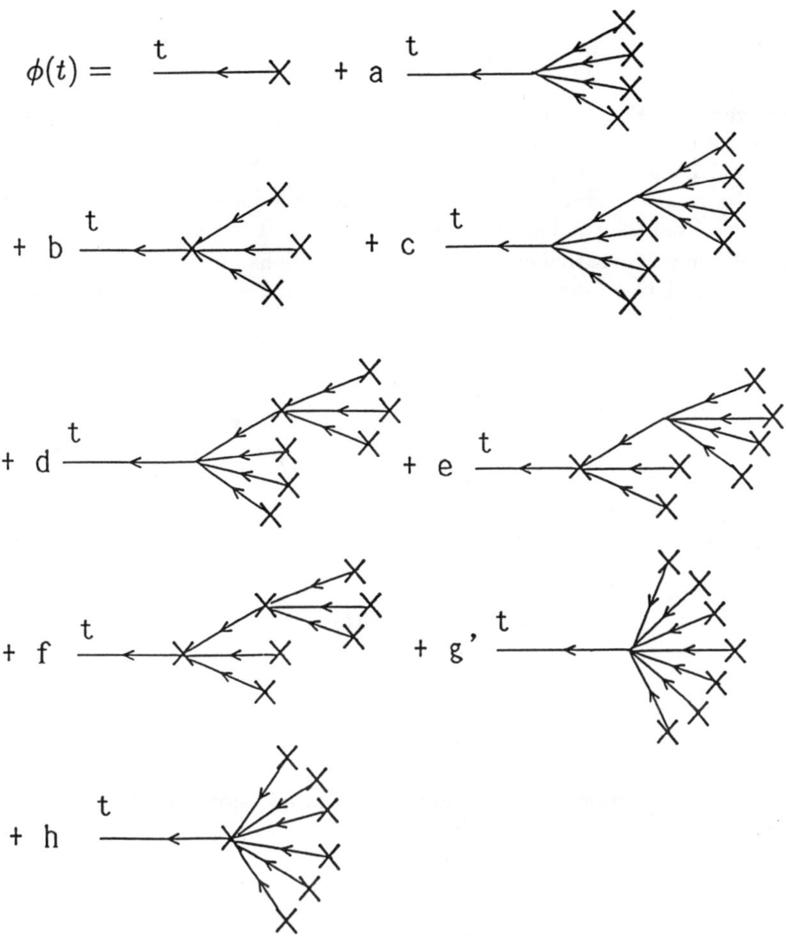

FIGURE 1: DIAGRAMATICAL EXPRESSION OF THE PERTURBATIVE SOLUTION OF (3.5) FOR THE ϕ^3-MODEL UP TO THE ORDER g^2. THE COEFFICIENTS $(a,\ b,\ \cdots,\ h)$ ARE:

$$a = \frac{m^2}{3} g\,, \quad b = -\frac{1}{6} g, \quad c = \frac{4m^4}{9} g^2, \quad d = -\frac{2m^2}{9} g^2$$
$$e = -\frac{m^2}{6} g^2, \quad f = \frac{1}{12} g^2, \quad g' = -\frac{m^2}{18} g^2, \quad h = \frac{1}{72} g^2\,.$$

On each vertex of Fig.1, internal fictitious-time integration is performed. Note that the Gaussian white noises, η's, expressed by the crosses also exist on vertices, being different from the usual Langevin perturbation in the naive SQM prescription where $\overline{K} = 1$. It is because the random term of (3.4) explicitly depends on the field ϕ, and interactions (vertices) come from the random term as well as the drift term.

We can obtain perturbative quantum n-point functions by calculating equal-fictitious-time n-point correlation functions using the solution in Fig.1. In performing the calculation, we use the formula,

$$< \phi_0(t)\phi_0(t') > = 2\int ds\ G(t-s)G(t'-s) \tag{3.8a}$$

$$= \frac{1}{m^2}\exp\left[-m^2 \mid t-t' \mid\right] \equiv D(t-t')\ , \tag{3.8b}$$

where $\phi_0(t)$ is the free part of the solution,

$$\phi_0(t) = \int_{-\infty}^{\infty} ds\ G(t-s)\eta(s)\ . \tag{3.9}$$

We also use the formula,

$$G(0) = \theta(0) = 0\ , \tag{3.10}$$

coming from the convention mentioned above. The propagator $D(t-t')$ is diagramatically expressed by the crossed line as in Table 2.

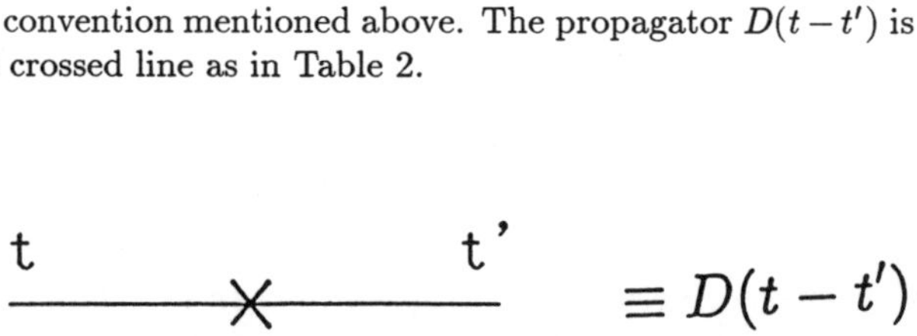

TABLE 2: DIAGRAMATICAL EXPRESSION OF THE FREE PROPAGATOR $D(t-t')$.

First, we calculate a three-point function up to the order of g^1. This calculation includes most of the interesting features of the perturbation based on our kerneled Langevin equation. Based on the solution in Fig.1 and using the formula (3.8), the three-point function is calculated as

$$\begin{aligned}
< \phi(t)^3 > = m^2 g \Big[3 \int ds\, G(t-s)D(s-s)D(s-s) \cdot D(t-t) \\
+12 \int ds\, G(t-s)D(s-s)D(s-t)D(s-t) \Big] \\
- \frac{1}{2} g \Big[6 \cdot 2 \int ds\, G(t-s)G(t-s)D(s-s)D(s-t) \\
+ 6G(0) \int ds\, G(t-s)D(s-t)D(s-t) \\
+3G(0) \int ds\, G(t-s)D(s-s) \cdot D(t-t) \Big] .
\end{aligned} \tag{3.11}$$

Each term in (3.11) corresponds to a diagram in Fig.2.

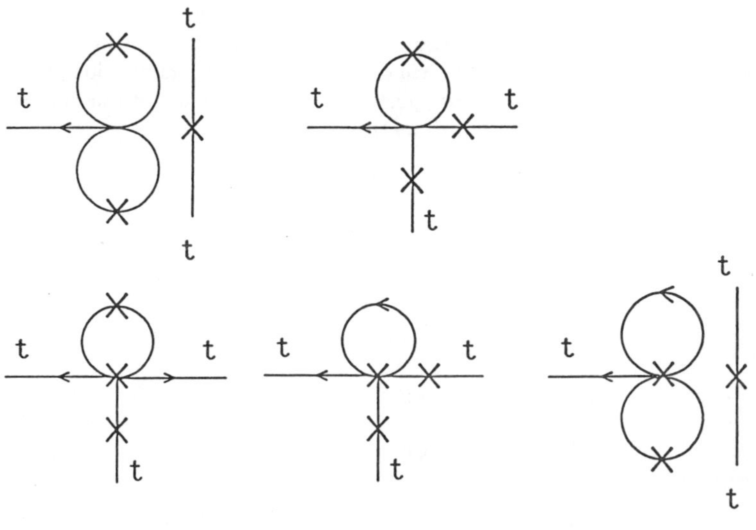

FIGURE 2: DIAGRAMS REPRESENTING PERTURBATIVE EXPANSION OF THE THREE-POINT FUNCTION (3.11) FOR THE ϕ^3-MODEL UP TO THE ORDER OF g^1. TO THE FIRST TWO DIAGRAMS, THE VERTEX GIVEN BY THE DRIFT TERM OF (3.4) CONTRIBUTES. TO THE LATTER THREE DIAGRAMS, THE VERTEX GIVEN BY THE RANDOM TERM OF (3.4) CONTRIBUTES.

The last two terms in (3.11), which vanish because of (3.10), correspond to diagrams with a closed directed line (closed G-line). Such terms do not appear in the usual Langevin perturbation from the beginning, but they do in our case because of the field-dependence of the random term in (3.4). In Fig.2, a vertex with a cross attached to two directed lines is evaluated as

$$2\int ds\ G(t-s)G(t-s)\times[\text{a function of } t \text{ and } s]\ , \tag{3.12}$$

where the two G's cannot be combined to make one D as in (3.8). In (3.12), the coefficient "2" comes from the diffusion constant in (1.1c). Conventionally, we can understand that a cross on a vertex represents a factor "2" in a perturbative expectation value (after taking ensemble averages).

Performing the fictitious-time integration in (3.11), we obtain the following result:

$$<\phi(t)^3>=3g\left(\frac{1}{m^2}\right)^3+2g\left(\frac{1}{m^2}\right)^3+\mathrm{O}(g^2)\ . \tag{3.13}$$

In (3.13), the first term of the right-hand side corresponds to the non-zero disconnected diagram in Fig.2 (the first diagram), and the second term to the sum of the non-zero connected diagrams (the second and third diagrams). The first term also coincides with the contribution of the disconnected diagram in the conventional path-integral quantization method (see Fig.3a), while the second term coincides with the contribution of the connected diagram (see Fig.3b). This fact shows perturbative consistency of our method with the conventional path-integral quantization method.

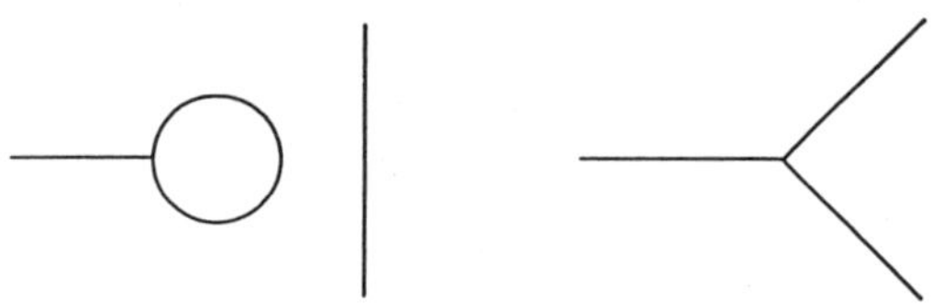

FIGURE 3: THE DISCONNECTED (A) AND CONNECTED (B) DIAGRAMS IN THE THREE-POINT FUNCTION OF THE ORDER g^1 FOR THE ϕ^3-MODEL, WHICH APPEAR IN THE CONVENTIONAL PATH-INTEGRAL QUANTIZATION METHOD.

Next, we consider terms of the order g^2 in the three-point function. When we make a product of three $\phi(t)$'s in Fig.1, terms of the order g^2 contain odd (nine) η's. Hence, these terms vanish because of the Gaussian property of the noise. (A correlation of odd noises is zero.) Then the three-point function becomes

$$<\phi(t)^3>=5g\left(\frac{1}{m^2}\right)^3+\mathrm{O}(g^3)\ . \tag{3.14}$$

This exactly coincides with the three-point function calculated with the conventional path-integral quantization method up to the order of g^2.

We can calculate one and two-point functions in the same way. The results are

$$< \phi(t) >= g \left(\frac{1}{m^2}\right)^2 + O(g^3) \ , \tag{3.15}$$

$$< \phi(t)^2 >= \frac{1}{m^2} + 5g^2 \left(\frac{1}{m^2}\right)^4 + O(g^3) \ . \tag{3.16}$$

These results agree with the ones obtained by the conventional path-integral quantization method. As for the one-point function in (3.15), it is clear from Fig.1 that each term of the order g^0 or g^2 contains odd (one for the former and seven for the latter) η's, and so such a term vanishes. As for the two-point function in (3.16), each term of the order g^1 contains five η's, and so it vanishes. To terms of the order g^2 in the two-point function, nineteen connected three-loop diagrams[2] and one disconnected diagram contribute. The disconnected diagram is shown in Fig.4a, whose contribution coincides with the contribution of the disconnected diagram in Fig.4b appearing in the conventional path-integral method.

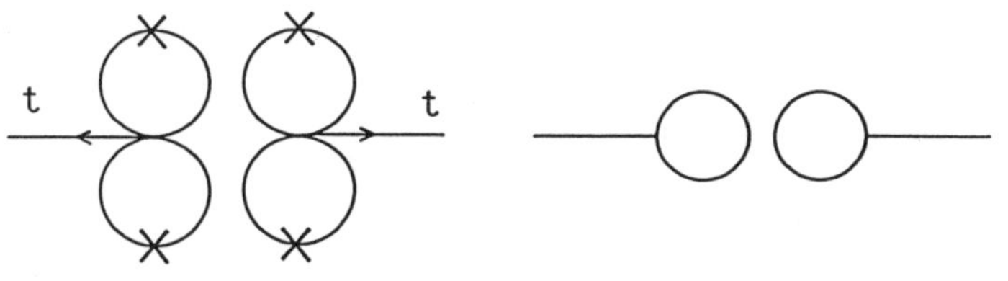

FIGURE 4: THE DISCONNECTED DIAGRAM FOR THE ϕ^3-MODEL APPEARING IN OUR LANGEVIN PERTURBATION OF THE TWO-POINT FUNCTION (A) AND ITS CORRESPONDING DIAGRAM IN THE CONVENTIONAL PATH-INTEGRAL QUANTIZATION METHOD (B).

Thus, for the ϕ^3-model, we have shown that our method gives the same perturbative results as the conventional path-integral quantization method for the one, two and three-point functions up to the second order of the coupling constant. Moreover we have confirmed that our disconnected diagrams exactly correspond to the disconnected diagrams in the conventional path-integral method.

Perturbative calculation in the bottomless ϕ^4-model (2.5) can be done in the same way. Based on the Langevin equation (2.8), the two-point function for a zero-dimensional theory is calculated as:

$$< \phi^2 >= \frac{1}{m^2} + \frac{3}{m^6}\lambda + O(\lambda^2) \ . \tag{3.17}$$

This result also coincides with the one obtained through perturbative calculation based on the conventional path-integral quantization method.

[2]Some of the nineteen diagrams look exactly the same but are made of different combination of vertices in Fig.1.

4. Numerical simulation

Next we demonstrate that our method is applicable to numerical calculation of nonperturbative expectation values. We also check validity of the general argument in Section 2 numerically.

First, we perform numerical simulation of the ϕ^3-model (3.1). We calculate the two-point function,

$$< \phi^2 >= \lim_{t\to\infty} < \phi^2(t) > , \tag{4.1}$$

numerically solving (3.3) for various values of g, keeping $m = 1$. The result is shown in Fig.5 (squares).

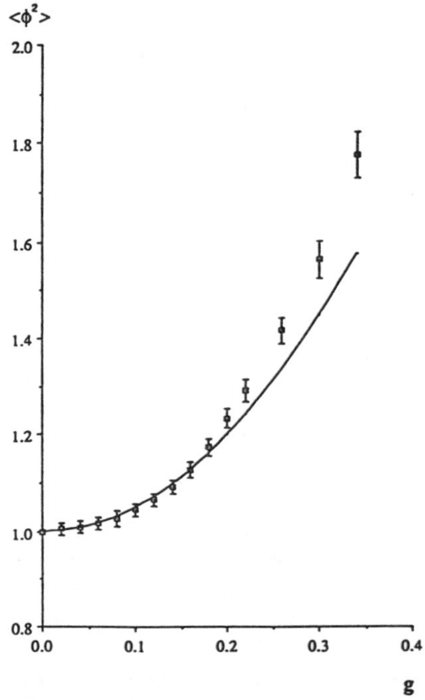

FIGURE 5: EXPECTATION VALUE $< \phi^2 >$ OF THE ϕ^3-MODEL FOR VARIOUS VALUES OF g WITH $m = 1$. $< \phi^2 >$ IS CALCULATED BY SOLVING (3.3) NUMERICALLY (OPEN SQUARES). WE CHOOSE A DISCRETIZED FICTITIOUS TIME STEP OF $\Delta t = 0.0025$, PERFORM 2,000,000 ITERATIONS, AND USE THE LAST 1,950,000 ITERATIONS TO CALCULATE THE EXPECTATION VALUES. THE SOLID LINE SHOWS THE PERTURBATIVE RESULT UP TO THE ORDER g^2.

The perturbative result (3.16) with $m = 1$ is represented by the solid line in Fig.5. We see that the numerically calculated nonperturbative expectation values are consistent with the perturbative result.

Next, Fig.6 shows the distribution of the numerical solutions of ϕ given by (3.3) for $g = 0.1$, $m = 1$ (open circles).

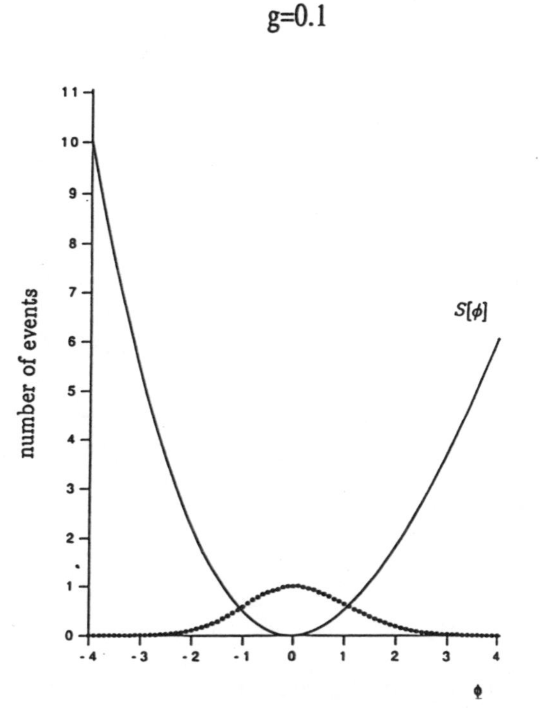

FIGURE 6: DISTRIBUTION OF NUMERICAL SOLUTIONS OF (3.3) (OPEN CIRCLES) FOR $g = 0.1$, $m = 1$ IN THE ϕ^3-MODEL. FOR COMPARISON, WE PLOT THE ACTION S OF (3.1) (SOLID LINE) FOR THE SAME PARAMETERS.

We see that the solution is indeed confined to a finite region, as is expected from the general argument in Section 2.

Fig.6 also shows that the form of the distribution is consistent with the form of the action $S[\phi]$ (solid line). To see this more clearly, we calculate $S_{\rm cal}[\phi] \equiv -\ln(\text{distribution in Fig.6})$. This should coincide with the action $S[\phi]$ apart from a constant shift, since the equilibrium Fokker-Planck distribution should be proportional to $\exp[-S]$. Fig.7 shows that this is indeed correct.

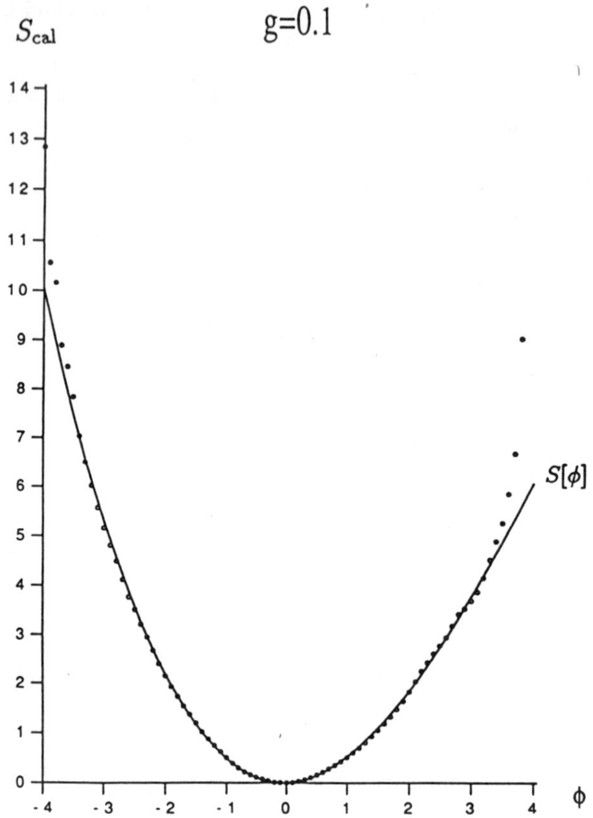

FIGURE 7: THE ACTION S_{cal} FOR THE ϕ^3-MODEL NUMERICALLY CALCULATED BY $-\ln$ (DISTRIBUTION IN FIG.6) (OPEN CIRCLES) IN COMPARISON WITH THE ORIGINAL ACTION S OF (3.1) (SOLID LINE).

We also perform numerical simulation of the bottomless ϕ^4-model (2.5) in zero dimension.

We numerically calculate the two-point function (4.1) for various values of λ, keeping $m = 1$. The preliminary result is shown in Fig.8 (circles).

The perturbative result (3.17) with $m = 1$ is represented by the solid line in Fig.8. We see that the numerically calculated nonperturbative expectation values are consistent with the perturbative result.

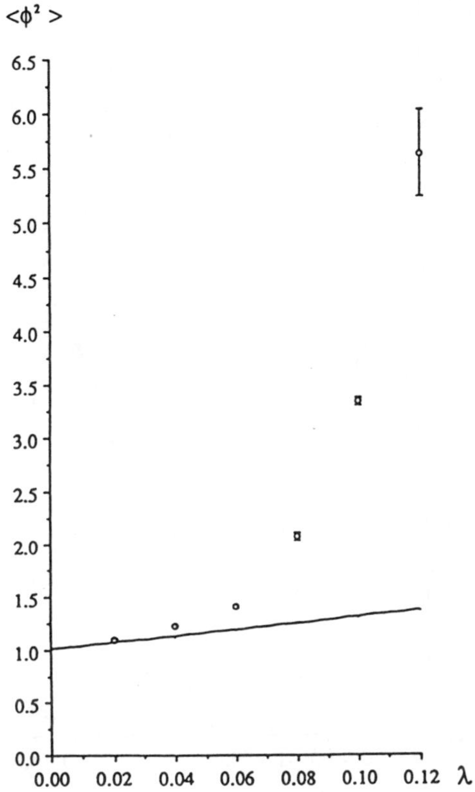

FIGURE 8: EXPECTATION VALUE $<\phi^2>$ OF THE BOTTOMLESS ϕ^4-MODEL (2.5) IN ZERO DIMENSION FOR VARIOUS VALUES OF λ WITH $m = 1$. $<\phi^2>$ IS CALCULATED BY SOLVING (2.8) NUMERICALLY (OPEN CIRCLES). WE CHOOSE A DISCRETIZED FICTITIOUS TIME STEP OF $\Delta t = 0.01$, PERFORM 1,500,000 ITERATIONS, AND USE THE LAST 1,450,000 ITERATIONS TO CALCULATE THE EXPECTATION VALUES. THE SOLID LINE SHOWS THE PERTURBATIVE RESULT UP TO THE ORDER λ^1.

Next, Fig.9 shows the distribution of the numerical solution of (2.8) for $\lambda = 0.1$, $m = 1$. We see that the solution is confined to a finite region, as in the ϕ^3-model.

Fig.10, in which we plot $S_{\rm cal} \equiv -\ln(\text{distribution in Fig.9})$, shows that the equilibrium Fokker-Planck distribution is proportional to $\exp[-S]$.

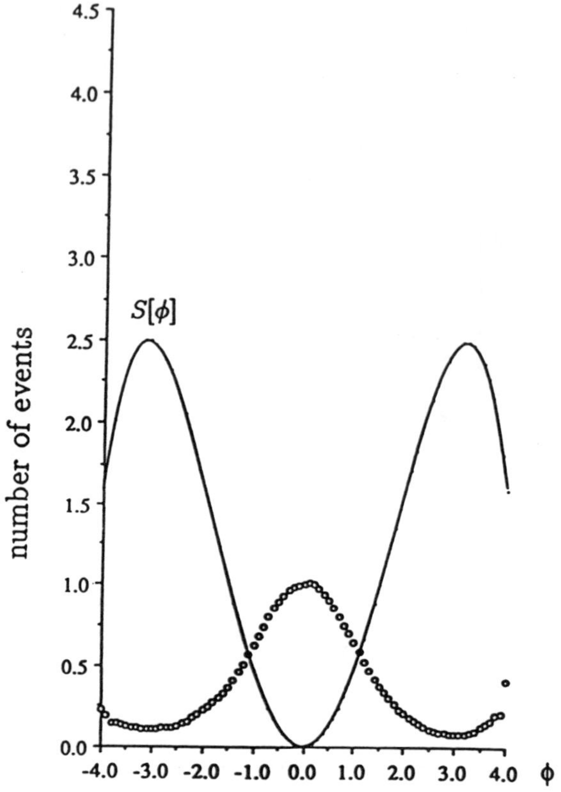

FIGURE 9: DISTRIBUTION OF NUMERICAL SOLUTIONS TO (2.8) (OPEN CIRCLES) FOR $\lambda = 0.1$, $m = 1$ IN THE BOTTOMLESS ϕ^4-MODEL. FOR COMPARISON, WE PLOT THE ACTION S OF (2.5) (SOLID LINE) FOR THE SAME PARAMETERS.

Finally, we check the validity and accuracy of the numerical simulation based on our kerneled Langevin equation. In Fig.11 we show how relative errors in the action calculated from the distribution decrease as the sampling number increases.

The relative error is defined by

$$\Delta S/S\ , \qquad \Delta S \equiv S_{\text{cal}} - S\ , \tag{4.2}$$

where $S_{\text{cal}}[\phi] \equiv -\ln(\text{distribution in Fig.9}) + \text{const}$, and the constant is chosen so that $S_{\text{cal}}[\phi = 0] = 0$. Fig.11 shows that the distribution generated by our Langevin equation correctly approaches the Feynman path-integral measure as the sampling number increases. We should note that a larger sampling number is necessary in areas beyond the maximum points of $S[\phi]$.

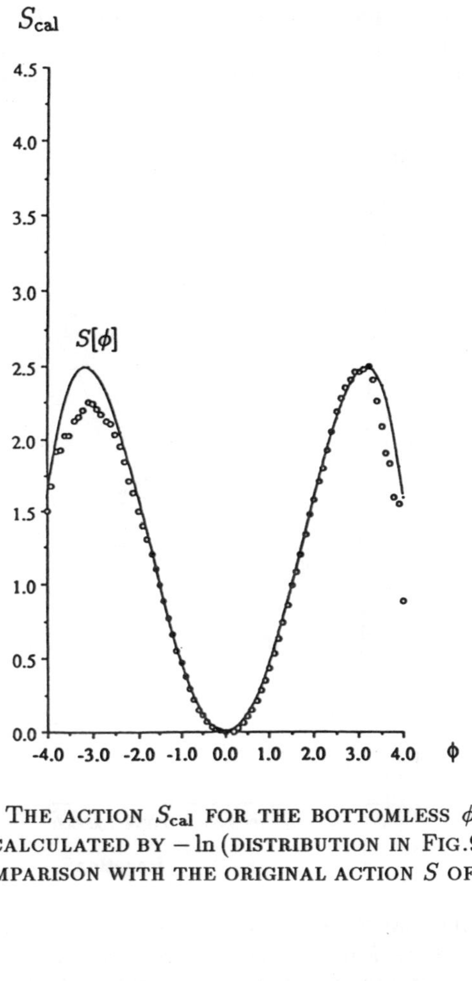

FIGURE 10: THE ACTION S_{cal} FOR THE BOTTOMLESS ϕ^4-MODEL NUMERICALLY CALCULATED BY $-\ln$ (DISTRIBUTION IN FIG.9) (OPEN CIRCLES) IN COMPARISON WITH THE ORIGINAL ACTION S OF (2.5) (SOLID LINE).

5. Conclusions

In this paper, we have proposed a new method to quantize systems with bottomless action functionals. Our method is based on a generalized stochastic quantization scheme, using a Langevin equation with a field-dependent kernel. We have shown that our kerneled Langevin equation describes a stochastic process with a thermal equilibrium state. This equilibrium state is described by the Feynman path-integral measure, $\exp[-S]$, with the boundary condition that fields are confined to a finite region. This argument makes the relation between our method and the conventional path-integral quantization method clear. In simple models, we have checked numerically that the desired equilibrium state is realized and that we can calculate nonperturbative quantum expectation values. We have also demonstrated that our method gives the same perturbative results as the conventional path-integral quantization method. We can also construct a $(d+1)$–dimensional path-integral version of our method that allows us to give a meaning to the ill-defined d–dimensional Feynman path-integral. For details, see [9]. Application of our method to 2-dimensional gravity is in progress.

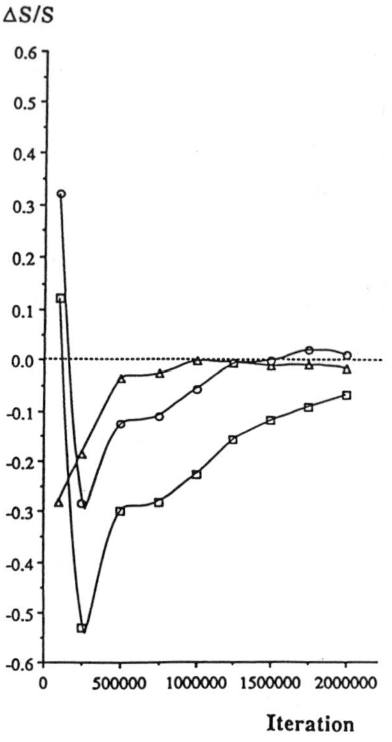

FIGURE 11: DECREASING RELATIVE ERRORS (4.2) IN THE ACTION FOR VARIOUS VALUES OF ϕ IN THE BOTTOMLESS ϕ^4-MODEL: TRIANGLES SHOW THE RELATIVE ERRORS FOR $\phi = 2.0$, CIRCLES FOR $\phi = 3.2$, AND SQUARES FOR $\phi = 3.5$.

REFERENCES

1. G. Parisi and Y. S. Wu, Sci. Sin. **24** (1981), 483; for a general review, see: M. Namiki, *Stochastic Quantization*, Springer, Heidelberg, 1992.
2. J. Greensite and M. B. Halpern, Nucl. Phys. **B242** (1984), 167.
3. E. Gozzi, Phys. Rev. **D28** (1983), 1922; Y. Nakano, Prog. Theor. Phys. **69** (1983), 361; C. M. Bender et al., Nucl. Phys. **B219** (1983), 61.
4. M. Namiki and Y. Yamanaka, Prog. Theor. Phys. **69** (1983), 1764.
5. H. Kawara, M. Namiki, H. Okamoto and S. Tanaka, Prog. Theor. Phys. **84** (1990), 749.
6. H. Nakazato, K. Okano, L. Schülke and Y. Yamanaka, Nucl. Phys. **B346** (1990), 611.

7. M. Karliner and A. Migdal, Mod. Phys. Lett. **A5** (1990), 2565; J. Ambjørn, J. Greensite and S. Varsted, Phys. Lett. **249B** (1990), 411; J. Ambjørn and J. Greensite, Phys. Lett. **254B** (1991), 66; see also: E. Marinari and G. Parisi, Phys. Lett. **240B** (1990), 375.
8. for example: B. Sakita, Quantum Theory of Many-Variable Systems and Fields, World Scientific, Singapore (1985); P.H. Damgaard and K. Tsokos, Nucl. Phys. **B235** (1984), 75; J.D. Breit, S. Gupta and A. Zaks, Nucl. Phys. **B233** (1984), 61; H. Okamoto, K. Okano, L. Schülke and S. Tanaka, Nucl. Phys. **B324** (1989), 684.
9. S. Tanaka, M. Namiki, I. Ohba, M. Mizutani, N. Komoike and M. Kanenaga, Waseda University Preprint WU-HEP-92-2.
10. S. Tanaka, I. Ohba, M. Namiki, M. Mizutani, N. Komoike and M. Kanenaga, Waseda University Preprint WU-HEP-92-4.
11. N. Komoike, Prog. Theor. Phys. **86** (1991), 575.
12. R. Mochizuki, Mod. Phys. Lett. **A6** (1991), 117.

DEPARTMENT OF PHYSICS, WASEDA UNIVERSITY, TOKYO 169, JAPAN

DYNAMICAL SYMMETRY OF THE $X-X$ CHAIN IN A MAGNETIC FIELD

T.T. Truong

Abstract. It is shown that for a very simple quantum chain a dynamical symmetry algebra may be realized by generators of corner transfer matrices recently studied in the literature of exactly soluble models in statistical mechanics. As by–products we obtain some new relationships for orthogonal polynomials and some extensions of the Yang–Baxter equations.

Contribution in honor of the 60th birthday of John R. Klauder

1. In the usual sense the concept of dynamical symmetry, where the Hamiltonian operator of a system is shown to be an element of a Lie–algebra, has concerned mainly quantum systems with finite number of degrees of freedom. The hydrogen atom and the harmonic oscillator are prime examples and have been widely discussed [1]. In most cases of interest this algebra is simple and different energy eigenstates may be connected by applying raising and lowering operators. Recently the hypothesis of conformal invariance in critical bidimensional systems has been shown to imply a dynamical symmetry for systems with infinitely many degrees of freedom which is realized by an infinite dimensional algebra: the Virasoro algebra (for a review, see [2]). Although a wealth of results has been obtained so far through the study of the representations of the Virasoro algebra, the explicit construction of this algebra from the building blocks of a definite solvable model has not been totally satisfactory. We would like to reconsider here this problem but restrict ourselves to the construction of the $SU(1,1)$ algebra, a subalgebra of the Virasoro algebra, for a simple 6–vertex model in lattice statistical mechanics.

Such a vertex system on a square lattice is completely specified by the Boltzmann weights:

$$\begin{array}{ll} w_1 = \sinh(\lambda+\alpha) & w_2 = \sinh(\lambda-\alpha) \\ w_3 = w_4 = \sinh\alpha & w_5 = w_6 = \sinh\lambda \end{array} \tag{1}$$

in the standard notations and parameterization of Baxter [3]. Moreover these weights also fulfill the free fermion condition:

$$w_1 w_2 + w_3 w_4 = w_5 w_6 \tag{2}$$

The row–to–row transfer matrix of this lattice commutes with the Hamiltonian

$$H = \sum_n H_{n\ n+1} \tag{3}$$

where

$$H_{n\ n+1} = \frac{1}{2\sinh\lambda}\left\{(\sigma_n^x\sigma_{n+1}^x + \sigma_n^y\sigma_{n+1}^y) + \cosh\lambda\,(\sigma_n^z + \sigma_{n+1}^z)\right\} \tag{4}$$

This is the so–called $X - X$ chain in the magnetic field $h = \cosh\lambda$ [4].

Another transfer matrix, relevant for the computation of one–point (correlation) functions is the corner transfer matrix introduced by Baxter. It has been shown that its strong anisotropic limit is the operator [5]:

$$L_0 = \sum_n nH_{n\ n+1} \tag{5}$$

which is suspected to be the "central" generator of the Virasoro algebra, the algebra generated by the $\{L_n, n \in \mathbb{Z}\}$ and closed under the rule:

$$[L_n, L_m] = (n-m)L_{n+m} + \frac{c}{12}n(n^2-1)\,\delta_{n+m,0} \tag{6}$$

c being the central charge of the algebra; it has the value $1/2$ in the case of free fermions. However if one considers only $n = 0, \pm 1$, the central charge c does not play a role in (6). (Note that we understand the $SU(1,1)$ symmetry in the loose sense of $SU(1,1) \simeq SL(2,\mathbb{R}) \simeq SO(2,1)$).

Recently we have shown that the L_0–operator may be diagonalized and its spectrum completely determined [6]. This is due to the fact that L_0 may be written as a bilinear form in fermions and that the associated matrix is a tridiagonal Jacobi matrix: $\mathcal{L}_0$, defined as follows. Let $|\psi>$ be a vector on which $\mathcal{L}_0$ operates, then if ψ_n is the nth component of $|\psi>$ one has:

$$\mathcal{L}_0\psi_n = \frac{1}{2\sinh\lambda}\{n\psi_{n+1} + (2n-1)\cosh\lambda\,\psi_n + (n-1)\psi_{n-1}\}. \tag{7}$$

In [6] it is shown that the components ψ_n are given by Gottlieb polynomials $l_n(x,\lambda)$, known also as Meixner polynomials of the first kind [7]

$$\psi_{n+1} = (-e^{\lambda})^n l_n(x,\lambda). \tag{8}$$

For an infinite lattice the spectrum of $\mathcal{L}_0$ is an oscillator spectrum

$$\varepsilon_\nu = (\nu + \frac{1}{2}) \quad , \quad \nu = 0, 1, 2, \ldots \tag{9}$$

and the component of the corresponding eigenvector is

$$\psi_{n+1} = (-e^{\lambda})^n l_n(\nu,\lambda). \tag{10}$$

As the Gottlieb polynomials satisfy the orthogonality and normalization relations:

$$\sum_{\nu=0}^{\infty} e^{-2\lambda\nu} l_n(\nu) l_m(\nu) = \delta_{mn}\frac{e^{-2\lambda n}}{1-e^{-2\lambda}} \tag{11}$$

and the discrete symmetry property [7]:

$$e^{2n\lambda} l_n(\nu) = e^{2\nu\lambda} l_\nu(n) \tag{12}$$

for both n and ν integers, it will be possible to compute the norm squared of the eigenstate $|\psi(\nu)>$:

$$\begin{aligned}(\text{Norm}|\psi>)^2 &= \sum_{n=0}^{\infty} e^{2\lambda n} l_n(\nu,\lambda) l_n(\nu,\lambda) \\ &= \frac{1}{(1-e^{-2\lambda})}.\end{aligned} \tag{13}$$

The normalized components of $|\psi>$ are then:

$$\psi_{n+1}(\nu) = \sqrt{1-e^{-2\lambda}}\,(-e^{\lambda})^n\, l_n(\nu,\lambda) \tag{14}$$

for $n = 0, 1, 2, \ldots$.

2. Now we may ask whether there also exist tridiagonal Jacobi matrices $\mathcal{L}_{\pm 1}$ which obey with $\mathcal{L}_0$ the commutation rules of the $SU(1,1)$ algebra. The answer was given in a special case by C. Itzykson [8] long ago and in the general case by A. S. Zhedanov [9] quite recently. Ityzkson started out with a group theoretical approach whereas Zhedanov begins with the Lax–pair of a Toda lattice. It is remarkable that the Lax–pair method for the Toda lattice already introduces the tridiagonal Jacobi matrices from the very start. In any case both $\mathcal{L}_{\pm 1}$ are defined by:

$$\mathcal{L}_{\mp 1}\psi_n = \frac{1}{2\sinh\lambda}\{e^{\mp\lambda} n\psi_{n+1} + (2n-1)\psi_n + e^{\pm\lambda}(n-1)\psi_{n-1}\}. \tag{15}$$

From the structure one may imply that the $\mathcal{L}_{\pm 1}$ matrices belong to generators of corner transfer matrices:

$$L_{\pm 1} = \sum_n n H^{\pm}_{n\ n+1} \tag{16}$$

where

$$H^{\mp}_{n\ n+1} = \frac{1}{2\sinh\lambda}\left\{2e^{\mp\lambda}\sigma_n^+\sigma_{n+1}^- + 2e^{\pm\lambda}\sigma_n^-\sigma_{n+1}^+ + (\sigma_n^z + \sigma_{n+1}^z)\right\} \tag{17}$$

clearly with a critical magnetic field $h = h_c = 1$.

Evidently $H^{\pm}_{n\ n+1}$ come from two free fermion 6–vertex systems defined by the Boltzmann weights with broken chiral symmetry:

$$\begin{aligned} w_1^{\pm} &= \sinh\alpha & w_2^{\pm} &= -\sinh\alpha \\ w_3^{\pm} &= e^{\pm\lambda}\sinh\alpha & w_4^{\pm} &= e^{\mp\lambda}\sinh\alpha \\ w_5^{\pm} &= 0 & w_6^{\pm} &= 0 \end{aligned} \tag{18}$$

Thus $L_{\pm 1}$ may be viewed as generators of corner transfer matrices for those lattice 6–vertex models.

Before we discuss the relationship between the three systems of vertices, let us come back to the $SU(1,1)$ structure which was studied by A. O. Barut and L. Girardello long ago [10]. This structure means that if $|\psi >$ is an eigenstate of L_0 then $L_\pm|\psi >$ is an eigenstate of L_0 with the eigenvalue $(\nu \mp 1)$

$$L_0(L_\pm|\psi >) = (\nu \mp 1)(L_\pm|\psi >). \tag{19}$$

An immediate consequence is therefore the relation (eq. (15))

$$(2\sinh\lambda)\,\psi_n(\nu \mp 1) = \left\{e^{\pm\lambda} n\psi_{n+1}(\nu) + (2n-1)\psi_n(\nu) + e^{\mp\lambda}(n-1)\psi_{n-1}(\nu)\right\} \tag{20}$$

which may be reflected on the Gottlieb polynomials through eq. (14).

As in [10] we may look for eigenstates $|\phi >$ of L_+ (the lowering operator): the "coherent states". The defining equation is:

$$\mathcal{L}_+|\phi >= z|\phi > \tag{21}$$

where z is a complex number. Projecting (21) on components ϕ_n we obtain the recursion relation:

$$e^{\lambda} n\phi_{n+1} + (2n-1)\phi_n + e^{-\lambda}(n-1)\phi_{n-1} = 2z\sinh\lambda\,\phi_n\,. \tag{22}$$

This recursion is in fact a recursion for the simple Laguerre polynomials with the substitution:

$$\phi_{n+1} = (-e^{-\lambda})^n L_n(2z\sinh\lambda) \tag{23}$$

namely:

$$(n+1)L_{n+1} = (2n+1-2z\sinh\lambda)\,L_n - nL_{n-1}. \tag{24}$$

Then using the following expansion [11],

$$\sum_{n=0}^{\infty} t^n L_n(x)L_n(y) = \frac{1}{1-t}\exp\left[-\frac{(x+y)t}{1-t}\right] I_0\left(\frac{2\sqrt{xyt}}{1-t}\right) \tag{25}$$

where I_0 is a modified Bessel function, the normalization factor of $|\phi >$ may be evaluated:

$$\sum_{n=0}^{\infty} \phi_{n+1}^2 = \frac{-1}{(e^{-2\lambda}-1)}\exp(2ze^{-\lambda})\,I_0(2z)\,. \tag{26}$$

Thus the normalized components ϕ_{n+1} are:

$$\phi_{n+1} = \frac{e^{-ze^{-\lambda}}}{\sqrt{I_0(2z)}}\sqrt{-e^{-2\lambda}+1}\,(-e^{-\lambda})^n\,L_n(2z\sinh\lambda)\,. \tag{27}$$

From (19) and (21) one may obtain an expansion of the coherent state $|\phi >$ in terms of the eigenstates $|\psi(\nu) >$ of L_0

$$|\phi >= |\phi(z) >= \sum_{\nu} z^{\nu}|\psi(\nu) > \,. \tag{28}$$

Comparison of components of both sides of (27) yields the following relationship between Laguerre and Gottlieb polynomials:

$$L_n(2z\sinh\lambda) = e^{(ze^{\lambda}-\lambda 2n)}\sqrt{I_0(2z)}\sum_{\nu=0}^{\infty}(z)^{\nu} l_n(\nu,\lambda)\,. \tag{29}$$

It seems that such expansions are new, and the overlap amplitude is simply:

$$<\phi(z')|\phi(z)>=(1-z'^{*}z)^{-1}\,. \tag{30}$$

3. As pointed out earlier this $SU(1,1)$ – dynamical symmetry relates free–fermion and 6–vertex models with different magnetic fields. One might seek then a kind of generalization of the Yang–Baxter equations which gives the relationship between vertex systems having the same invariants; e.g. Lieb parameter, external fields as given by Krinsky and Felderhof [12]. To this end we follow Sutherland's procedure [13] by giving a "weak form" of the Yang–Baxter equations

$$\left[H^{\pm}_{n\ n+1}, R_n(\alpha)R_{n+1}(\alpha)\right] = \left[R_n(\alpha)B^{\pm}_{n+1}(\alpha) - B^{\pm}_n(\alpha)R_{n+1}(\alpha)\right]\,. \tag{31}$$

Here $R_n(\alpha)$ is the Baxter standard vertex operator at site n [3] with the weights (1) and $B^{\pm}_n(\alpha)$ are derived vertex operators obtained from the weights (18) in the following way:

$$(\cosh\alpha)\,B^{\pm}_n(\alpha) = e^{\mp\alpha}\frac{d}{d\alpha}R^{\pm}_n(\alpha)\,. \tag{32}$$

Following Krinsky [12] we conclude that the Hamiltonians

$$H^{\pm} = \sum_n H^{\pm}_{n\ n+1} \tag{33}$$

commute with the row–to–row transfer matrix of the 6–vertex system (1): $T(\alpha) = Tr\left(\prod_n R_n(\alpha)\right) = Tr\mathcal{T}(\alpha)$

$$[H^{\pm}, T(\alpha)] = 0 \tag{34}$$

where $\mathcal{T}(\alpha)$ is the so–called monodromy matrix of the model (see e.g. [14]). However (29) leads to

$$[L_{\pm 1}, \mathcal{T}(\alpha)] = -\sum_n \mathcal{T}^{\pm}_n(\alpha) \tag{35}$$

where $\mathcal{T}^{\pm}_n(\alpha)$ is the monodromy matrix of a Kondo–like system [17]: $\mathcal{T}^{\pm}_n(\alpha)$ is obtained from $\mathcal{T}(\alpha)$ by replacing at site n the operator $R_n(\alpha)$ by an operator $R^{\pm}_n(\alpha)$. When one consider the effect of L_0 on $\mathcal{T}(\alpha)$ [5]

$$[L_0, \mathcal{T}(\alpha)] = -\frac{d}{d\alpha}\mathcal{T}(\alpha) \tag{36}$$

one realizes that the three differential operators:

$$\left(\frac{d}{d\alpha}, e^{\mp\alpha}\frac{d}{d\alpha}\right) \tag{37}$$

also form a representation of the $SU(1,1)$ algebra [8].

4. To conclude, we would like to indicate some possible directions for investigation. It is known that the vertex system (1) admits a quantum symmetry described by a "quantum" group [4] which is itself obtainable from the Yang–Baxter algebra in some asymptotic limit [15]. A feature not yet explored is the relationship between such a quantum group and the dynamical symmetry group $SU(1,1)$.

Second, the L_0 operator through (34) was shown to be a "boost" operator for a lattice Poincaré algebra [16] generated by the "conserved charges" of the theory. We do not know yet what would be the action of $L_{\pm 1}$ on this lattice Poincaré algebra.

Third, within the regime of $\lambda = i\mu$ where criticality of the system exists [6] it should be desirable to extend the $SU(1,1)$ symmetry algebra to the whole Virasoro algebra. This is presumably possible with an improved type of Mellin–transform technique as in [8].

Finally, an extension of this construction for dynamical symmetry groups should be made for vertex systems without the free fermion condition in general.

References

1. M. J. Englefield, *Group Theory and the Coulomb Problem*, Wiley Interscience, New York, 1972.
2. J. L. Cardy, *in Phase transitions and critical phenomena, Vol 11*, (C. Domb and J. Lebowitz, eds.) Academic Press, New York, 1986.
3. R. J. Baxter, *Exactly solved models in statistical mechanics*, Academic Press, New York, 1982.
4. H. Saleur, in Proceedings of the 8th Symposium on Theoretical Physics, Sokcho, Korea (July 1989); Saclay Preprint SPhT/89-157.
5. H. Frahm and H. B. Thacker, J. Phys. A: Math. Gen. **24** (1991), 5587.
6. T. T. Truong and I. Peschel, Int. Jour. Mod. Phys. **B4** (1990), 895; B. Davies and P. Pearce, J. Phys. A : Math. Gen. **23** (1990), 1295.
7. M. Gottlieb, Am. Jour. Math **60** (1938), 453; T. S. Chihara, *An introduction to Orthogonal Polynomials*, Gordon and Breach, New York, 1978.
8. C. Itzykson, Jour. Math. Phys. **10** (1969), 1109; H. Bacry, Jour. Math. Phys. **31** (1990), 2061.
9. A. S. Zhedanov, Theor. Math. Phys. **82** (1990), 11.
10. A. O. Barut and L. Girardello, Comm. Math. Phys. **21** (1971), 41.
11. E. A. Rainville, *Special Functions*, MacMillan, New York, 1960.
12. S. Krinsky, Phys. Lett. **39A** (1972), 169; B. U. Felderhof, Physica **66** (1973), 279 and 579.
13. B. Sutherland, Jour. Math. Phys. **11** (1970), 3183.
14. L. A. Takhtadzhan and L. D. Faddeev, Russian Math. Surveys **34** (1979), 186.
15. H. J. de Vega, Int. Jour. Mod. Phys. **B4** (1990), 755.
16. M. G. Tetelman, Sov. JETP **55** (1982), 306; H. B. Thacker, Physica **D18** (1986), 348; K. Sogo and M. Wadati, Prog. Theor. Phys. **69** (1983), 431.
17. M. F. Weiss and K. D. Schotte, Nucl. Phys. **B257** (1985), 599.

Université de Tours, Faculté des Sciences et Techniques, Laboratoire de Modèles de Physique Mathématique, Parc de Grandmont, F-37200 Tours, FRANCE

QUANTUM NOISE AND THERMAL NOISE

H. UMEZAWA

ABSTRACT. A mechanism is discussed for the creation of quantum correlations in a pure state of a system consisting of two independent parts sharing vacuum. This mechanism of vacuum correlation is studied for the cases of a two mode squeezed state, a black hole and thermal physics. The physical role of the second freedom in thermo field dynamics (TFD) is analyzed. The phase operator and time operator in TFD are reviewed.

Contribution in honor of the 60th birthday of John R. Klauder

1. Introduction

My first meeting with Dr. John Klauder took place in the middle of the sixties in a workshop in Germany, although I had known his name before then through reading his papers. Since the first meeting we became good friends, although his mathematics is so deep that my mathematics has never been able to match his. His work on ultralocal theories, representation theories in classical and quantum theories, stochastic quantization and many papers on functional integral methods and quantum optics have enlightened my mathematics and physics in many ways. Therefore, it is my great pleasure to contribute this article to the Festschrift for celebrating the sixtieth birthday of John. The subject of this article overlaps in some way with a problem in quantum optics in which John is one of the greatest figures.

2. Mixed– or Pure–State–Noise; Doubling Degrees of Freedom

The main topic in this paper is the doubled degrees of freedom which were introduced for purely technical reasons in the beginning of the seventies [1,2]

Let us consider two commuting sets of oscillators, a and $\tilde{a}$:

$$[a, a^\dagger] = [\tilde{a}, \tilde{a}^\dagger] = 1\,, \tag{2.1}$$

$$[a, \tilde{a}] = [a, \tilde{a}^\dagger] = 0\,. \tag{2.2}$$

We denote the eigenvalues of $a^\dagger a$ and $\tilde{a}^\dagger \tilde{a}$ by m and $\tilde{m}$ respectively. Here $\tilde{m}$ means the eigenvalue m of $\tilde{a}^\dagger \tilde{a}$. We then choose the eigenvectors $|m, \tilde{m}'\rangle$ for a basis of the Hilbert space for the two oscillators.

Now consider the pure states in this doubled space defined as

$$|0\rangle = c_1 \sum_{m=0}^{\infty} \rho^{\alpha}|m, \tilde{m}\rangle, \tag{2.3}$$

$$\langle 0| = c_2 \sum_{m=0}^{\infty} \langle m, \tilde{m}|\rho^{1-\alpha}, \tag{2.4}$$

where α is any real constant, and c_1 and c_2 are real normalization constants constrained by

$$\langle 0|0\rangle = 1. \tag{2.5}$$

The operator ρ is the density operator defined by

$$\rho = f^{a^\dagger a}, \tag{2.6}$$

where f is a c-number. The normalization condition (2.5) gives

$$c_1 c_2 = 1 - f. \tag{2.7}$$

Note that, when we wish to have a unitary representation in which $\langle 0|$ is the Hermitian conjugate of $|0\rangle$, we have only one parameter f because then $\alpha = 1/2$ and $c_1 = c_2$; otherwise, we have three parameters α, c_1/c_2 and f.

The equations (2.3) and (2.4) define the vacuum with a condensation of $a\tilde{a}$-pairs. We then find for any operator A consisting of a and $a^\dagger$ the relation

$$\langle 0|A(a, a^\dagger)|0\rangle = c_1 c_2 \sum_{m=0}^{\infty} f^m \langle m|A(a, a^\dagger)|m\rangle \tag{2.8}$$

$$= Tr[\rho A(a, a^\dagger)]/Tr\rho. \tag{2.9}$$

The right hand side of (2.9) is the statistical average of A with respect to the density operator ρ. The relation [2,3]

$$\begin{aligned} n &= \langle 0|a^\dagger a|0\rangle \\ &= c_1 c_2 \sum_{m=0}^{\infty} m f^m \\ &= \frac{f}{1-f} \end{aligned} \tag{2.10}$$

relates f to the particle number parameter n. We thus find the equality of the vacuum expectation value (a pure state expectation value in a doubled space) to the statistical average (mixed state expectation value).

According to (2.3) the vacuum $|0\rangle$ is a superposition of the $a\tilde{a}$-pair states, $|m, \tilde{m}'\rangle$ with $m = m'$. This implies that the vacua, $\langle 0|$ and $|0\rangle$, are not those of a- and $\tilde{a}$- operators. Let us denote the vacuum operators ξ and $\tilde{\xi}$ which satisfy

$$\xi(\theta)|0(\theta)\rangle = \tilde{\xi}(\theta)|0(\theta)\rangle = 0, \tag{2.11}$$

$$\langle 0(\theta)|\xi^\dagger(\theta) = \langle 0(\theta)|\tilde{\xi}^\dagger(\theta) = 0, \tag{2.12}$$

where θ stands for the three parameters $(\alpha, c_1/c_2, f)$ which signify each choice of the vacuum. To simplify the notations, we make use of the doublet notations:

$$a^1 = a, \quad a^2 = \tilde{a}^\dagger, \tag{2.13}$$

$$\bar{a}^1 = a^\dagger, \quad \bar{a}^2 = -\tilde{a}, \tag{2.14}$$

and

$$\xi^1 = \xi, \quad \xi^2 = \tilde{\xi}^\dagger, \tag{2.15}$$

$$\bar{\xi}^1 = \xi^\dagger, \quad \bar{\xi}^2 = -\tilde{\xi}. \tag{2.16}$$

Then it can be shown that the $(a, \tilde{a})$-operators and $(\xi, \tilde{\xi})$-operators are related to each other through the Bogoliubov transformation:

$$\xi(\theta)^\mu = B(\theta)^{\mu\nu} a^\nu, \tag{2.17}$$

$$\bar{\xi}(\theta)^\mu = \bar{a}^\nu B^{-1}(\theta)^{\nu\mu}, \tag{2.18}$$

with the Bogoliubov matrix of the form [4,5]

$$B(\theta)^{\mu\nu} = (1+n)^{1/2} e^{s\tau_3} \begin{bmatrix} 1 & -f^\alpha \\ -f^{1-\alpha} & 1 \end{bmatrix}. \tag{2.19}$$

Here, the parameter s is determined by c_1/c_2. Henceforth, we use the three parameters (α, s, f). The number n is related to f through the relation (2.10). It is known that this Bogoliubov transformation forms a three–parameter group associated with SU(1,1).

The above consideration finds an immediate application in statistical mechanics. Such a formalism of statistical mechanics has been applied to quantum field systems and is called thermo field dynamics (TFD). An obvious benefit of TFD is that it can employ most of the quantum field theoretical techniques including both the Green's function method and the operator method in analysis of problems in thermal physics. It has been shown that the usual Feynman diagram methods can be used in TFD with any α and s when the number n has the equilibrium form so that $f = \exp(-\beta\omega)$ with quantum energy ω. When n does ***not*** have the equilibrium form, the Feynman diagram method requires that $\alpha = 1$ or $\alpha = 0$ [6,7].

However, this merit is somewhat compensated by the fact that TFD needs double the number of degrees of freedom. Some people called the tilde operators ghost operators. If the tilde operators were redundant operators, they would not be welcome. Are they really redundant?

The equation (2.9) shows that expectation value of an observable $A(a, a^\dagger)$ of the system a is influenced by the presence of $\tilde{a}$, even when there is no interaction between a and $\tilde{a}$. Furthermore, this effect acts as statistical noise in the a-state due to $\tilde{a}$. This is called the TFD mechanism of vacuum correlation. This is something we can test in a realistic situation. Since this noise effect appears in a pure state, it acts as a correlation through the vacuum; since the vacuum contains condensate of $a\tilde{a}$-pairs, each a-quantum is inevitably correlated with its partner $\tilde{a}$. In such a situation, $\tilde{a}$ is not a redundant freedom.

3. Two Mode Squeezed Waves

One example of realistic examples of the kind mentioned above is the two mode squeezed states.

To make our discussion concrete, let us work with a simple model in quantum optics. Consider two independent incident modes of light denoted by the two sets of oscillator operators introduced in the last section: $(a, a^\dagger)$ and $(\tilde{a}, \tilde{a}^\dagger)$. Suppose that these modes interact through a crystal medium for a short time. Before the interaction the vacuum of the system is the one annihilated by the operators a and $\tilde{a}$, i.e. $|0\rangle\rangle$ with $a|0\rangle\rangle = 0$. When one considers a type of crystal interaction [8] inducing the Bogoliubov transformation in (2.17-18), the vacuum changes into $|0(\theta)\rangle\rangle$. In other words, the state just at the time when the two waves come out of the crystal is $|0(\theta)\rangle\rangle$. This time is called t^1. The time when the waves enter the crystal is denoted by t^0. Thus, before t^0 the vacuum is the one associated with a and $\tilde{a}$, while the vacuum at t^1 is $|0(\theta)\rangle\rangle$ associated with ξ and $\tilde{\xi}$. The parameter θ which identifies the vacuum at time t^1 depends on the coupling constant of the interaction through the crystal media.

A remarkable observation in this mechanism is that, as a result of the fact that the two systems share the common vacuum $|0(\theta)\rangle\rangle$, a measurement of a manifests noise due to the presence of $\tilde{a}$. This is despite the fact that the a- and $\tilde{a}$-systems have no mutual interactions after t^1.

A useful comment here is on the temporal behavior of the state after t^1. Let ω and ω' denote the quantum energy of a and $\tilde{a}$ respectively. Since $|0(\theta)\rangle\rangle$ contains the pair $a^\dagger\tilde{a}^\dagger$ according to the consideration in the last section, the vacuum state after t^1 is dependent on time unless $\omega+\omega' = 0$, because the pair term changes in time as $\exp[i(\omega+\omega')(t-t^1)]a^\dagger\tilde{a}^\dagger$; the final vacuum is a state with oscillating pairs. This becomes significant when we consider the TFD mechanism of vacuum correlation in thermal physics.

This noise effect motivates us to study the uncertainty relations. To do this we introduce the canonical operators through the relations

$$q = \frac{1}{\sqrt{2}}(a + a^\dagger), \qquad p = \frac{1}{\sqrt{2}i}(a - a^\dagger), \tag{3.1}$$

$$\tilde{q} = \frac{1}{\sqrt{2}}(\tilde{a} + \tilde{a}^\dagger), \qquad \tilde{p} = -\frac{1}{\sqrt{2}i}(\tilde{a} - \tilde{a}^\dagger). \tag{3.2}$$

We introduce also $\{q(\theta), p(\theta), \tilde{q}(\theta), \tilde{p}(\theta)\}$ by replacing a by $\xi(\theta)$. We then find the uncertainty relation

$$\langle(\Delta q)^2\rangle\langle(\Delta p)^2\rangle \geq \frac{\hbar^2}{4} + \langle\Delta q\Delta\tilde{q}\rangle\langle\Delta p\Delta\tilde{p}\rangle, \tag{3.3}$$

where the brackets mean the expectation values with respect to the vacuum $|0(\theta)\rangle\rangle$. This shows that the uncertainty in the state $|0(\theta)\rangle\rangle$ is enhanced by the second term [9] which is

$$\langle\Delta q\Delta\tilde{q}\rangle\langle\Delta p\Delta\tilde{p}\rangle = \frac{\hbar^2}{2}\sqrt{n(1+n)}. \tag{3.4}$$

A similar noise effect shows up in the Wigner function [10].

4. Black Hole

There is a dynamical system which creates thermal effects in our world through the TFD mechanism of vacuum correlation. That is the black hole. A black hole creates a horizon. The world beyond the horizon is completely dissociated from our world. However, these two dissociated worlds share a common vacuum. Let a stand for dynamical variables of our world, while those in the world beyond the horizon are represented by $\tilde{a}$. It has been shown that the vacuum is not the vacuum $|0\rangle\rangle$ associated with a and $\tilde{a}$, but is $|0(\theta)\rangle$, which is related to $|0\rangle\rangle$ through the thermal Bogoliubov transformation. The parameter θ of the Bogoliubov transformation is determined by the mass of the black hole, while it determines the temperature of radiation around the black hole (the Hawking temperature) [11,12]. Thus the analogy to the case of a two mode squeezed state is perfect. Furthermore, it has been shown that the Hamiltonian which controls the entire system has the structure $\hat{H} = H - \tilde{H}$, in which H is the Hamiltonian for dynamical phenomena in our world and $\tilde{H}$ is the one for the other world. For a free oscillator model this means $\omega + \omega' = 0$, implying the stationary nature of the vacuum $|0(\theta)\rangle$.

An extension of this picture motivates the idea that the origin of the thermal degree of freedom in our world could be due to the presence of other universes which are totally dissociated from ours, though they share the same vacuum with our world [12]. This description of the thermal situation around a black hole is exactly the same as the TFD formalism for equilibrium states.

5. TFD Mechanism of Vacuum Correlation in QFT

So far we described the TFD mechanism of vacuum correlation in terms of quantum mechanics for oscillators. However, its appearance becomes more dramatic in quantum field theory (QFT), because in quantum field theory two vacua, $|0\rangle\rangle$ and $|0(\theta)\rangle$ do not belong to the same Hilbert space. As a matter of fact, $|0(\theta)\rangle$ with different θ do not belong to the same Hilbert space. Thus, we have an uncountable set of Fock spaces $\mathcal{H}(\theta)$, each being associated with a choice of θ. This means that different choices of θ correspond to different physical situations, each manifesting different amount of vacuum correlations.

6. Thermal Physics

It is obvious that the above idea of the TFD mechanism of vacuum correlation can be used in reformulating statistical mechanics. Such a reformulation leads to a real-time formalism of thermal quantum theory and is called thermo field dynamics (TFD). In TFD, to each operator A is associated another operator $\tilde{A}$. The rule for obtaining $\tilde{A}$ from A has been formulated and is called the tilde conjugation rule [1,2,3,4]. Recall now that we need the relation $\omega + \omega' = 0$ in order to have a stationary vacuum. Therefore, the TFD Hamiltonian has the following structure. Suppose a dynamical Hamiltonian H is given. Then its tilde conjugate $\tilde{H}$ is obtained from it through the tilde conjugation rule. Then the total Hamiltonian $\hat{H}$ is not the addition of H and $\tilde{H}$, but the difference:

$$\hat{H} = H - \tilde{H}. \tag{6.1}$$

This is because TFD should be able to accomodate stationary equilibrium states.

This implies that the tilde quanta carry negative energies. This might enhance the impression that the tilde degrees of freedom would be ghost freedom and lead many people to the question: do we need ghost freedoms? However, we will find that this negative sign in front of $\tilde{H}$ in the above structure of $\hat{H}$ plays the most significant role in deriving many basic laws of thermal physics. *The fact that the eigenvalue spectrum of $\hat{H}$ is unbounded below is the most significant feature of TFD.* This unboundedness originates from the presence of $\tilde{H}$ in (6.1) with a negative sign. In this sense the tilde degree of freedom is not a ghost freedom. We will explain this in the next section.

7. Degenerate Thermal Vacua and Heat Energy

To pursue our question we need to recall a fundamental feature of thermal physics. It is known that there are two kinds of energies; dynamical energy and heat energy. To move from one thermal situation to another usually consumes heat energy. Furthermore, although the dynamical energy is non negative, the heat energy is inexhaustible. The stability of a system is not supported by the energy conservation but by the entropy law. This can be well taken care of by the unboundedness (from below) of hat-energy, where hat-energy means the eigenvalue of $\hat{H}$. This will be briefly touched upon in this section.

Consider a set of stationary thermal states. They should be the zero hat-energy states

$$\hat{H}|0(\theta)\rangle = 0, \tag{7.1}$$

because they are stationary. A state satisfying this relation is called the thermal vacuum.

Since the hat-energy spectrum is unbounded below, there is an uncountable set of thermal vacua $|0(\theta)\rangle$ satisfying this stationarity condition [13]. We use the parameter θ to classify such vacua. As was pointed out in section 2, θ is known to have three components associated with SU(1,1) algebra. The dynamical energy E is given by

$$E(\theta) = \langle 0(\theta)|H|0(\theta)\rangle, \tag{7.2}$$

which depends on θ. This means that the dynamical energy changes when we move from one thermal vacuum to another. This change is compensated by the heat energy: the first law of thermodynamics. In this way the concept of heat naturally emerges in TFD [14]. This is an example of the general statement that thermal quantities such as heat, entropy, etc. need the tilde degree of freedom when we make use of a pure state formalism. The heat is inexhaustible because we have a continuous set of thermal vacua.

Summarizing, the unboundedmess of hat-energy from below gives rise to a continuous set of hat-energy degenerate thermal vacua, which in turn gives rise to the concept of heat.

The presence of hat-energy degenerate thermal vacua suggests that thermal physics in TFD has a structure of spontaneous breakdown of a certain symmetry. Since the parameter θ is associated with transformation parameters of the group associated with SU(1,1) algebra, it is not difficult to show that the hat-energy degenerate vacuum is associated with spontaneous breakdown of the SU(1,1) symmetry [13].

8. Time-Dependent TFD

Being equipped with the set of thermal vacua it is easy to move to time dependent thermal changes. There are two ways to describe time dependence: to introduce the time dependence either through state vectors or through operators. We take the latter approach. When a thermal situation changes in time, we make use of a state $|0(\theta)\rangle$ at time, say t_0, as the thermal vacuum and all the thermal changes are taken care of by the operator changes. This thermal vacuum will be denoted by $|0, t_0\rangle$. This means that we make use of the Heisenberg representation. Then, the Hamiltonian cannot be $\hat{H}$, because the dynamical energy is not conserved.

Let $\hat{H}_Q$ denote the new Hamiltonian and write it as

$$\hat{H}_Q = \hat{H} - \hat{Q}. \tag{8.1}$$

Here, $\hat{Q}$ should be proportional to $\dot{n}$ (time derivative of n), because $\hat{Q}$ should vanish when $\dot{n}$ is zero.

Here is a comment on the choice of $|0, t_0\rangle$. Since we have an uncountable number of thermal vacua $|0(\theta)\rangle$, a particular choice for $|0, t_0\rangle$ should be specified. Since we wish to have a well-defined picture of quasi-particles, and since the quasi-particles have been so far introduced in a stationary situation, we require that $|0, t_0\rangle$ be a stationary state; $\dot{n} = 0$. Then, t_0 cannot be the initial time, because a time dependent process starts at the initial time. Thus, we choose $t_0 = \infty$. Furthermore, this stationarity requirement for $|0, t_0\rangle$ means that $\hat{H}_Q(t_0)|0, t_0\rangle = 0$, which gives

$$\hat{Q}(t_0)|0, t_0\rangle = 0. \tag{8.2}$$

This is not a trivial condition, because there is no obvious reason that this condition should be consistent with other laws such as the entropy law. Therefore, it is remarkable that this condition has a solution which leads to the entropy law.

Consider the case of a free field. Since we are considering a time dependent thermal process, the Bogoliubov matrix in equation (2.17-18) becomes time dependent. Since the thermal vacuum under consideration is $|0, t_0\rangle$, its vacuum operators are $\xi(t_0)$ and $\tilde{\xi}(t_0)$. They are denoted simply by ξ and $\tilde{\xi}$, respectively. Though θ stands for three parameters (α, s, f), only f is physically significant. Therefore, we make f dependent on time; $f = f(t)$. It means that the number depends on time as $n(t) = f(t)/[1 - f(t)]$. With $t_0 = \infty$ the Feynman diagram method is applicable only to $\alpha = 0$. For the sake of easy calculation we specify the parameter s by the requirement that all the matrix elements of the Bogoliubov matrix be linear in $n(t)$. With this choice of the parameters, the relation (2.17-18) becomes [7,15] :

$$\xi_k^\mu = B_k(t)^{\mu\nu} a_k(t)^\nu, \tag{8.3}$$

$$\bar{\xi}_k^\mu = \bar{a}_k(t)^\nu B_k^{-1}(t)^{\nu\mu}, \tag{8.4}$$

with the Bogoliubov matrix of the form

$$B_k(t) = \begin{bmatrix} 1 & -1 \\ -n_k(t) & 1 + n_k(t) \end{bmatrix}. \tag{8.5}$$

Here we introduced the momentum suffix because we are treating fields. It was shown in [7] that calculations of Feynman diagrams and manipulation of Green's functions are tremendously simplified with this choice of the $B_k(t)$-matrix. The change of the Hamiltonian, $\hat{Q}$, is obtained from this knowledge of the $B_k(t)$-matrix.

To describe a temporal behavior, we need an equation for the number $n_k(t)$. This equation was obtained in [7] from a self–consistent renormalization of the self–energy diagram. It is expected that the usual definition of the on-shell self–energy is hardly applicable to time dependent processes. Therefore, it is worth mentioning that in this formalism the usual on-shell definition for the self–energy can be used without any modification. The result with the self–energy loop with two vertices turns out to be the Boltzman equation, which gives rise to the law of entropy. In this way, the second law of thermodynamics is accommodated by TFD. Furthermore, it opens the possibility of calculating the nonequilibrium entropy even when we consider more complicated diagrams. This Boltzman equation should be solved under a given initial condition for n_k. This is where the initial condition enters; the time t_0 has nothing to do with this initial condition.

In the above we considered only one parameter dependent on time. This formalism has been extended to an Abelian gauge theory [16]. However, from a very general point of view we are dealing with the three parameters associated with SU(1,1) algebra. From this view point there has emerged an SU(1,1) gauge theory for TFD [17]; in this paper the authors present a preliminary analysis of topological problems in this gauge theory.

9. Phase Operator and Time Operator in TFD

In quantum theory observables are usually represented by certain operators. However, there are some exceptions such as the phase and time operators. It is well known that time does not have its own operator, though the generator of time translation does exist and is called the Hamiltonian. We also have the generator of the phase transformation of fields and it is the number operator. However, the phase operator has not been constructed properly [18,19,20,21,22,23]. For some physicists this might appear to be an unsatisfactory aspect of the quantum theory. For those physicists it may be an interesting remark that the phase operator and time operator do exist in TFD.

The main cause of the difficulty in the usual theory is that the number spectrum and energy spectrum are confined to the non-negative domain, and therefore, it is bounded below. It was shown by Newton [21] that, when we introduce a number operator that is unbounded from below by doubling the degrees of freedom, a phase operator appears. This consideration has been elaborated by Ban in his recent article [23], in which the consideration was applied to the TFD mechanism. The time operator was also discussed in this article. In this section we study the phase and time operators, closely following this article.

For sake of simplicity we omit the momentum suffix and consider free oscillators without interaction. Then, the Hamiltonian in TFD has the form

$$\hat{H}_0 = H_0 - \tilde{H}_0 \tag{9.1}$$

$$= \omega \hat{N} \tag{9.2}$$

with the number operator

$$\hat{N} = a^\dagger a - \tilde{a}^\dagger \tilde{a}. \tag{9.3}$$

Although the dynamical energy operator of the a-system is H_0 whose eigenvalue is non-negative, the eigenvalue (hat-energy) of the Hamiltonian $\hat{H}_0$ is unbounded below. The same is true for the TFD number operator $\hat{N}$. It is due to this aspect of TFD that the canonical conjugate of $\hat{H}_0$ exists as an operator and acts as the time operator. In the same way the canonical conjugate of the number operator $\hat{N}$ exists and it acts as a phase operator.

Let us begin with the phase operator. To do this we use the basis vectors $|n, \tilde{m}\rangle\rangle$, in which n and $\tilde{m}$ stand for the eigenvalues of N and $\tilde{N}$, respectively. Let $\hat{n}$ denote the eigenvalues of $\hat{N}$. Then, each member, $|n, \tilde{m}\rangle\rangle$, of this base set can be specified by $\hat{n}$ together with $\tilde{m}$. We denote this state by $|\hat{n}, \tilde{m}\rangle\rangle$; that is $|\hat{n}, \tilde{m}\rangle\rangle = |n, \tilde{m}\rangle\rangle$ with $\hat{n} = n - \tilde{m}$. Therefore, the base set can be written as $(|\hat{n}, \tilde{m}\rangle\rangle : -\infty < \hat{n} < \infty, 0 \leq \tilde{m} < \infty)$. We then have the orthonormalization relation

$$\langle\langle \hat{n}, \tilde{m} | \hat{n}', \tilde{m}' \rangle\rangle = \delta_{nn'} \delta_{m,m'} \tag{9.4}$$

and the completeness relation

$$\sum_{m=0}^{\infty} \sum_{n=-\infty}^{\infty} |\hat{n}, \tilde{m}\rangle\rangle\langle\langle \hat{n}, \tilde{m}| = 1, \tag{9.5}$$

where 1 stands for the identity operator.

Introduce the operator D by the relation

$$D = \sum_{m=0}^{\infty} \sum_{n=-\infty}^{\infty} |\hat{n} - 1, \tilde{m}\rangle\rangle\langle\langle \hat{n}, \tilde{m}|. \tag{9.6}$$

We can then prove the relations

$$DD^\dagger = D^\dagger D = 1, \tag{9.7}$$

$$D|\hat{n}, \tilde{m}\rangle\rangle = |\hat{n} - 1, \tilde{m}\rangle\rangle, \tag{9.8}$$

$$D^\dagger|\hat{n}, \tilde{m}\rangle\rangle = |\hat{n} + 1, \tilde{m}\rangle\rangle. \tag{9.9}$$

These relations lead us to

$$[D, \hat{N}] = D\,, \tag{9.10}$$

which shows that D is the exponential of $i \times$ (the phase operator).

To clarify this point, we introduce the states

$$|\theta, \tilde{m}\rangle\rangle = \frac{1}{\sqrt{2\pi}} \sum_{n=-\infty}^{\infty} \exp(-in\theta)|\hat{n}, \tilde{m}\rangle\rangle. \tag{9.11}$$

Then, making use of the relation

$$\sum_{n=-\infty}^{\infty} \exp(-in\theta) = 2\pi\delta(\theta), \tag{9.12}$$

with θ being confined to the domain $-\pi \leq \theta \leq \pi$, we can derive the orthonormalization relations

$$\langle\langle \theta, \tilde{m} | \theta', \tilde{m}' \rangle\rangle = \delta_{m,m'}\delta(\theta - \theta') \tag{9.13}$$

and the completeness relation

$$\sum_{m=0}^{\infty} \int_{-\pi}^{\pi} d\theta \; |\theta, \tilde{m}\rangle\rangle\langle\langle\theta, \tilde{m}| = 1. \tag{9.14}$$

Now we can derive the relation

$$D|\theta, \tilde{m}\rangle\rangle = \exp(-i\theta)|\theta, \tilde{m}\rangle\rangle. \tag{9.15}$$

We can write

$$D = \sum_{m=0}^{\infty} \int_{-\pi}^{\pi} d\theta \; |\theta, \tilde{m}\rangle\rangle \exp(-i\theta) \langle\langle\theta, \tilde{m}|. \tag{9.16}$$

We thus obtain the *phase operator* conjugate to $\hat{N}$.

By replacing $\hat{N}$ with $\hat{H}_0$, we obtain the *time operator*. Existence of phase and time operators gives us a modest comfort, but it remains to be seen how important they can be.

Acknowledgement

The author would like to thank Dr. Y. Yamanaka and Dr. H. Chu for stimulating discussions and valuable comments. This work was supported by NSERC, Canada and the Dean of Science, the University of Alberta.

References

1. L. Leplae, H. Umezawa and F. Mancini, Physics Reports **10C** (1974), 153.
2. Y. Takahashi and H. Umezawa, Collect. Phenom. **2** (1975), 55.
3. H. Umezawa, H. Matsumoto, and M. Tachiki, *Thermo Field Dynamics and Condensed States*, North-Holland, Amsterdam, 1982.
4. H. Umezawa and Y. Yamanaka, Advances in Physics **37** (1988), 531.
5. T.S. Evans, I. Hardman, H. Umezawa and Y. Yamanaka, *A time-dependent nonequilibrium calculational scheme towards the study of temperature fluctuations*, Univ. of Alberta preprint (1990).
6. T.S. Evans, I. Hardman, H. Umezawa and Y. Yamanaka, *Heisenberg and Interaction Representations in Thermo Field Dynamics*, J. Math. Phys. **33** (1992), 370.
7. Y. Yamanaka, H. Umezawa, K. Nakamura and T. Arimitsu, *Thermo Field Dynamics in Time Representation*, Univ. of Alberta preprint (1991).
8. A. Mann and M. Revzen, Phys. Lett. **134A** (1988), 531.
9. H. Umezawa and Y. Yamanaka, Physica **170A** (1991), 291.

10. M. Berman, Phys. Rev. **A40** (1989), 2057.
11. W. Israel, Phys. Lett. **57A** (1976), 107.
12. R. Laflamme, Physica **158A** (1989), 58.
13. H. Umezawa and Y. Yamanaka, Phys. Lett. **155A** (1991), 75.
14. I. Hardman, H. Umezawa and Y. Yamanaka, Phys. Lett. **146A** (1990), 293.
15. H. Umezawa and Y. Yamanaka, *Temporal Description of Thermal Quantum Fields*, Univ. of Alberta preprint (1991).
16. E. Ahmed, Int. Jour. Theor. Phys. **28** (1989), 1351.
17. P. A. Henning, M. Graf and F. Matthäus, *The Geometry of the Thermal Quasi-Particle Transformation*, Gesellschaft für Schwerionenforschung, Darmstadt, preprint (1990).
18. L. Susskind and J. Glogower, Physics **1** (1964), 49.
19. P. Carruthers and M. M. Nieto, Rev. Mod. Phys. **40** (1968), 411.
20. J.M. Levy-Leblond, Ann. Phys. **101** (1976), 319.
21. R. Newton, Ann. Phys. **124** (1980), 327.
22. D.T. Pegg and S.M. Barnett, Phys. Rev. **A41** (1990), 3427.
23. M. Ban, J. Math. Phys. **32** (1991), 3077.

THE THEORETICAL PHYSICS INSTITUTE, UNIVERSITY OF ALBERTA, EDMONTON, ALBERTA T6G 2J1, CANADA